W9-BKX-819

Birds of Missouri

Birds of Missouri

Their Distribution and Abundance

Mark B. Robbins and
David A. Easterla

University of Missouri Press
Columbia and London

University of Missouri Press, Columbia, Missouri 65201
Printed and bound in the United States of America

5 4 3 2 1 96 95 94 93 92

Library of Congress Cataloging-in-Publication Data

Robbins, Mark B., 1954–
Birds of Missouri : their distribution and abundance / Mark B. Robbins and David A. Easterla.
p. cm.
Includes bibliographical references and index.
ISBN 0-8262-0791-X
1. Birds—Missouri. 2. Birds—Missouri—Geographical distribution. 3. Bird populations—Missouri. I. Easterla, David A. II. Title.
QL684.M8R6 1991
598.29778—dc20 91-19284
CIP

∞™ This paper meets the requirements of the American National Standard for Permanence of Paper for Printed Library Materials, Z39.48, 1984.

Designer: Rhonda Gibson
Typesetter: Connell-Zeko Type & Graphics
Printer: Edwards Brothers, Inc.
Binder: Edwards Brothers, Inc.
Typeface: Korinna

To the past, present, and future bird-watchers of Missouri.

And we especially take great pleasure in dedicating this book to our late friend and field companion, Floyd R. Lawhon. We deeply regret that Floyd did not live to see the fruition of his contributions to this work.

Contents

Acknowledgments

The foundation of a book of this kind is the result of many dedicated field observers, who not only spend countless hours afield, but accurately record their observations. We greatly acknowledge their contributions. Without the invaluable contribution of Otto Widmann at the turn of the century, there would have been very little historical perspective to our work. More recently, the late Floyd Lawhon's voluminous notes were indispensable. Up to his untimely death, Floyd indefatigably answered our seemingly endless queries. Richard "Dick" Anderson, like Lawhon, was inundated with a constant barrage of questions from us. Dick helped immensely with the clarification of many St. Louis observations. JoAnn Garrett graciously aided in tracking down details of sightings in the Kansas City region. Much valuable unpublished material was also provided by Randy Korotev and Jim Haw. Korotev also made helpful comments on several of the Passerine Species Accounts. The Gruiformes and Charadriiformes accounts benefitted from comments by Ron Goetz,sds did the Falconiform section by Mark Peters' critique.

We are deeply indebted to David Plank for his extraordinary color paintings and line drawings which greatly enhance this work. Dan Brauning, George "Skip" Glenn and Fred Sheldon were very helpful with suggestions on the use of computer software for data analysis and graphics. We thank Frank Gill and the Philadelphia Academy of Natural Sciences' Ornithology Department for use of computer software and hardware. Appreciation is also extended to Northwest Missouri State University's Department of Biology and the College of

Agriculture Science and Technology for providing assistance during the course of our work.

A number of people at the Missouri Department of Conservation have provided information. In particular, we especially want to thank James D. Wilson, Brad Jacobs, Norb Giessman, John W. Smith, John B. Lewis, Diana L. Hallett, and Jim Rathert.

The following people aided in one or more ways to our work: Ike Adams, James Dinsmore, Bill Eddleman, William Elder, Mike Flieg, Leo Galloway, Bill Goodge, Paul Heye, Chris Hobbs, David Jones, Tom Kennedy, Ben King, Alan Knox, Kevin Kritz, Rebecca Matthews, David Mead, Russell Mumford, Bill Reeves, Van Remsen, Jr., Bill Rudden and Phoebe Snetsinger.

We are most grateful to Sam Droege and the U.S. Fish and Wildlife Service for providing data from the North American Breeding Bird Survey. Doug Wechsler and Steve Holt of project VIREO at the Academy of Natural Sciences, Philadelphia, aided in the deposition of documentary photographic material, as did Greg Budney and the Library of Natural Sounds, Cornell Laboratory of Ornithology, with the deposition of vocal material.

We are very thankful to the following curators and museums for aiding us with locating and examining Missouri material (listed in alphabetical order by institution): Mary LeCroy (American Museum of Natural History), Kenneth Parkes (Carnegie Museum of Natural History), Joseph Geist (Central Methodist College, Fayette), John Hess (Central Missouri State University), Vicki Byre (Chicago Academy of Sciences), Betsy Webb (Denver Museum of Natural History), Jim Ware (Gaylord Laboratory, University of Missouri), James Dinsmore (Iowa State University), Van Remsen, Jr. (Louisiana State Museum of Natural Science), Raymond Paynter, Jr. (Museum of Comparative Zoology, Harvard), Gary Schnell (Oklahoma Museum of Natural History), Richard Nolf and David Mead (St. Joseph Museum), James Houser (St. Louis Science Center), Gary Graves (U.S. National Museum, Smithsonian), Richard Stiehl (Southeast Missouri State University), Marion Mengel (University of Kansas Museum of Natural History), William Elder and Mark Ryan (University of Missouri), and Lloyd Kiff (Western Foundation of Vertebrate Zoology).

*   *   *

I would like to express my sincerest appreciation and love to my wife, Kathy Robbins, who unselfishly tolerated all too many evenings and weekends as a “widow” during the preparation of this book. She was also very supportive and understanding when I disappeared in the field for extended periods of time. I am also very deeply indebted to my parents, Patricia and the late Norman Robbins, whose endless support and love allowed me the freedom to pursue such a unique career. I owe special thanks to David Easterla who, through inviting me to join his ornithology class field trip to Squaw Creek over twenty years ago, opened the door to the world of birds.

—Mark Robbins

* * *

I would like to thank my parents, Gene and Hattie Easterla, for tolerating my numerous absences (since I was 10 years old) to chase those "crazy" birds. Other relatives and friends have also tolerated those often, untimely field endeavors, including my "neglected" brother Richard. I am indebted especially to my two sons, David J. and Todd B. for their field companionship in earlier years; for Todd, this birding companionship has continued at a high level of fun, competition, and enjoyment. To Patti, the mother of these two sons, I express my appreciation and gratitude for her eighteen years of time and tolerance to my ornithological pursuits.

To my wife Debbie Easterla, I say "thank you" for being so tolerant and allowing me to be my "own person," which usually meant never being at home, working late hours at the office, or being in the field and returning at any unexpected, unimaginable hour! Appreciation is also extended to Lee R. Crail, who first taught a twelve-year-old Ozark boy how to keep a field notebook and introduced him to R. T. Peterson's *Eastern Field Guide to the Birds.* Previously Chester Reed's bird guide had been my constant field companion until Peterson's overruled.

I am also indebted to Dr. William H. Elder, Emeritus Professor of Zoology, University of Missouri–Columbia, for introducing me to the more scientific aspects of birding. Lisle and Alice Jeffrey and Kay and Bob Eastman were always tolerant of an often overenthusiastic birder. I want to thank my Versailles High School agriculture teacher, Andy White, for personally taking me to the University of Missouri–Columbia and encouraging my attendance there, and I would like especially to single out and thank Richard Anderson for his original encouragement, cooperation, and enthusiasm in the compilation of our first 1967 *Checklist of Missouri Birds* for the Audubon Society of Missouri.

Many of my field trips would not have been as pleasant if it had not been for the wit and companionship of such birders as: David Snyder, Wally George, Kay Stewart, Jim Gilmore, Gary Jackson, Nick Holler, Duane Kelly, Mike Flieg, Jim Rising, Elizabeth Cole, Ted Pucci, Ted Anderson, Walter Cunningham, Steve Hilty, Jack Hilsabeck, Floyd Lawhon, John Hamilton, Si Rositsky, Leo Galloway, Richard Rowlett, Drew Thate, Jim Grace, Doyle Damman, Richard Anderson, Jim F. Comfort, J. Earl Comfort, Earl Hath, Bertha Massie, David Mead, Ken Jackson, and all the others not mentioned but greatly appreciated. Students in my numerous ornithology classes cannot be forgotten; I hope they enjoyed all of the field trips as much as I did. Certainly, Kenny and Carolyn Fisher must be mentioned in our numerous "boat pursuits" of birds at Lake Viking.

Last, but not least, is my very close and talented friend and birding companion Mark Robbins, who is always a pleasure to work with. His field enthusiasm is synonymous with health and youth; may it always remain on the planet Earth.

—David Easterla

Birds of Missouri

Introduction

The avifauna of Missouri is relatively rich, even though the state is landlocked and has only a moderate amount of relief. A total of 385 species have been positively recorded since the arrival of European man, and another 20 species are placed on the hypothetical list; thus, a total of 405 species are treated in this work.

The relatively high diversity is the result of the state's boundaries encompassing several ecological zones. The fact that Missouri lies at the suture of the eastern deciduous forest and the eastern limits of the Great Plains is the most important factor contributing to its diversity. Many eastern deciduous forest species (e.g., several warblers) reach, or nearly reach, their westernmost point of distribution here; the same is true for western species at the eastern edge of their range (e.g., Swainson's Hawk, Western Kingbird, Great-tailed Grackle). Moreover, only in a handful of states can one witness the impressive eastern passerine migration, as well as have the opportunity to observe spectacular numbers of species whose main migratory path is through the interior of the country (e.g., American White Pelican, Greater White-fronted and Snow geese, and a number of shorebird and sparrow species).

A southern flavor is added to the state's fauna as a result of the southern border extending to 36° 30′ N latitude and even farther in the "bootheel" to 36° 0′ N. The southeastern corner is especially important to the state's avian diversity, since it is here where another unique ecological zone is encompassed. Species such as Anhinga, Glossy Ibis, Mississippi Kite, Purple Gallinule, and Fish Crow are principally found in this area of the state. In addition, there are a few species that reach their northeastern-most limit in the southwestern corner of the state, such as Greater Roadrunner, Scissor-tailed Flycatcher, and the interior population of the Painted Bunting. Finally, complementing the southern component are species of more northerly climes that irregularly visit the state in winter. These include Northern Goshawk, Snowy Owl, Northern Shrike, and northern fringillid finches.

Not since Widmann's treatise (1907) has there been a thorough summary of the distribution and abundance of the birds in the state. Our book is the result of over twenty years of involvement with Missouri birds by Robbins and some forty years by Easterla. We have attempted to couple the literature with much unpublished information to give a historical perspective as well as the current status for each species. This book is not intended to be a field identification guide or displayed on the "coffee table." Given the number of excellent field guides that are now available, we feel it would be redundant to include descriptions of all the state's birds. This space can be better utilized by presenting more data on the status and distribution of Missouri's avifauna. We hope that this work will not only be used as a reference book but will be taken in the field and used in concert with one or more of the thorough field guides. One of our goals is to make the field observer cognizant of the significance of his observations—*while still in the field*—to improve the documentation of unusual sightings. In addition, this book readily emphasizes areas where more observations are needed to clarify the status and relative abundance of species in the state.

Photographs included in this work were obviously not chosen for their aesthetic value (although several are of good quality). We primarily chose photographs that document the occurrence of a species in the state or during a particular season. In a couple of instances we used photographs to illustrate points of identification of species that have been a problem for birders in the state. We encourage observers to document their unusual observations with photographs to ensure a permanent record for future workers.

We chose subjects for David Plank's lovely watercolor paintings that would highlight some of the uniqueness of Missouri's avifauna and a few of the significant North American ornithological discoveries that were first made in the state. Plank's shorebird painting portrays four of the species whose primary spring migration route passes through at least part of the state. The Greater Roadrunner, Scissor-tailed Flycatcher, Painted Bunting, and Bachman's Sparrow all reach the northern limit of their distribution in Missouri. Missouri's early contribution to North American ornithology is underscored by the paintings of the three species first discovered in the state and the pair of Bachman's Warbler. See the appropriate species accounts for details of these discoveries.

Although we have attempted to check as many as possible of the published records against original field notes, it was impossible to check them all. In those records where the original data were compared to the published accounts, we found approximately 1 percent transcriptional rate of error—undoubtedly a few errors of this type still remain. Furthermore, additional errors are likely to have been generated in the transcription of observations during the preparation of this book. We would be most grateful to anyone who brings any errors to our attention. Likewise, we welcome comments on the format of the book for future revisions. This work includes information made known to us through late November 1990.

History of Missouri Ornithology

During the eighteenth century and the first decade of the nineteenth century, explorers' accounts were limited to game and the more conspicuous avian species. Missouri, however, was fortunate to have a number of expeditions, with competent observers, pass through the state before Europeans significantly modified the presettlement environment. Because the rivers were the main avenues of travel, the earliest accounts are confined to the Missouri and Mississippi rivers. The first trained ornithologist to visit the state was Thomas Say. He accompanied the Long Expedition in 1819 that traveled up the Mississippi River from the mouth of the Ohio and then across the state on the Missouri River. Between 21 and 23 June, while the expedition was stationed at Bellafontaire in St. Louis County, about four miles from the mouth of the Missouri River, Say collected the first Lark Sparrow known to science (Say 1823). Say also made several references to the Common Raven during this journey, a species that is now extirpated from the state.

In the spring of 1834, Thomas Nuttall and John Townsend passed upstream on the Missouri to the Kansas City area. On 28 April of that year, as they traveled overland from Independence to Westport in Jackson County, they collected science's first Harris' Sparrow (Harris 1919a). The next significant contribution to Missouri's ornithology came when the renowned John J. Audubon, Edward Harris, and John G. Bell passed through the state on the Missouri River in 1834. While at St. Joseph in Buchanan County on 6 May 1843, Bell collected a new vireo. Audubon later named the bird the Bell's Vireo, in honor of the collector (Cooke 1910). Two days prior to the discovery of the vireo, Audubon and his colleagues collected several Carolina Parakeets just below St. Joseph. The collection is now deposited at the Academy of Natural Sciences, Philadelphia (ANSP).

The exploration of western Missouri in 1854, led by P. R. Hoy, under the auspices of the Smithsonian Institution, produced the first list of birds for the state. Hoy (1865) reported such unusual species as the Purple Sandpiper on this trip; however, the deposition of his collection remains a mystery because it was never cataloged at the Smithsonian. Scott (1879) published a list of birds recorded by him in the spring and early summer of 1874 in the Warrensburg, Johnson County, area.

Certainly the most fortuitous event to shape Missouri ornithology was the arrival of Otto Widmann to St. Louis in 1867. During the first few years of Widmann's residence he was occupied with business matters, but by early 1875 he began to make serious observations of the state's avifauna. Over the next half century, he made detailed observations of the birds throughout the state, with particular emphasis on the St. Louis area. One of his most important contributions was the discovery and description of the first Bachman's Warbler's nest and eggs from the Mississippi lowlands (Widmann 1897). He had nearly completed a manuscript on the state's birds when, in 1902, while he was away visiting in Germany, a fire destroyed the manuscript and all of his notes. It took Widmann a few years to recover from this disaster, but he nonetheless rewrote his indispensable book, *A Preliminary Catalog of the Birds of Missouri,* which finally appeared in 1907. Widmann's treatise on the state's birds is surprisingly complete and accurate, given the state of ornithology at that time. His work provides not only a solid account of the pre–1900 status and distribution of Missouri's birds but much of the information still depicts an accurate picture of today's avifauna. At the time Widmann was putting the final touches to his book, E. Woodruff made several important observations of the little-known fauna of the Shortleaf Pine region in Shannon and Carter counties. His valuable collection is deposited at the American Museum of Natural History, New York (AMNH).

As early as 1901 the Audubon Society of Missouri was established by thirteen people. Shortly after its formation, there was a hiatus in meetings and business until 1933 when the society was rejuvenated; since that time, it has been continuously active. The society's publication, *The Bluebird,* is the principal journal where bird observations for the state are published. Beginning in 1962, observations from across the state were incorporated into an edited summary for each season. Until then, observations were independently published by the observer or by subdivisions within the state. Over the past seventy years, many important observations from the St. Louis area, which were not published elsewhere, have appeared in *Nature Notes,* the journal of the Webster Groves Nature Study Society.

Another careful observer and compiler was Harry Harris of Kansas City. *Birds of the Kansas City Region* (1919b) was the culmination of his work. In 1920 A. E. Shirling wrote *Birds of Swope Park,* which presents the seasonal distribution and ecology of the birds in Kansas City's largest park. J. A. Neff (1923) summarized observations from the then little-known area of the southwestern Ozarks. Rudolf Bennitt's 1932 annotated *Check-list of the Birds of Missouri* was the next compilation of the state's avifauna. Much of the information contained in this work was based on Widmann's and Harris' publications. The status he presented for some species unfortunately was dated (e.g., see his American Swallow-tailed Kite account), and some of the new information was questionable (see crossbill accounts) with many subspecific determinations based on invalid taxa that were widely recognized at that time. Bennitt (1937) also produced the first field card checklist, *Pocket List of Missouri birds,* for the Audubon Society of Missouri.

The first modern summary of the status of birds in the St. Louis region was completed by Eugene Wilhelm in 1958. Wilhelm's detailed work, *Birds of the St. Louis Area,* unfortunately includes some records that subsequent workers feel are questionable. A decade later Richard Anderson and Paul Bauer used a bar graph format in their publication, *A Guide to Finding Birds in the St. Louis Area,* to present the status and temporal pattern of the birds of that region.

In 1966 the Audubon Society of Missouri "charged" David Easterla and Richard Anderson with researching and producing an "official" *Checklist of Missouri Birds.* This list, which designates major regional differences in the status and distribution for each species, was published in 1967 and was revised in 1971 and 1979. In 1986, Mark Robbins joined the authors in producing the current checklist, *Annotated Check-list of Missouri Birds,* published under the auspices of the Audubon Society of Missouri.

Rising et al.'s *Birds of the Kansas City Area* (1978) updates the status of birds in the Kansas City region. *The Status, Distribution, and Habitat Preferences of the Birds of Missouri* (1982), by Richard Clawson, is a summary of the birds recorded in the state between 1970 and 1980. The information is primarily based on a questionnaire sent out to the Missouri birding community.

While the above summarizes most of the major publications on Missouri birds, it does not herald the importance of all the observers who have contributed over the past few decades to *The Bluebird, American Birds,* and *Nature Notes.* Of course, these observations form the backbone of our knowledge of the distribution and abundance of birds in the state over the past half century. Certainly foremost, both in duration and in detail, among contributors was the late Floyd Lawhon of St. Joseph. His daily notes spanned thirty-five years (1952–87). Rivaling Lawhon in duration of observations was James Earl Comfort of the St. Louis area. His contributions covered over three decades (late 1930s through the early 1970s). Other major contributors from the St. Louis area: Wayne Short (1930s and 1940s), James F. Comfort (1930s through 1970s), and Richard Anderson (1950s–). During the past decade the St. Louis area has been fortunate in having a host of outstanding observers. Some of these observers' names (Barksdale, Goetz, Korotev, Peters, Rudden) are seen repeatedly in the species accounts that follow.

One of the premier contributors from the Kansas City area was Walter Cunningham. His contributions were not, however, confined to that region but encompassed several areas of the state between the 1920s and 1950s. Other important observers from the Kansas City region have included John Bishop, Elizabeth Cole, Harold Hedges, Ben King, and Jim Rising; and more recently, Robert Fisher, JoAnn Garrett, Chris Hobbs, Mick McHugh, and Sebastian Patti. Peers of Lawhon's in the northwestern corner who have made significant contributions include Harold Burgess, Leo Galloway, John Hamilton, Simon Rositsky, and Jack Hilsabeck.

Until very recently the southwestern corner of the state had been largely neglected from an ornithological standpoint. However, Pat Mahnkey, Jeff Hayes, and company have been indefatigably rectifying that deficiency. The status and distribution of birds in the eastern Ozarks is rapidly being clarified by Steve Dilks, Bob Lewis, Bill Reeves, and others. Ike Adams, Bill Goodge, and Jim Rathert have been instrumental in gathering information on the status and distribution of birds in the Columbia area. Over a decade ago, the addition of James D. Wilson as the Missouri Department of Conservation's (MDC) ornithologist has greatly facilitated the coordination and cooperation of bird-watchers statewide. More recently, Brad Jacobs joined the MDC staff to help coordinate the "Breeding Bird Atlas."

From the standpoint of breadth and relatively large nongame series, the material accumulated by Easterla at Northwest Missouri State University, Maryville, beginning primarily in the early 1960s, is the most important Missouri collection. This collection contains the state's only extant specimen for over twenty species. The University of Missouri–Columbia collection has the largest holdings of game species. Unfortunately, some specimens have misleading information on the specimen tags; in particular, several preparators dated the tag when the specimen was prepared instead of when the bird was captured. The American Museum of Natural History, New York, the University of Kansas, Lawrence, and the Stevens Natural History Museum at Central Methodist College, Fayette, Missouri, also house valuable Missouri material.

Over the past fifteen years the MDC has been actively involved in determining the status and distribution of nongame species, especially those that are rare and endangered. Studies that they have sponsored range from ascertaining the breeding status of the state's heron populations to determining if a viable habitat still persists for the Red-cockaded Woodpecker (Eddleman and Clawson 1987). An integral part of MDC's sponsored avian studies has been the identification and preservation of critical habitats of these rare and endangered species; large tracts of land have been set aside for ensuring that these species will be present for future generations of bird-watchers.

In order to get a better understanding of the status and distribution of the breeding birds of the state, the MDC initiated a seven-year "Breeding Bird Atlas" in 1986. This ambitious project's goal is to assess which species nest in each of about twelve hundred sampling blocks that are uniformly distributed throughout the state. So far, the project has relied on nearly four hundred volunteers to collect the information.

The Missouri Rare Bird Records Committee (MRBRC), under the auspices of the Audubon Society of Missouri, was formed in October 1987 for the purpose of evaluating current and future rare and unusual records. The Committee's decisions are published in *The Bluebird* as an annual report.

Missouri Birds and Their Environment

The boundaries of the state encompass 68,925 square miles, making it the largest state east of its western border but the third smallest to the west. Map 1 shows the state's 114 counties. The principal localities mentioned in the text are depicted in Map 7 (see p. 31). Relief and climate information are summarized from Collier (1955) and Rafferty (1982).

Map 1.
The 114 Missouri counties.

Relief

Even though Missouri is removed from any of the major mountainous regions of North America, it nevertheless has a fair amount of relief (see Collier 1955, or Rafferty 1982, for an excellent color, graphic representation of relief in Missouri). The most dramatic changes in elevation occur in the southern third of the state. Elevations range from 250 feet (76 m) in the extreme southwestern corner of Dunklin County in the bootheel to 1,772 feet (537 m) on Taum Sauk Mountain in Iron County in the eastern Ozarks. The most extensive area of lowest elevation is the Mississippi Lowlands where elevations generally do not exceed 300 feet (90 m). The greatest area of highest elevation is the broad, rolling plateau in southeastern Webster County where most elevations exceed 1,600 feet (485 m). As one moves from east to west in the northern half of the state, the elevation increases gradually from about 700 feet (212 m) along the Mississippi River to over 1,200 feet (365 m) in the northwestern corner.

Climate

Warm, moist air from the Gulf of Mexico; cold, dry air from the Arctic; and warm, dry air from the west drive the rather dramatic differences in precipitation and temperatures in the state. The Gulf air mass is the primary source of precipitation with as much as two-thirds of the annual moisture coming during the spring and summer. A gradient from the northwest to southeast of increasing rainfall exists. The northwestern corner receives an annual average of about 34 inches (86 cm), whereas the southeastern corner receives as much as 50 inches (125 cm). Although a gradient exists at all seasons, it is steepest in the winter months (December through February) when the northwest may receive an annual average of as little as 2 inches (5 cm), while the southeast may receive an average of up to 12 inches (30 cm). The direction of the gradient is reversed and less pronounced in the summer months when, on average, the northwest receives about 4 inches (10 cm) more than the southeast.

Temperature differences are equally dramatic. In the northern tier counties, the average date of the first killing freeze occurs in mid-October, but the average date does not occur until almost mid-November in the southeastern corner. This disparity is also true for the last killing freeze in the spring: it averages as late as the third week of April in the extreme north but not past the third week of March in the southeast. The average daily maximum and minimum temperature in the north during January is 40° F and 16° F, respectively, whereas the southeastern corner averages a maximum and minimum of 48° F and 28° F, respectively. The average difference between the north and the southeast is less during the hottest month of the year, July. The July daily average maximum and minimum for the extreme north are 90° F and 65° F, respectively. The extreme southeastern part of the bootheel averages 92° F and 71° F.

The Natural Communities

Although various workers have subdivided the state based on one or more features (e.g., geology or flora), Thom and Wilson (1980) were the first to partition the state into natural communities that reflected the ecology and natural history of the fauna. These divisions were superbly illustrated with photographs by Nelson (1985). We have adopted the six major components of these communities that accurately reflect differences in the status and distribution of birds. Below we briefly summarize Thom and Wilson's (1980) divisions; Map 2 depicts the distribution of each. The natural vegetation, as it existed at the time of European settlement, is presented in Map 3.

Glaciated Plains: About one-third of the state is represented by this division. The gently rolling hills, which characterize this entire zone, were formed during the Nebraskan and Kansan Glacial periods (between about 1,800,000 and 600,000 years ago) of the Pleistocene epoch ("Ice Age"). This region is dissected by narrow river drainages that are mostly oriented north to south, except for the extreme eastern section where the orientation is northwest to southeast. Soils are largely composed of loess (especially in the western section), glacial till, and alluvium.

The tallgrass prairie dominated the presettlement landscape of this zone. Well over half of the state's original 18,474 square miles of prairie was found here. Now, with the exception of a few small, isolated patches, the prairie has been completely converted to agricultural land. Likewise, much of the woodland and forest of this division, which was primarily restricted to riparian areas, has diminished since presettlement. Although the overall acreage of forest has decreased since presettlement, many counties in the northwestern and north-central sections now have a larger percentage of forest cover than at presettlement (Rafferty 1982). Despite these dramatic changes over the past 150+ years, a gradient from west to east of increasing rainfall and woodland and forest (Oak-Hickory is the predominate association) occurs today, as it did in presettlement times.

Although the climatic and vegetation gradients described above are subtle, the composition of the avifauna reflects these slight differences. The northwestern corner of this region is the most xeric in the state, and it is here where a number of species, such as the Western Meadowlark and Lark Sparrow, are most common. This relatively dry, open area has been the only site where vagrant Great Plains' species have bred, such as the

Map 2.
The Natural Divisions of Missouri. Adopted from Thom and Wilson (1980).

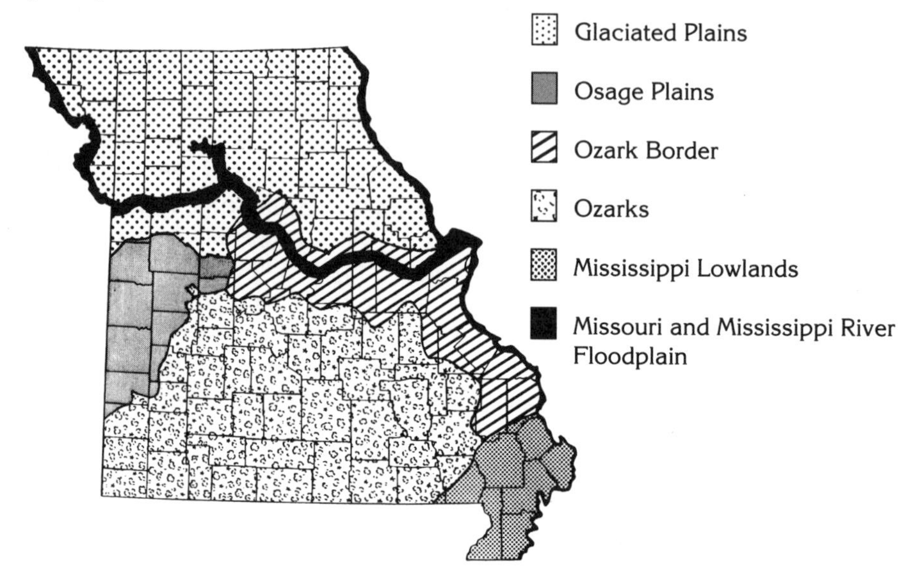

Map 3.
The presettlement vegetation of Missouri. Courtesy of Clair Kucera (1961).

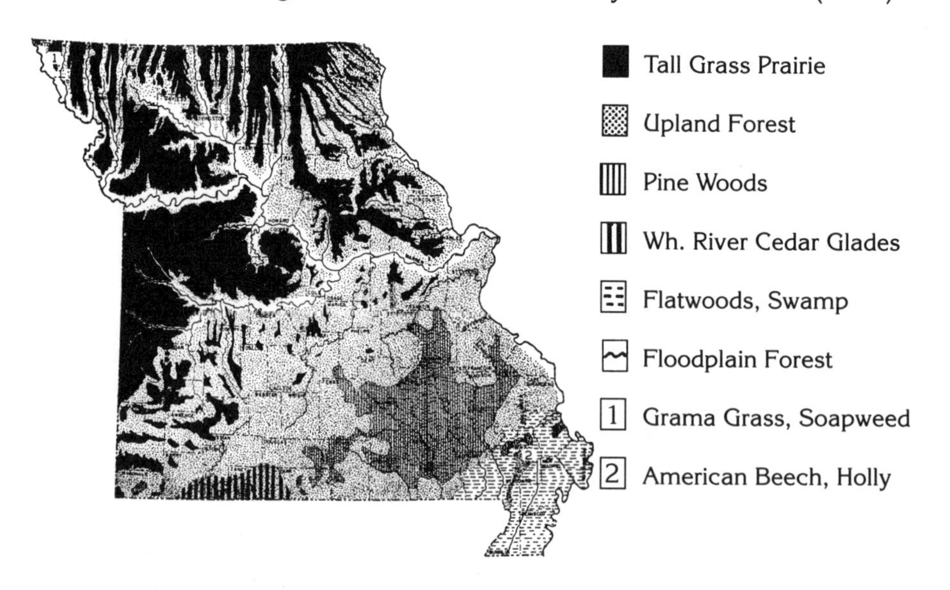

Burrowing Owl and Lark Bunting. Eastward, as the precipitation and vegetation gradually increase, open country birds become less common (e.g., Loggerhead Shrike, Dickcissel, Brown-headed Cowbird), and woodland-edge and forest-inhabiting species increase in abundance (e.g., Red-bellied Woodpecker, Pileated Woodpecker, Blue Jay, Carolina Wren, Rufous-sided Towhee). The following species are primarily restricted to this region either seasonally or year-round: Ring-necked Pheasant, Rose-breasted Grosbeak, Vesper Sparrow, and Bobolink. A number of species that are widespread during the summer in Missouri reach their greatest abundance in this zone: Red-headed Woodpecker, Northern Flicker, House Wren, American Robin, Gray Catbird, Brown Thrasher, American Redstart, and American Goldfinch.

Another micro habitat that frames the western and eastern borders of this division is the bluffs above the Missouri and Mississippi rivers. These are composed of windblown loess from the floodplains. In presettlement times, the slopes above the Missouri River floodplain were principally covered with tallgrass prairie, but today, with the containment of fire, they are largely forested. The vegetation along the Missouri River bluffs is more xeric than that above the Mississippi, with a number of Great Plains floral and faunal species reaching their easternmost distribution here. The Mississippi bluffs continue to be forested, though less extensively, than they were during presettlement. The bluffs along both rivers are highly dissected by deep forested ravines. These ravines are a haven for a number of species that are more localized or even nonexistent a few miles away. For example, the Chuck-will's-widow is known to breed regularly only along these bluffs in the Glaciated Plains. The Acadian Flycatcher, Carolina Wren, Ovenbird, Louisiana Waterthrush, and Summer Tanager are all more prevalent and common here than in most other areas of this division.

Osage Plains: Vegetatively and faunally this division is very similar to the Glaciated Plains, but unlike that zone it escaped glaciation. The 8 percent of the state encompassed within this region is relatively flat. The tallgrass prairie comprised over 70 percent of this division during presettlement. Its forested areas were primarily restricted to riparian areas and along the extreme eastern and southern edges where this zone interdigitates with the Ozarks. Like the Glaciated Plains most of the prairie has been plowed and converted into agricultural land. Virtually all of the remaining 70,000 acres of prairie in Missouri are found in this division.

Most of the species that are indicative of the Glaciated Plains are also found in this region. An exception is the Scissor-tailed Flycatcher, which is primarily restricted to this division. Besides the prairie specialists (Table 1), the following species reach their greatest breeding abundance in this region: Red-tailed Hawk, Northern Mockingbird, Loggerhead Shrike, Bell's Vireo, Dickcissel, and Eastern Meadowlark.

Ozark Border: As the name implies this community is a transition in relief and biota from the abutting natural divisions. It composes roughly 13 percent of the state. Only the Ozarks surpass this division in relief. The dominant vegetation was, and still is, upland deciduous forest (Oak-Hickory). In presettlement Missouri, less than 10 percent of this region was prairie, with most of it located in isolated patches in the western and northern sections. There are no avian species that are indicative of this area, and the relative abundance of many species is intermediate of that in adjacent divisions. This intermediate effect in the relative abundance of species is probably less pronounced now than it was in presettlement times: extensive clearing of the Ozarks has made its vegetational phenology (less forest and more ecotonal habitat) more similar to that of the original Ozark Border. If data concerning relative abundance of breeding were available for this region, they would probably indicate that species most abundant in the forest-edge ecotones (Yellow-breasted Chat, Northern Cardinal) would reach their greatest densities in this zone.

Ozarks: The topography of this division, which was never glaciated, is the most rugged and dramatic in the state, with the highest elevations (highest about 1,800 feet [545 m]) found here. Nearly 40 percent of the state is included in this section. Soils are very thin throughout. In presettlement Missouri, sections of the western portion were interspersed with deciduous forest and prairie, whereas the eastern sections were more uniformly forested and more heavily dissected by streams and rivers. Generally, as one would predict, forest-inhabiting species reach their greatest densities in this division. A few specific examples include: Red-bellied Woodpecker, Pileated Woodpecker, Eastern Wood-Pewee, Tufted Titmouse, Wood Thrush, Red-eyed Vireo, Kentucky Warbler, and Summer Tanager.

Two unique habitats within this division are worth highlighting because they are important to a host of avian species. The first, the Shortleaf Pine (*Pinus echinata*) section, is principally found in the southeastern area of the Ozarks (Map 3). Like most of the state's forest, it was largely cleared around the turn of the century. The last sizable stand (near Round Spring, Shannon County) of virgin pine was cut in 1946; the last colony of Red-cockaded Woodpeckers disappeared with this stand. Species that are unique or reach (or once did) their greatest abundance in this habitat are listed in Table 1.

The second unique subdivision within the Ozarks is what is commonly referred to as the White River Cedar Glades (see Map 3). This area is characterized by extensive limestone glades. The glades are dominated with herbaceous flora and with scattered Eastern Red Cedar (*Juniperus virginiana*); as a result, this habitat is often referred to as "Cedar Glades." Soils are very thin throughout this area. Although no avian species are restricted to this subdivision, several species appear to reach their highest densities here: Greater Roadrunner, Prairie Warbler, Blue Grosbeak, Painted Bunting, and Field Sparrow.

Mississippi Lowlands: The vegetation of this division has perhaps been modified more than any other in the state since presettlement. At one time the entire area (about 5 percent of the state), except for the isolated Crowley's Ridge, was covered with tall, luxurious swampy forest. Now, except for Big Oak Tree State Park (about 1,000 acres), the area has almost completely been converted to agricultural land for crops such as cotton and soybeans. Even with the loss of most of the vegetation, it still has the highest average precipitation and temperatures in Missouri. The area is largely flat, except for Crowley's Ridge, and low in elevation (mostly below 300 feet [90 m]).

In presettlement times, when extensive sections consisted of bald cypress, tupelo, and mixed deciduous bottomland forest, many avian species were restricted to or reached their greatest breeding abundance here (see Table 1 for a few examples). As a result of this entire area being largely denuded, a number of open-habitat-preferring species now obtain their highest relative abundance in the Mississippi lowlands: American Kestrel, Killdeer, Horned Lark, European Starling, Common Yellowthroat, Red-winged Blackbird, Common Grackle, and House Sparrow.

Missouri and Mississippi River Floodplain: Like all the other regions, the floodplains of these two large rivers have been extensively modified since presettlement. At one time, tall riparian forest bordered the rivers, and many sloughs, marshes, and wet prairie were found in the floodplains. Along the upper Missouri, especially in Atchison and Holt counties, wet prairie was the dominant vegetation in the floodplain; however, farther downstream, forest predominated (Schroeder 1981). Forest has always been the dominant vegetation along the banks and in the floodplain of the Mississippi.

Channelization has greatly affected the width and the flow of these two rivers. For example, the Missouri River has been estimated to have lost 50 percent of its original surface area between 1879 and 1972 (Funk and Robinson 1974). Furthermore, greater than 90 percent of the surface area of islands was eliminated between 1879 and 1954. Locks and dams have greatly modified the Mississippi River. A few of the riparian species that reach their greatest relative abundance along these two rivers include the Warbling Vireo, Yellow Warbler, and American Redstart. The only area of the state where the Least Tern is now know to breed is along the lower part of the Mississippi River. Mississippi Kites and Fish Crows are also primarily restricted to the lower Mississippi River floodplain. The upper Missouri floodplain (upstream from Kansas City) is the only area where Western Kingbirds and Yellow-headed Blackbirds regularly breed in the state. Moreover, the only recorded nesting records of a few waterfowl, like the Northern Pintail and Northern Shoveler, are from the latter area.

Changes in Natural Communities and the Avifauna since Presettlement

As mentioned briefly under each Natural Community, the natural vegetation has changed considerably since presettlement. At the time European man first arrived, about 70 percent of the state was covered in forest, and the remaining 30 percent was tallgrass prairie (Map 3). Most of the state's forest was relentlessly cleared between 1890 and 1920, leaving only an estimated 50 percent of the original forest by 1937 (Mayes 1937). At present, about 30 percent of the state is covered with forest (Map 4; Giessman et al. 1986). Although this latter figure demonstrates how extensive the clearing has been, it does not reveal the magnitude of the change in the structure of today's forest. Much of the remaining forest is highly dissected, and virtually none of it is virgin.

The loss of continuous, mature tracts of forest has drastically influenced the status and distribution of many of the state's birds (see Table 1). It has been postulated that the opening of the eastern forests has contributed to the decline in some eastern deciduous forest species (Robbins 1979; Whitcomb et al. 1981; Blake and Karr 1987). Furthermore, the containment of fire, which aids in keeping the understory of forest relatively open, has affected several avian species. Mature stands of Shortleaf Pine are now quite rare, and those without a relatively tall deciduous understory are even scarcer (Eddleman and Clawson 1987; pers. obs.). The lack of fire in such areas has undoubtedly aided in the decline of several species (see Table 1).

The Europeans nearly annihilated Missouri's prairie, most of which has been plowed since the Civil War, reducing the original acreage from an estimated 15 million to now less than 70,000. North of the Missouri River, where about 10,400 square miles of prairie existed prior to presettlement, the only remaining traces of this once vast sea of grass are limited to railroad right-of-ways and a handful of very small, isolated patches. Today, most of the prairie is located in the westernmost counties of the Osage Plains. Even there, the three largest tracts are islands surrounded by thousands of cultivated acres: Taberville Prairie, Osage Prairie (each about 1,400 acres), and Prairie State Park (about 2,500 acres). It is not surprising that a number of species have seriously declined as result of this conversion (see Table 1).

The loss of 90 percent of the state's wetlands since settlement has had a devastating effect on the waterbird community (Reffalt 1985). As a result some species, like the Black Tern, are now extirpated as breeders, and

Map 4.
Current forest cover of Missouri (forested tracts of greater than 160 acres). Courtesy of Giessman et al. (1986).

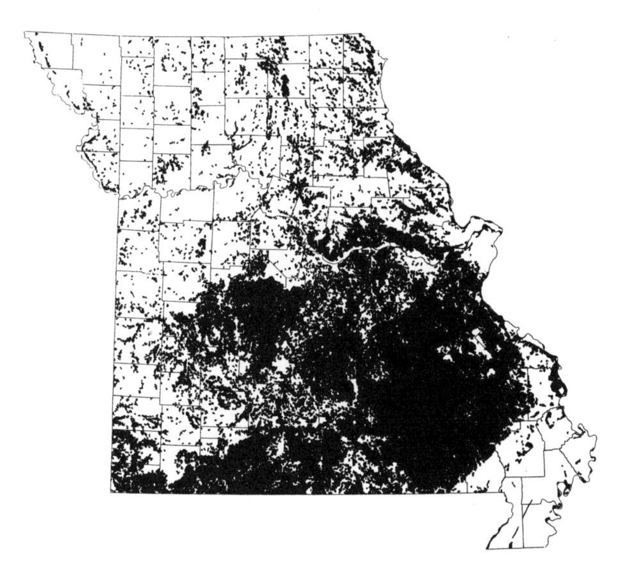

others only occupy a small fraction of their former distributions within the state (King Rail, Least Tern). The draining of marshes and the channelization of rivers not only has negatively affected the breeding community but has significantly reduced key feeding and resting sites for migratory species such as waterfowl and shorebirds.

Since European settlement a total of 191 species have been recorded breeding in Missouri. Four of these are now either extinct or nearly so, and another 28 species are either extirpated through the loss of habitat or only accidentally nest in the state. Over the past decade 159 species have been documented breeding. Approximately 150 species now breed on an annual basis within the state's border. This latter figure is probably slightly higher during wet years when good marsh habitat is available. A breakdown of the top five natural groups (at or below the family level) that annually breed in the state is as follows:

Parulinae (wood-warblers) . . . . 18 species
Ardeidae (herons) . . . . 10 species
Icterinae (blackbirds) . . . . 10 species
Accipitrinae (kites, eagles, hawks) . . . . 9 species
Emberizinae (towhees, sparrows) . . . . 9 species

Although Widmann's (1907) relative abundances for species are not

Table 1.

Species that have declined in Missouri since 1900 due to man's alteration of their primary breeding habitat.

Species marked with an asterisk (*) once had their primary breeding range restricted to riparian areas in the Mississippi Lowlands. Species marked with a plus sign (+) are primarily restricted to riparian habitat but are found (or were found) elsewhere in the state. Unmarked species are either principally found in upland forest or commonly occur in upland and riparian forest.

Deciduous Woodland and Bottomland/Riparian Forest

Double-crested Cormorant*
Anhinga*
Black Vulture
Hooded Merganser+
Osprey+
American Swallow-tailed Kite*
Mississippi Kite*
Bald Eagle+
Red-shouldered Hawk+
Barred Owl+
Pileated Woodpecker
Ivory-billed Woodpecker*
Eastern Wood-Pewee
Acadian Flycatcher+
Brown Creeper*
Wood Thrush
Warbling Vireo+
Red-eyed Vireo
Bachman's Warbler*
Yellow-throated Warbler+ (also see Shortleaf Pine)
Cerulean Warbler+
Black-and-white Warbler
American Redstart+
Prothonotary Warbler+
Worm-eating Warbler
Swainson's Warbler*
Ovenbird
Kentucky Warbler
Hooded Warbler+

Shortleaf Pine

Red-cockaded Woodpecker
Brown-headed Nuthatch
Yellow-throated Warbler
Pine Warbler
Bachman's Sparrow
Chipping Sparrow

Prairie

Northern Harrier
Greater Prairie-Chicken
Upland Sandpiper
Short-eared Owl
Sedge Wren
Dickcissel
Grasshopper Sparrow
Henslow's Sparrow

Marshes

Pied-billed Grebe
American Bittern
Least Bittern
King Rail
Common Moorhen
American Coot
Black Tern
Marsh Wren
Yellow-headed Blackbird

empirically supported, they nonetheless serve as a baseline against which comparisons can be made between the state's avifauna prior to 1900 and now. Those breeding species that have clearly either declined or increased considerably since before 1900, as a result of man's modification of the environment, are indicated in Tables 1 and 2.

Format and Relative Abundance Definitions for the Species Accounts

Taxonomy and nomenclature follow the *A.O.U. Check-list of North American Birds* (1983) and subsequent supplements: 35th (1985), 36th (1987), and 37th (1989) published in the *Auk.* We deviate from the above in only a single instance (see vultures). The Order and Family names proceed all groups. Each species account begins with the Common and Scientific names. Immediately below these names the Status is given. The following definitions have been used to depict the status of species.

Temporal Occurrence

Permanent Resident: Individuals of a species are present throughout the year, although different populations may be involved at different times of the year. If status varies throughout the year, it is indicated. Example: Blue Jay.
Summer Resident: Present as a breeder. Example: Bell's Vireo.
Summer Visitant: Irregularly present in small numbers; breeding has not been documented; includes postbreeding species (especially Ciconiiformes) and possibly some late spring and early fall migrants. Examples: Purple Gallinule and White Ibis.
Winter Resident: Present for most of the winter. Example: White-throated Sparrow.
Winter Visitant: Generally refers to species that appear infrequently and then only for a short period; typically involves late fall migrants that have lingered into the winter. Example: Indigo Bunting.
Transient: A migrant that passes through the state during spring and/or fall. Example: White-rumped Sandpiper.
Regular: Predictable; seen annually in the state and/or season. Example: Ruddy Turnstone.
Irregular: Not recorded annually; usually refers to cyclic and erratic species. Examples: Northern Goshawk, Snowy Owl, northern fringillid finches.

Table 2.

Breeding species that have increased in Missouri since 1900 due to man's alteration of the state's natural communities.

Alteration of Natural Habitats	*Artificial Structures*	*Introductions*
Blue Jay	Common Nighthawk	Gray Partridge
Northern Mockingbird	Chimney Swift	Ring-necked Pheasant
Northern Cardinal	Purple Martin	Rock Dove
Great-tailed Grackle	Northern Rough-winged Swallow	European Starling
Common Grackle	Bank Swallow	House Finch
Red-winged Blackbird	Cliff Swallow	House Sparrow
Brown-headed Cowbird	Barn Swallow	Eurasian Tree Sparrow
Orchard Oriole	House Wren	

Relative Abundance Criteria

Common: Encountered daily in relatively large numbers, more than 10 individuals/day. Example: Northern Cardinal.
Uncommon: Observed on most days in relatively low numbers, 1–10 individuals/day; sporadically seen in relatively large numbers. Example: Scarlet Tanager.
Rare: More than 15 records; infrequently recorded and usually found in small numbers. Examples: Whimbrel and Sharp-tailed Sparrow.
Casual: 5–15 records; only occasionally reported. Examples: Red-throated Loon; Sedge Wren in winter.
Accidental: 1–4 records; a vagrant. Examples: Slaty-backed Gull; Clay-colored Sparrow in winter.
Extirpated: Formerly occurred in the state and/or at a particular season, but no longer recorded. Examples: Red-cockaded Woodpecker and Brown-headed Nuthatch.
Extinct: Species no longer exists. Example: Carolina Parakeet.

The Documentation line follows the Status statement. Where a specimen is available it takes precedence over the other categories of documentation. The following criteria were used for selection of specimens and photographs: the significance of the record, its physical condition, and the likelihood that it will still be extant in the future (specimens deposited in institutions that have a good curatorial legacy have been given higher priority). Categories for Documentation are as follows:

1. **Specimen:** An extant specimen is deposited at a museum or university. See Abbreviations at the beginning of the Species Accounts for institutions' initials. The catalog number is given after the institution's abbreviation; for the Northwest Missouri State University (NWMSU) collection, the preparator's number is included.
2. **Photograph:** An extant photograph is deposited in either VIREO (a photographic, archival collection at ANSP; see Myers et al. 1986), or the Missouri Rare Bird Records Committee File, or has been published.
3. **Sight Record(s) Only:** Since the MRBRC was not established until October of 1987, we have established criteria for acceptance of a species on the state list based only on sight records that predate the records committee's formation.
   a. All single party, first state sightings have automatically been relegated to the Hypothetical list. Thus, a minimum of at least two separate reliable parties must see the bird(s) or there must be at least two separate, reliable sightings before a species is added to the list.
   b. First state observations that involve more than one party have been individually evaluated based on the following criteria: credibility of the observer(s); observer(s) prior experience with the species; the level of difficulty in the species identification; the length of time the species was under observation; the amount of detail recorded at the time of observation; the likelihood of occurrence; and the circumstances of the observation(s), such as weather conditions.
4. **Hypothetical:** Does not meet the above criteria, but the record has some reasonable cause for being considered. With the exception of two highly questionable species (Lesser Prairie-Chicken and Northern Hawk Owl), which we include herein because they have been repeatedly listed in prior works, all other questionable records are excluded. This category also includes species definitely recorded, but whose origins are questionable; they may be escapees (e.g., Greater Flamingo and Barnacle Goose). Brackets denote a species of hypothetical status.

The sex and age (if known), date, locality, observer(s), and the published reference (if any) are given for all documentations. If the species has been documented with a specimen or photograph, then the institution of deposition is given. Abbreviations for institutions and observers are included in the Abbreviation and Observer lists that immediately precede the Species Accounts. The following boldfaced abbreviations are used for references to the primary literature. The volume and page numbers follow each of these references. *The Bluebird* citations include the issue number (placed in brackets) because the number of issues per year has varied for this publication. Pagination has also been inconsistent for this journal—at times each issue started with page 1.

American Birds- **AB**
Audubon Field Notes- **AFN**
Audubon Magazine- **AM**
The Bluebird- **BB**
Bird Lore- **BL**
Nature Notes- **NN**

A brief statement on Habitat follows the Documentation line. This statement gives the primary habitat in which the species is found. If breeding or wintering habitat differs from that of migration, it is so stated.

The Records section is presented next. The four seasons, beginning with spring and ending with winter, have arbitrarily been divided as follows (following Easterla et al. 1986):

Spring Migration: 21 February through 31 May.
Summer: 1 June through 31 July.
Fall Migration: 1 August through 14 December.
Winter: 15 December through 20 February.

Under each season a synopsis is given of any major movements that occur. This includes information on the known earliest and latest dates, as well as high counts (if available) for the period. The most significant observation is listed first under each of these categories, the earliest date given first followed by the second. When specific records are listed, the number of birds observed, date, location, observer(s), and reference (if published, see above for boldface reference abbreviations) are given. When there are more than one observation for the same season and year, a single citation is given per species account. If the two earliest spring dates for a species occurred during the same year, only one of the records is referenced.

If the species is migratory, and there is a decided difference in arrival to various regions of the state, it is so stated. For those species whose overall or seasonal status is accidental or casual, the specific records (up to and including five observations) are listed in chronological order (by day within month) for each season.

For some species a Comments section follows the Records summary. Remarks on subspecies, identification, behavior, taxonomy, and distribution patterns are included in this section. For a few species, such as those that are extinct or extirpated, or for species whose position on the state list is based on only a single record, the Comments section may be the only section following the Documentation line.

The Data

As mentioned above, since the MRBRC was not established for reviewing rare records until October 1987, the burden of critiquing the pre–1987 body of literature fell upon us. Unfortunately, not until the past couple of decades have observers been encouraged to include written details for unusual sightings. In those cases in which observations lacked any details, we relied primarily on the observers' reputation for evaluation. A person's reputation was based on our personal experience with the observer, or on conversations with reliable people who had field experience with the observer, or on published observational patterns of the observer. Observations of a person who repeatedly reported unusual sightings that no one else saw have been excluded.

Literature

Beginning in November of 1984, a systematic review of the literature began in preparation for this book. Initially the two most important publications for bird sightings in the state were perused, *The Bluebird* and the more regional (St. Louis area) *Nature Notes.* These were followed by *Audubon Field Notes,* and *American Birds,* as well as their precursor, *Bird Lore.* In addition, the three major American ornithological journals, *Auk, Condor,* and *Wilson Bulletin,* were reviewed. Other publications thoroughly scanned include *Journal of Field Ornithology* (formerly known as *Bird-Banding*), *Transactions of the Missouri Academy of Science,* and the *Oologist.* Naturally, Widmann (1907), Harris (1919b), Bennitt (1932), Anderson and Bauer (1968), and Rising et al. (1978) were consulted continually during the writing phase. Format and abbreviations for citations are given above under the Documentation section.

Unpublished Field Notes

Field notes from various observers throughout the state extensively supplemented (and corrected) published records. As mentioned in the "History of Missouri Ornithology," Floyd Lawhon's *daily,* detailed notes (primarily of birds in the northwestern corner) were a keystone of this book. Our unpublished material was used to amplify Lawhon's northwestern observations and to supplement published information on other regions of the state. Patterns in the Mississippi lowlands were elucidated with Jim Haw's notes, along with those of Paul Heye and Bill Reeves. Randy Korotev provided 12 years of data on passerine migration through Forest Park, St. Louis. In defining the range of dates chosen for each species, the early and late dates were excluded; this limited the birds/hour analysis to each species' main migration period. For example, although Korotev recorded the Blackpoll Warbler from 19 April through 31 May, we restricted the analysis to the dates, 30 April through 26 May, as when he consistently observed this species. We hope that observers in other sections of the state will follow Korotev's lead and record their observations in a similar, detailed manner (see Korotev 1990 for methods). A number of other people extracted observations from their notes (see Acknowledgments).

Breeding Bird Survey Data

The U.S. Fish and Wildlife Service's North American Breeding Bird Survey (BBS) data were used to determine the relative abundance of species among the natural divisions in Missouri and to track changes in the state's breeding populations. These roadside counts, which involve making a total of fifty, 3-minute stops every half mile along a predetermined route (see Bystrak 1981), were initiated in 1967 in Missouri. A total of 37 routes (Map 5) have been established in Missouri, representing all major physiographic regions of the state. There has been considerable variability in the number of times each route has been conducted, ranging from only 4 to 21 (through 1987; Table 3).

All routes were used (through 1987; maps were generated before the 1988–89 data were received) for ascertaining the relative abundance of a species among the natural divisions. Means for each species/route were determined and then pooled with means for all other routes in the same natural division. Table 3 depicts which runs were used for each of these divisions. Ozark and Ozark Border data were combined because the three routes that traverse the Ozark Border also spill over into adjacent natural divisions. Thus, these three have been lumped with the Ozark region (n=18 routes). A total of 14 routes represents the Glaciated Plains region, and for those species whose relative abundance differed significantly within this zone, these data have been further subdivided into east (routes 19, 25, 26, 27, 28, 33, 34) vs. west (routes 29, 30, 31, 32, 35, 36, 37) and/or north (routes 33, 34, 35, 36, 37) vs. south (routes 19, 25, 26, 27, 28, 29, 30, 31, 32). Four routes (16, 22, 23, 24) were used to express the relative abundance of each species in the Osage Prairie. Only a single route exists for the Mississippi Lowlands.

The above data are presented in map form, using unique shade patterns to represent differences in the average mean (expressed in birds/route) for all routes in the same natural division. When the relative abundance for a species is similar in two natural divisions, a range is given and the same shading pattern is used for both.

Map 5.
Missouri Breeding Bird Survey Routes.

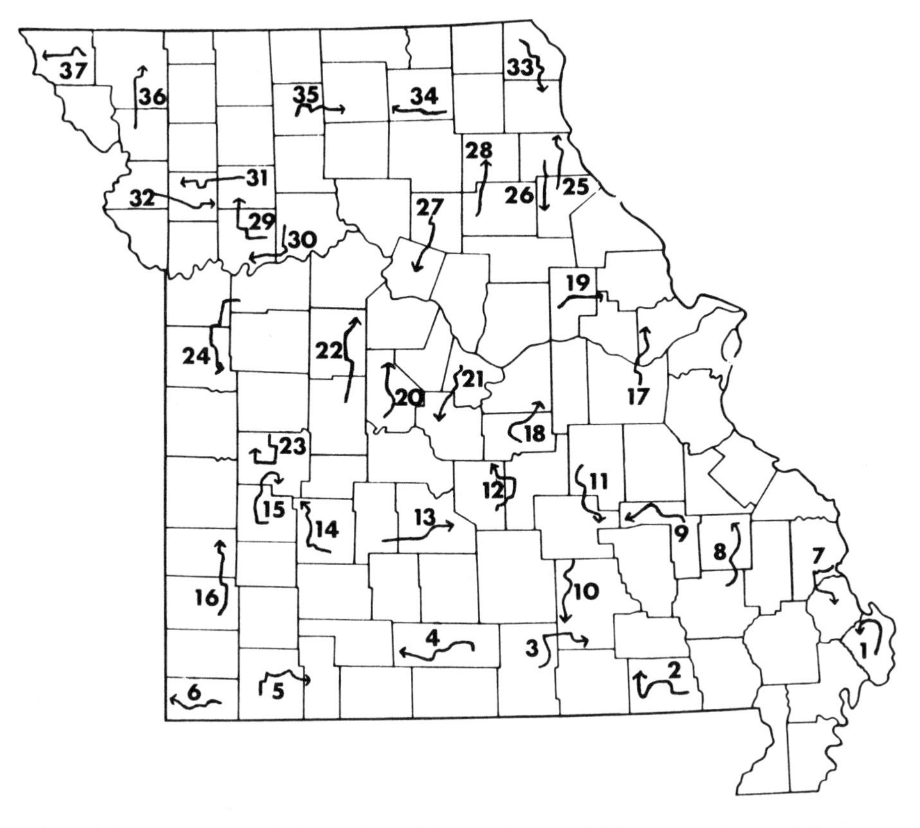

In addition, we used these data (through the 1989 season) to exhibit, in graph form, population trends for several species. The number of birds/route, using only those routes that ever recorded this species, was used for generating an annual statewide mean. Even though generating means based on only those censuses in which the species were recorded is more robust than calculating means for all routes, there are, nevertheless, some biases with using this data subset.

These biases include differences in the number of times a route has been run. During the early years (mid-1960s to early 1970s) a number of Missouri routes were not conducted. Additionally, some routes were consistently run by the same observer, while others were done by a number of people. Recently, the U.S. Fish and Wildlife Service has attempted to address these problems by applying more complex statistics to these data referred to as the Annual Index (see Appendix E in Robbins et al. 1986). We have elected to use the more simplified data analysis (average number of

Table 3.

Missouri Breeding Bird Survey Routes. Routes are listed under the appropriate natural division along with the number of years (through 1987) that each has been conducted. *See Map 6 for location of routes.*

Glaciated Plains

Route #	19	25	26	27	28	29	30	31	32	33	34	35	36	37
Years run	21	4	6	14	18	16	15	10	21	7	16	8	18	16

Ozarks

Route #	2	3	4	5	6	7	8	9	10	11	12	13	14	15	17	18	20	21
Years run	8	12	17	16	8	15	15	9	20	16	12	20	19	10	11	13	18	15

Ozark Plains

Route #	16	22	23	24
Years run	12	13	14	17

Mississippi Lowlands

Route #	1
Years run	16

birds/route that ever recorded this species) for two reasons: the same population fluctuations are exhibited by both analyses, and the majority of this book's audience will find the average number of birds/route that ever recorded this species statistic more understandable than the Annual Index.

Users of these data should remember that the BBS technique involves censusing from roadsides, so it inherently contains an "edge effect" bias. Those species most common in ecotones are oversampled, whereas strictly forest-inhabiting species are underrepresented. Moreover, the sampling occurs primarily during June, when some species have already completed the nesting cycle or are vocalizing very little, such as woodpeckers and titmice. Furthermore, this is a poor technique for censusing colonial (herons, swallows) and nocturnal species (owls, goatsuckers). Finally, at least with the Missouri data, riparian species are especially underrepresented.

Christmas Bird Count Data

We used Christmas Bird Count (CBC) data to determine early winter relative abundance for species among natural divisions and population trends in the state. All counts used in these analyses are listed, along with the number of years each count was conducted, under the appropriate natural division (Table 4). The location of each is depicted in Map 6. Two criteria were used in selecting counts for these analyses. First and foremost, the reliability and consistency of the results were considered. All count results were reviewed by us because the quality of editing CBCs has varied considerably. Three counts, which are located in key areas of the state, have been persistently plagued with errors and were omitted from these analyses. Second, to reduce bias from uneven coverage of counts, we only used counts run for a minimum of 15 years in the period from 1968–69 through 1986–87. Thus, the number of years counts were run ranged from 15 to 19. Counts prior to 1968 were excluded from the analyses because the variance in years run and effort per count greatly increased prior to the 1968–69 counts. (Note: information from counts conducted prior to this date was incorporated in the species accounts. For example, high counts and unusual species).

Using the above criteria, one of the natural regions, the Osage Plains, was not represented, and the relative abundance for species in the Ozark Border and Ozark areas was derived from only three counts each. Furthermore, the Glaciated Plains region is highly biased toward the western half (west vs. east, 5 to 1 counts). Finally, the Mississippi lowlands is represented by only a single count. In generating the maps with the abundance results, the mean for the Glaciated Plains was also used for the Osage Plains because the Osage Plains is most like the Glaciated Plains in terms

Table 4.

Missouri Christmas Bird Counts. Only counts that were used in generating winter relative abundance figures are included. Only data from the 1968–69 through 1986–87 counts were used. See Map 7 for location of counts.

Glaciated Plains

Count Name	Years run
Kansas City North	15
Kansas City Southeast	19
Maryville	19
Orchard Farm	19
St. Joseph	19
Squaw Creek NWR	18

Ozark Border

Count Name	Years run
Columbia	19
Gray Summit	19
Weldon Springs	19

Ozarks

Count Name	Years run
Mingo NWR	18
Springfield	19
Sullivan	17

Mississippi Lowlands

Count Name	Years run
Big Oak Tree SP	19

of vegetation and birds. This is especially true because the Glaciated Plains results are heavily biased toward the western half (see above), which abuts with the Osage Plains. However, we suspect that if data were available, a slight cline would exist in the winter abundance for some species from the northern edge of the Glaciated Plains to the southernmost Osage Plains. For example, we believe that the following species are probably more common at the southern end of the Osage Plains than at the northern end of the Glaciated Plains: Eastern Bluebird, Carolina Wren, Northern Mockingbird, and Field Sparrow.

The relative abundance for a species in each region was derived by taking a mean (expressed in birds/party hour; Raynor 1975; Falk 1979; Confer et al. 1979) for each count and then pooling the means across counts within a region. For example, a mean was calculated for the

Map 6.
Location of Missouri Christmas Bird Counts used in expressing early winter relative abundance of species.

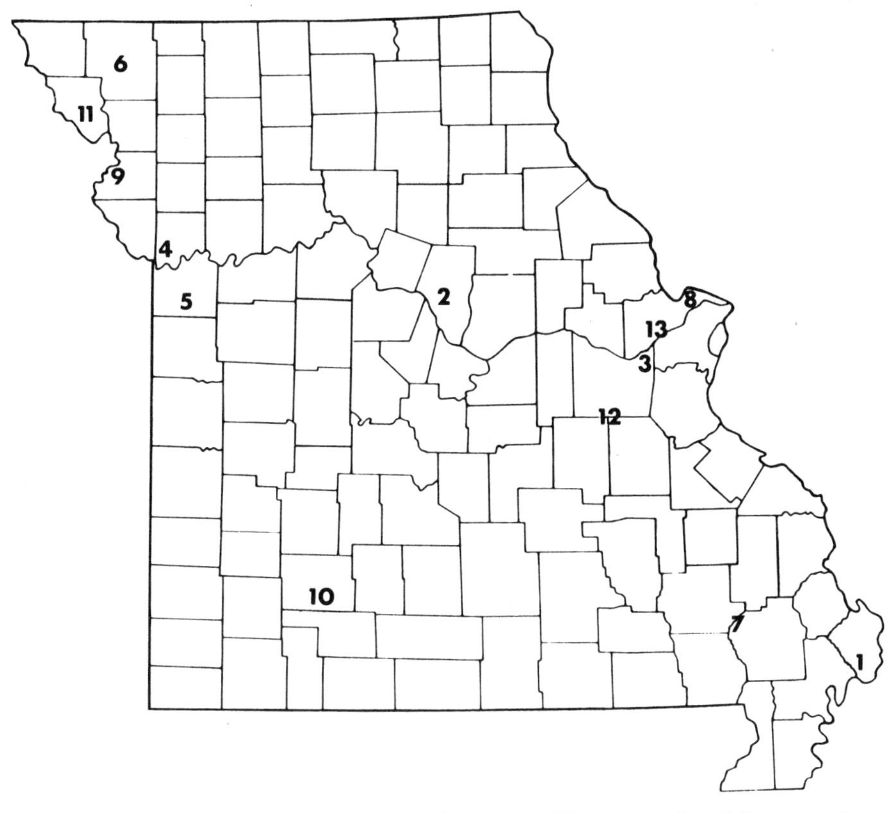

number of Blue Jays recorded on the Maryville count (n=19 years; between 1968 and 1986). This mean (2.0 birds/party hour) was then pooled with the means for the other five counts within the Glaciated Plains to give an average figure (2.3 birds/party hour) for that division (Map 33). If the relative abundance mean for a species is similar for two natural divisions, a range is given and included under the same shading pattern (see White-breasted Nuthatch, Map 37).

Specimen Data

All major U.S. natural history museums, as well as most Missouri universities, colleges, and museums, were either contacted or visited by us for assessing their holdings of Missouri material (see Acknowledgments). Computerization of the Academy of Natural Sciences, Philadelphia

(ANSP), Carnegie Museum of Natural History (CM), Chicago Academy of Sciences (CAS), United States National Museum (Smithsonian, USNM), Iowa State University (ISU), University of Kansas (KU), University of Missouri–Columbia (MU), University of Nebraska, and the Oklahoma Museum of Natural History collections facilitated specimen searches. The authors visited the following institutions' collections: ANSP; American Museum of Natural History (AMNH); Central Methodist College, Fayette, Missouri (CMC); Central Missouri State University (CMSU); Delaware Museum of Natural History; Missouri Department of Conservation (Fish and Wildlife Unit, MDC), Northwest Missouri State University (NWMSU); Southeast Missouri State University (SEMO); Gaylord Memorial Lab, Puxico, Missouri (GL); and the University of Missouri–Columbia.

Information on the collection at the St. Louis Science Center (STSC, formerly known as the St. Louis Academy of Science) was provided by Richard Anderson, who scrutinized this collection in 1973.

Abbreviations Used in Species Accounts

A. Observers' initials and corresponding names that appear five or more times in the Species Accounts.

A

IA-Ike Adams
KA-Kathryn Arhos
RA-Richard Anderson

B

BB-Bob Brown
RBE-Ron Bell
BBR-Bob Bright
CB-Catherine Bonner
FB-Felicia Bart
HB-Harold Burgess
JB-John Bishop
JVB-Jack Van Benthuysen
MB-Mike Braun
PB-Paul Bauer
RB-Robert Brundage
TB-Tim Barksdale

C

DC-Dick Coles
EC-Elizabeth Cole
JC-James F. Comfort
JEC-Jim Earl Comfort
WC-Walter Cunningham

D

GD-Gerald Dobbs
SD-Steve Dilks

E

BE-Bill Eddlemann
DAE-David A. Easterla
JE-Joey Eades
TE-Todd Easterla

F

HF-Hal Ferris
NF-Nathan Fay
RF-Robert Fisher

G

BG-Bill Goodge
JG-JoAnn Garrett
KG-Ken Granneman
LG-Leo Galloway
RG-Ron Goetz
WG-Wally George

H

CH-Chris Hobbs
HH-Henry Harford
JH-Jim Haw
JHA-John Hamilton
JHI-Jack Hilsabeck
KH-Kelly Hobbs
PH-Paul Heye
SHA-Steve Hanselmann
SH-Steve Hilty

J

BJ-Brad Jacobs
DJ-David Jones
JJ-Joan Jefferson
NJ-Nanette Johnson

K

BK-Ben King
RK-Randy Korotev

L

BL-Bob Lewis
CL-Chris Lundberg
EL-Eugenia Larson
FL-Floyd Lawhon
WL-Wade Leitner

M

KM-Keith McMullen
MMH-Mick McHugh
PM-Pat Mahnkey
PMC-Paul McKenzie

P

CP-Carmen Patterson
MP-Mark Peters
TP-Tom Parmeter
SP-Sebastian Patti

R

BR-Bill Rudden
BRE-Bill Reeves
BRO-Bill Rowe
JR-Jim Rathert
JRI-Jim Rising
MBR-Mark B. Robbins
RAR-Richard A. Rowlett
SR-Simon Rositsky
SRU-Skip Russell

S

CS-Claudia Spener
JS-Jerry Sowers
MS-Mildred Schaeffer
PS-Phoebe Snetsinger
SS-Sherman Sutter
WS-Wayne Short

T

DT-Drew Thate

W

JW-Jim Wilson
RW-Rea Windsor

Z

JZ-Jim Ziebol

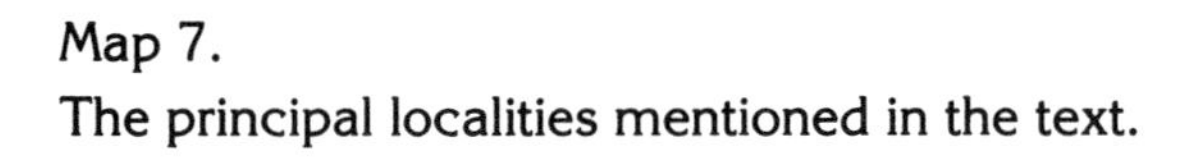

Map 7.
The principal localities mentioned in the text.

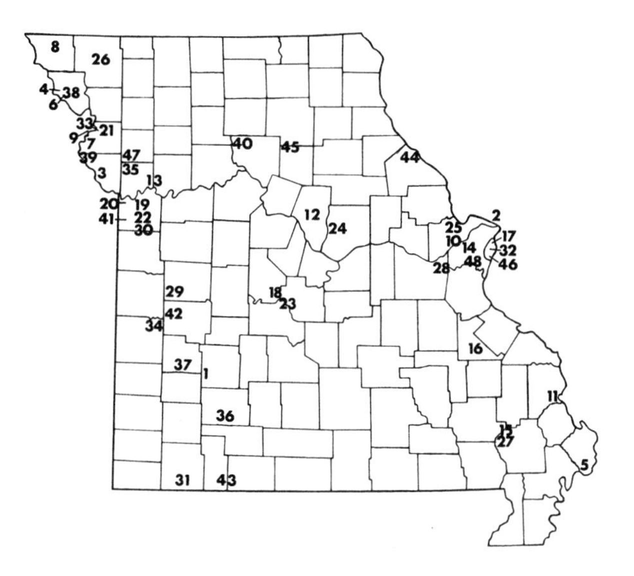

B. List of abbreviated localities that are frequently mentioned in the Species Accounts. Localities are listed in alphabetical order. Numbers correspond to Map 7.

1. Aldrich, Polk Co., southeastern arm of Stockton Lake.
2. Alton Dam, St. Charles Co., on Mississippi River. As of 1990 this dam is no longer functional and has been replaced by another dam, the Mel Price Dam, a few miles down river.
3. Beverly L., Platte Co., Missouri River bottom.
4. Big L.SP, Holt Co., Missouri River bottom.
5. Big Oak Tree SP, Mississippi Co.
6. Bigelow Marsh, Holt Co., located just northwest of Squaw Creek National Wildlife Refuge and Bigelow.
7. Bluffwoods SF, Buchanan Co.—state forest located south of St. Joseph along the Missouri River bluffs.
8. Brickyard Hill WA, Atchison Co.
9. Browning L., Buchanan Co., along Missouri River, at St. Joseph.
10. Busch WA, August A. Busch WA, St. Charles Co.
11. Cape Girardeau, Cape Girardeau Co.
12. Columbia, Boone Co.
13. Cooley L.—wildlife area located in the extreme southeastern corner of Clay Co.

14. Creve Coeur L., St. Louis Co.
15. Duck Creek—wildlife area that straddles Bollinger and Stoddard counties.
16. Farmington, St. Francois Co.
17. Forest Park—within St. Louis city limits.
18. Gravois Mills. Almost all references are to an area ca. 8–10 miles south of Gravois Mills, Morgan Co., on or adjacent to Lake of the Ozarks, Mill Creek Arm.
19. Independence, Jackson Co.
20. Kansas City, Jackson Co.
21. L. Contrary, Buchanan Co., located on southern outskirts of St. Joseph.
22. L. Jacomo, Jackson Co.
23. L. of the Ozarks SP, Miller Co.
24. Little Dixie L., Callaway Co., east of Columbia.
25. Marais Temps Clair, St. Charles Co., near Orchard Farm.
26. Maryville SL—sewage lagoons located on the eastern outskirts of Maryville, Nodaway Co.
27. Mingo—national wildlife refuge which straddles Bollinger and Stoddard counties.
28. Missouri Botan. Arboretum, Franklin Co., near Pacific.
29. Montrose, Henry Co.—wildlife area located north of the town of Montrose.
30. Reed WA, James A. Reed WA, Jackson Co.
31. Roaring R. SP, Barry Co.
32. St. Louis, St. Louis Co.
33. St. Joseph, Buchanan Co.
34. Schell-Osage—wildlife area located in northeastern corner of Vernon Co.
35. Smithville L., Clay Co.
36. Springfield, Greene Co.
37. Stockton L.—straddles Cedar and Dade counties.
38. Squaw Creek—national wildlife refuge, Holt Co.
39. Sugar L., Platte Co.—Missouri River bottoms.
40. Swan L.—national wildlife refuge, Chariton Co.
41. Swope Park—within Kansas City city limits, Jackson Co.
42. Taberville Prairie, St. Clair Co.—located north of Taberville.
43. Table Rock L.—straddles Barry, Stone and Taney counties; most references in text are to the dam area in Stone Co.
44. Ted Shanks—wildlife area along Mississippi River in Pike Co.
45. Thomas Hill Res.—straddles Macon and Randolph counties. Most shorebird references are to the north end of the reservoir, while most others are to the dam area and the southern end.
46. Tower Grove Park—located in St. Louis city limits.
47. Trimble—wildlife area that was formerly located in extreme south-

western Clinton Co., until the formation of Smithville Lake. Now located in the extreme northwestern corner of Clay Co.
48. Tyson Valley RC—research center located in St. Louis Co.

C. Abbreviations for museums. Listed alphabetically by initials.

AMNH—American Museum of Natural History, New York.
ANSP—Academy of Natural Sciences, Philadelphia.
CAS—Chicago Academy of Sciences, Chicago.
CM—Carnegie Museum of Natural History, Pittsburgh.
CMC—Central Methodist College, Fayette, Missouri.
CMSU—Central Missouri State University, Warrensburg.
DMNH—Denver Museum of Natural History, Denver.
GL—Gaylord Memorial Lab, Puxico, Missouri.
ISU—Iowa State University, Ames.
KU—University of Kansas Museum of Natural History, Lawrence.
LSUMNS—Louisiana State University Museum of Natural Science, Baton Rouge.
MCZ—Museum of Comparative Zoology, Harvard University.
MU—University of Missouri-Columbia.
NWMSU—Northwest Missouri State University, Maryville.
SEM0—Southeast Missouri State University, Cape Girardeau.
SJM—St. Joseph Museum, St. Joseph, Missouri.
STSC—St. Louis Science Center, formerly known as the St. Louis Academy of Science, St. Louis, Missouri.
USNM—United States National Museum, Smithsonian, Washington, D.C.
WFVZ—Western Foundation of Vertebrate Zoology, Los Angeles.

D. Miscellaneous Abbreviations

fide—according to
m.ob.—multiple observers; used when the reference did not include the primary observers
n—sample size
no.?—number of individuals unknown
o?—sex of specimen undetermined
pers. comm.—personal communication

Species Accounts

Order Gaviiformes

Family Gaviidae: Loons

Red-throated Loon (*Gavia stellata*)

Status: Casual transient in east; accidental in west.
Documentation: Specimen: o?, Dec 1937, Laclede Co. (SEMO 32).
Habitat: Rivers, lakes, and reservoirs.
Records:
Spring Migration: Only three records: 1, 7 Mar 1981, Thomas Hill Res. (SS, TB-**BB** 48[3]:16); 1, nuptial plumage, 30 Mar 1958, Sugar L. (WC-Rising et al. 1978); 1, nuptial plumage, 16 May 1971, Kirksville, Adair Co. (R. Luker).

Fall Migration: There are no fewer than eight records, all between late Oct and early Dec. Seven of the records have been obtained since 1970, and five of these are from the St. Louis area. Whitman (1907) mentions two basic plumaged specimens that were taken on 3 Nov 1902, near New Haven, Franklin Co., but these specimens apparently are no longer extant. Latest record: adult, 21 Oct–2 Dec 1990, Thomas Hill Res. (TB, PMC et al.).

Pacific Loon (*Gavia pacifica*)

Status: Accidental transient.
Documentation: Specimen: male, 23 Oct 1969, Browning L. (Easterla and Lawhon 1971; NWMSU, CLJ 149).

Habitat: Rivers, lakes, and reservoirs.

Records:

Spring Migration: Only one record for this season, which fits the general pattern for this loon in the interior of the country; it is much rarer in spring than in fall. An immature was first discovered on 23 Feb 1990 at Table Rock L. (CL, H. Hedges); it was joined by another bird on 7 Mar (Lundberg 1990). At least one of the birds remained until 9 May when it was still in basic plumage (CL et al.-**BB** 57[2]:101).

Fall Migration: Only four records: The first sighting was of 8 individuals south of Gravois Mills, L. of the Ozarks, 27–29 Nov 1964 (Easterla 1965b). The second was the above mentioned specimen that was present 19–23 Oct (Easterla and Lawhon 1971). Three were observed 15–17 Nov 1983, Thomas Hill Res. (PS, DJ, TB et al.-**BB** 51[1]:39). The most recent observation was of a single (possibly two birds were involved) that was present 13–23 Nov 1987, Big L. SP (B. Rose, DAE et al.-**BB** 55[1]:9; photos in MRBRC file).

Comments: The A.O.U. Check-list Committee (1985) now treats the former circumpolar Arctic Loon as two species. The population that primarily breeds in North America is now called the Pacific Loon (*G. pacifica*), whereas the Eurasian populations, consisting of two subspecies, are referred to as the Arctic Loon (*G. arctica*). The eastern population of the Arctic Loon (*G. arctica viridigularis*) does breed in western Alaska (Seward Peninsula) and could conceivably appear as a vagrant in Missouri. Walsh (1988), along with clarifications by Roberson (1989) and Schulenberg (1989), address how to distinguish these loons.

Common Loon (*Gavia immer*)

Status: Uncommon migrant; casual summer visitant; rare winter resident in south.

Documentation: Specimen: female, 29 Nov 1968, Gravois Mills, L. of the Ozarks (NWMSU, DLD 89).

Habitat: Rivers, lakes, and reservoirs.

Records:

Spring Migration: Birds begin appearing by mid-Mar and become more common in Apr. The largest single day counts have been recorded in late Apr and early May, with occasional stragglers through the end of May. High counts: 50, 4 May 1937, Mississippi R., St. Charles Co. (WS-**BB** 4[6]:62); 6, 4 May 1975, Springfield (NF-**BB** 43[1]:17).

Summer: Most of the records for this period are during the first two weeks of June and are of single, basic plumaged birds; however, there are records of birds lingering throughout the season: 4, 21 June–31 July (with two until 6 Aug) 1976, Fellows L., Springfield (CB, NF-**BB** 44[2]:19); 2, July–Aug 1977, Springfield (CB-**BB** 45[1]:23).

Fall Migration: Although there is an additional Aug record besides those listed above, migrants usually do not appear until mid-Oct; the earliest record is of 1, 10 Sept 1977, Maryville SL (DAE). The peak is generally during early Nov: 51, 10 Nov 1990, Thomas Hill Res. (PMC, TB et al.); 50, 8 Nov 1986, Smithville L. (RF).

Winter: The construction of reservoirs not only has increased the detectability of this species during migration, but also has dramatically increased the numbers wintering in the southern half of the state during relatively mild winters. The bird is now annually recorded on CBCs; some individuals linger throughout the winter. High counts: 20, Jan–Feb 1990, Table Rock L. (CL et al.-**BB** 57[2]:101); 11, 28 Dec 1965, Mingo CBC; 11, 23 Dec 1965, Springfield CBC.

Yellow-billed Loon (*Gavia adamsii*)

Status: Accidental winter resident and spring transient.

Documentation: Sight record only: see below.

Habitat: Large reservoirs, lakes, and rivers.

Records: A single record, although this species has probably been overlooked in the past as it closely resembles the Common Loon. A lone bird was first found on 14 Feb 1990, Table Rock L. (Lundberg 1990). By the beginning of May it was in full nuptial plumage. The loon was last seen on 9 May (CL et al.). Over one hundred seventy-five people saw the bird during its stay at Table Rock L.

Comments: The Yellow-billed Loon also has been recorded recently in two states that border Missouri: Oklahoma (S. Tomer et al.-**AB** 43:232) and Illinois (Bohlen and Zimmerman 1989). Now that observers are cognizant of the field marks of this species, it probably will prove to be of casual occurrence in the interior of North America.

Caution should be exercised in identifying this species; the only diagnostic field mark, in all ages and plumages, is the color of the culmen (ridge of upper mandible). The culmen is entirely dark in the Common, whereas the distal (outer) half, commonly the distal two-thirds, is pale in the Yellow-billed (Binford and Remsen 1974; Kaufman 1990). This feature can be very difficult to observe under some field conditions. There are other characteristics, such as a pale head and nape and small dark face patch, which basic plumage Yellow-billed Loons often exhibit.

Order Podicipediformes

Family Podicipedidae: Grebes

Pied-billed Grebe (*Podilymbus podiceps*)

Status: Common migrant; rare and local summer resident statewide; uncommon winter resident, primarily in south.

Documentation: Specimen: female, 15 Sept 1972, Maryville SL (NWMSU, DLD 175).

Habitat: Found on virtually every type of body of water during migration; primarily breeds in marshes with a relatively high water level.

Records:

Spring Migration: Migrants begin appearing as soon as the water opens. Numbers usually peak in mid-Apr. High count: 380, 17 Apr 1980, near St. Joseph (FL-**BB** 47[3]:10).

Summer: Although this species has been recorded breeding statewide, the principal nesting localities are in marshes north of the Missouri R. Breeding populations fluctuate considerably from year to year as a result of water levels; larger numbers occur during years of high water.

The spring of 1990 was especially wet and provided optimal nesting conditions at Squaw Creek. By the end of May, a total of 65 nests were located at the refuge (RBE-**BB** 57[3]:135).

Fall Migration: Relatively large numbers begin appearing in early Sept: 115, 4 Sept 1975, Maryville SL (DAE). Peak usually occurs in late Sept and early Oct: 540+, 28 Sept 1974, St. Joseph/Maryville areas (FL, MBR); 210, 1 Oct 1975, Busch WA (JC-**BB** 44[1]:18). Large concentrations may still be encountered through early Nov but rapidly drop off thereafter. High count for a single area: ca. 500, 25 Oct 1947, Creve Coeur L. (JEC et al.-**NN** 19:47).

Winter: Status varies considerably with the severity of the winter, ranging from almost no birds present during very cold winters to locally uncommon in milder years. Generally found on the larger bodies of water in the southern third of the state. High counts, both on the Mingo CBC: 64, 28 Dec 1965; 42, 27 Dec 1967.

Horned Grebe (*Podiceps auritus*)

Status: Uncommon migrant; rare to locally common winter resident in south.

Documentation: Specimen: female, 28 Apr 1973, Maryville SL (NWMSU, DLD 150).

Habitat: Larger and deeper rivers, lakes, and reservoirs.

Records:

Spring Migration: In average springs, birds begin augmenting the wintering population in late Feb and early Mar with numbers building rapidly: 100, 9 Mar 1985, Table Rock L. (M. Goodman-**BB** 52[3]:11). Peak usually occurs in late Mar and early Apr, with a few individuals still trickling through in May. High counts: 250, 11–25 Mar 1990, Table Rock L. (CL, JH-**BB** 57[3]:135); 152, 13 Apr 1969, Table Rock L. (JHA-**BB** 38[1]:2). Latest dates: 2, 18 May 1968, Table Rock L. (NF-**BB** 36[1]:3); 2, 15 May 1979, Maryville SL (DAE, MBR, TB-**BB** 46[3]:5).

Fall Migration: Casual in Aug with small numbers seen through Sept. Numbers increase in Oct and peak during the first half of Nov. Earliest records: 1, 10 Aug 1968, Table Rock L. (NF-**BB** 36[2]:15); 2, 14–21 Aug 1977, Maryville SL (DAE). High counts: 110, 18 Nov 1990, Table Rock L. (PM); 105, 25 Nov 1979, Stockton L. (TB-**BB** 47[1]:14).

Winter: During average winters it is a rare winter resident on the larger lakes and reservoirs in southern Missouri. Numbers vary according to the severity of the winter. High counts: an extraordinary concentration of 400+ were counted on 18 Feb 1990, Table Rock L. (DAE, PM, JH-**BB** 57[2]:101). High counts on CBCs: 148, 18 Dec 1988, Taney Co. CBC; 42, 27 Dec 1967, Mingo CBC (originally incorrectly identified as Eared Grebes).

Red-necked Grebe (*Podiceps grisegena*)

Status: Casual transient; accidental winter resident.

Documentation: Sight records only: see below.

Habitat: Rivers, lakes, and reservoirs.

Records:

Spring Migration: Only five definite records, all of single birds. Dates range from mid-Mar to late May: 19–30 Mar 1983, Springfield (CB et al.-**BB** 50[3]:3); 20 Mar 1985, L. Jacomo (m.ob.-**BB** 52[3]:11); 17–18 Apr 1967, Alton Dam (KA-**BB** 34[2]:5); 22 Apr 1942, Creve Coeur L. (JEC-**BB** 9[5]:28); 27 May 1984, Marais Temps Clair (L. Barber, JZ-**NN** 56:62).

Widmann (1907) states that a bird was collected by P. Hoy in the spring of 1854 in western Missouri; however, examination of Hoy's account (1865) mentions no such specimen. Nevertheless, in Hoy's summary list, he does mention *Podiceps cristatus,* a name that refers to the Great Crested Grebe of the Old World. The Great Crested Grebe does not resemble the Red-necked in any nonjuvenile plumage, but *cristatus*'s basic plumage is similar to that of the Western Grebe. It is conceivable that Hoy mistook a Western Grebe for *cristatus;* but without a specimen we will never know, so we have treated this record as suspect.

Fall Migration: Only five definite records, dates ranging from Sept through Nov: 1, early Sept 1954, near Buckner, Jackson Co. (FB-Rising et al. 1978); 1, 3 Oct 1978, Springfield (CB-**BB** 46[1]:21); 2, 9 Oct 1960, Busch WA (m.ob.-**BB** 28[1]:17); 1, 15–16 Nov 1976, Mississippi R., St. Charles Co.

(JE, PS et al.-**BB** 44[3]:19); 1, 29–30 Nov 1968, L. of the Ozarks, Morgan Co. (DAE-**BB** 36[4]:11).

Winter: Three records: 2, 18 Dec 1987, L. Wappapello, Wayne Co. (BRE, SD-**BB** 55[2]:59); 1, 1–3 Jan 1982, L. Jacomo (H. Gregory et al.-**BB** 49[2]:15); 1, 1–17 Jan 1983, Fellows L., Springfield (CB et al.-**BB** 50[2]:22).

Eared Grebe (*Podiceps nigricollis*)

Status: Uncommon migrant in west, rare east; casual summer visitant in west; hypothetical in winter in southwest.

Documentation: Specimen: male, 1 Dec 1974, Maryville SL (NWMSU, DLD 144).

Habitat: Marshes, ponds, lakes, and reservoirs.

Records:

Spring Migration: This species is encountered far more frequently in the western half of the state; nonetheless, it is regular in the east where observations almost invariably involve single birds. The first individuals begin appearing by mid-Mar, but the bulk of the migrants do not appear until the end of Apr; peak normally occurs in mid-May. A few birds linger through the end of May. Earliest records: 1, 14 Mar 1961, Columbia (DAE); 2 in nuptial plumage, 14 Mar 1974, Creve Coeur L. (R. Laffey, JEC et al.-**NN** 46:55). The largest number encountered followed several days of strong WSW winds, when this species was seen on many medium to large-sized bodies of water in the northwestern corner of the state. On 12 May 1978, J. Brady counted no less than 130 at Squaw Creek alone, and two days later 23 were present on the Maryville SL (MBR, TB-**BB** 45[3]:14). A high count under more normal conditions was 39, 15 May 1981, Squaw Creek (FL, MBR-**BB** 48[3]:16).

Summer: At least seven June (no July) records from the northwestern corner; the latest being of a single bird at Maryville SL, 17 June 1973 (MBR-**BB** 40[4]:2).

Fall Migration: Birds begin reappearing in late Aug and appear to peak in late Sept, but smaller numbers can continue through early Nov. An occasional individual may linger into early Dec. Earliest record: 1, 14–21 Aug 1977, Maryville SL (DAE). High count: 23, 25 Sept 1985, Maryville SL (DAE). Latest record: 1, 6 Dec 1982, L. Jacomo (RF-**BB** 50[2]:22).

Winter: To date, there still is no satisfactory winter record.

Western Grebe (*Aechmophorus occidentalis*)

Status: Rare transient in west, casual in east; accidental summer and winter visitant.

Documentation: Specimen: male, 21 Oct 1976, Nodaway Co. L. (DAE, NWMSU, DT 87).

Habitat: Marshes, ponds, lakes, and reservoirs.
Records:
Spring Migration: All records (12+) but two are from the western half of the state, and all but one are from late Apr through May. The latter exception is of a single bird at Schell-Osage on 30 Mar 1977 (G. Seek-**BB** 44[4]:24). High count: 4, 14–16 Apr 1990, Thomas Hill Res., Randolph Co. (BG, IA-**BB** 57[3]:135).
Summer: Only three summer records, all of single birds in June: most of June 1974, Pony Express L., Dekalb Co. (R. Lainy et al.); 11 June 1967, Squaw Creek (HB, FL, JHA-**BB** 6[7]:61); 26 June 1939, Alton Dam (L. Ernst, WS-**BB** 6[7]:61).
Fall Migration: There are more records in fall (ca. 20) than in spring, with at least five observations from the eastern half of the state. All records are from early Oct to early Dec, but the majority of sightings are between 21 Oct and 20 Nov. Earliest date: 1, 2 Oct 1977, Maryville SL (TB-**BB** 45[1]:23). High count: 6, 21 Oct–7 Nov 1967, Forest L., Kirksville, Adair Co. (R. Luker). Latest dates: 1, 10 Dec 1971, Browning L. (FL-**BB** 39[2]:10); 3, 10 Dec 1988, Harbor Point, St. Charles Co. (G. and T. Barker-**BB** 56[2]:52).
Winter: A single acceptable record: 1, 1–3 Jan 1988, Trimble CBC (RF et al.).

Clark's Grebe (*Aechmophorus clarkii*)

Status: Accidental transient and winter resident.
Documentation: Sight record only: see below.
Habitat: Marshes, ponds, lakes, and reservoirs.
Comments: There is a single observation of this grebe that was recently elevated to full species status (A.O.U. 1985). A single bird was present at Smithville L. 27 Nov–10 Jan 1988–89 (RF, MMH, L. Moore, DAE et al.-**BB** 56[1]:11; **BB** 56[2]:52). The above bird wintered with 2,000–3,000 Common Mergansers. Some of the records listed above under the Western Grebe may actually pertain to this species. Most states that border Missouri now have records of Clark's Grebe. Kaufman (1989) discusses how these two grebes can be distinguished from each other.

Order Procellariiformes

Family Hydrobatidae: Storm-petrels

Band-rumped Storm-Petrel (*Oceanodroma castro*)

Status: Accidental vagrant.
Documentation: Specimen (see below): 1, 3 Sept 1950, near Defiance, St. Charles Co. (K. Wesseling-**BB** 17[10]:2).

Comments: This extraordinary record apparently was the result of a hurricane. The bird was found injured beneath a farm windmill in St. Charles Co. by K. Wesseling. A description and measurements were sent to a USNM curator, who confirmed that it was indeed a Band-rumped Storm-Petrel. The specimen was mounted and deposited at the St. Louis Science Center (then known as the St. Louis Academy of Science), where it was subsequently misplaced or thrown out.

We failed to determine what hurricane was responsible for the bird's appearance, and who at the Smithsonian verified the original identification.

Order Pelecaniformes

Family Pelecanidae: Pelicans

American White Pelican (*Pelecanus erythrorhynchos*)

Status: Locally common transient in west, rare in east; rare summer visitant; casual winter resident.

Documentation: Specimen: male, found dead 8 Apr 1975, Big L. SP (NWMSU, DLD 160).

Habitat: Marshes, rivers, lakes, and reservoirs.

Records:

Spring Migration: This species is decidedly less common in spring than in fall statewide; nevertheless, relatively large numbers annually congregate during spring at some of the larger marshes in the western half of the state (Squaw Creek and Schell-Osage). Sightings in the east are sporadic and consist of small-sized flocks (usually < 10 birds/flock). The following generalizations pertain to western Missouri: Relatively small-sized flocks (20–100 birds/flock) begin appearing in late Mar, with numbers increasing until mid-Apr when the peak usually occurs (normally 2,500–3,500 birds). However, solitary birds and even relatively small-sized flocks are frequently seen through the end of May. Earliest records: 70+, 29 Feb 1988, Swan L. (KM-**BB** 55[2]:59); 50, 20 Mar 1982, L. Contrary (FL-**BB** 49[3]:15). Highest count: 5,000, 9–15 Apr 1965, Squaw Creek (HB-**BB** 32[1–2]:9).

Summer: Although some of the larger flocks seen in early June undoubtedly represent late migrants, e.g., 219, seen flying north, 8 June 1983, Squaw Creek (FL-**BB** 50[4]:30), smaller flocks encountered in mid- to late summer probably represent nonbreeding or failed breeders that have returned south.

Fall Migration: As in spring, this species is significantly more common in the western half of the state than in the east. Most of the eastern observations are in Oct, and flock size generally is relatively small (< 50 individuals). The following primarily refers to western Missouri: migrants begin appearing in the latter half of Aug and numbers increase until they peak (3,000–5,000 birds) in mid- to late Sept. Large-sized flocks (ranging from the low hundreds to a few thousand) may be seen through Oct: 2,250, 29 Oct 1981, Swan L. (SS-**BB** 49[1]:13). Very few individuals are encountered past mid-Nov. High counts, both at Squaw Creek: ca. 8,000, 11 Sept 1960 (FL); 6,000, mid-Sept 1962 (HB-**BB** 29[4]:22).

Winter: Most records are of late lingering migrants that normally remain only until the end of Dec; a number (all?) of these sightings are of sick or crippled birds. There are at least a dozen observations, with a few pertaining to overwintering birds: 4, 26 Dec–15 Mar 1963–64, Montrose (SH-**BB** 31[2]:9); 2, winter of 1984–85, Montrose (m.ob.-**BB** 52[2]:13); 1, 27 Dec–18 Jan 1981, Thomas Hill Res. (BG).

Brown Pelican (*Pelecanus occidentalis*)

Status: Accidental transient and summer visitant.
Documentation: Sight records only: see below.
Habitat: Lakes and reservoirs.
Records:

There are three old observations, two of which occurred in 1929: 3, 1 Apr, Parkville, Platte Co. (J. Jackson et al.; Bennitt 1932); 5, 28 June, L. Taneycomo, near Hollister, Taney Co. (G. Greenwell-**BB** 1[2]:10). A single adult bird was present for about two weeks, either in early July or mid-Aug 1950, at Gentle Slopes Resort, south of Gravois Mills, L. of the Ozarks (Easterla 1952). The latter bird was occasionally observed perched on boat dock roofs by G. Easterla, R. Easterla, C. Crenshaw, DAE et al.

Now that this species is making a comeback from the population lows (due to pesticides) of the 1950s and 1960s, it may reappear in the state.

Family Phalacrocoracidae: Cormorants

Double-crested Cormorant (*Phalacrocorax auritus*)

Status: Uncommon to locally common transient; rare summer visitant; former summer resident in Mississippi Lowlands; rare winter resident.
Documentation: Specimen: female, 5 Oct 1925, Fayette, Howard Co. (CMC 68).
Habitat: Lakes, reservoirs, and rivers.
Records:

Spring Migration: The first individuals begin appearing in mid- to late Mar, and relatively small-sized flocks start arriving in early Apr. Peak is usually during the third week of Apr, with numbers dropping off thereafter; however, a few individuals invariably linger until the end of May. High counts: 500+, 20 Apr 1990, Duck Creek/Mingo (BRE-**BB** 57[3]:135); 305, 29 Apr 1990, Squaw Creek (DAE).

Summer: Formerly, this species apparently bred in "considerable numbers" in the Mississippi Lowlands (Widmann 1907). For this same region, Bennitt (1932; 1940) listed the bird as an uncommon breeder, but it is unclear whether this statement was based on current information at that time or whether it was just a reiteration of Widmann's observations. The last documented nesting was at Mingo, where a small colony comprised of 8 nests in Bald Cypress was captured on film on 23 June 1956 (R. Coy, D. Reynolds; "Swampeast Missouri," deposited at SJM; Fig. 1).

Currently, this species is recorded nearly annually (primarily immature birds) during this period, and most of the records come from the western half of the state. Nesting possibly occurs along the Mississippi R., since it is known to breed along this river in Illinois (Bohlen and Zimmerman 1989) and Iowa (Dinsmore et al. 1984). Flocks seen in early to mid-June, undoubtedly represent late migrants.

Fall Migration: This species is much more common statewide in the fall

Fig. 1. This adult female Anhinga was photographed at a Double-crested Cormorant nesting colony in Bald Cypress (Mingo NWR, Stoddard and Bollinger counties, 23 June 1956). Photograph by Roy Coy and Don Reynolds. Courtesy of the St. Joseph Museum.

than in the spring. A few individuals begin appearing by late Aug, but larger numbers do not arrive until the latter half of Sept. Relatively large-sized flocks may be seen from the end of Sept through early Nov; they generally peak in early Oct. Numbers gradually taper off, and only a few individuals can be seen past late Nov. High counts: 2,000+, 9 Oct 1990, Thomas HIll Res. (IA); 1,800, 3 Nov 1983, Schell-Osage (JW-**BB** 51[1]:39).

Winter: As with the American White Pelican, most winter observations are concentrated in late Dec, with about 25 CBC records. During mild winters, a few individuals may overwinter on some of the larger, southernmost lakes: 54, 17 Dec 1989, Stockton L., Dade Co. (JH, PM-**BB** 57[2]:101); 8, 15 Jan 1985, Taney Co. (PM-**BB** 52[2]:13); 8, 20 Feb 1988, Springfield (CB-**BB** 55[2]:59). Other noteworthy records: 1, 1–11 Jan 1976, Alton Dam area (JE-**BB** 44[1]:24); 1, 8 Jan 1983, Swan L. (T. Miller-**BB** 50[2]:22).

Olivaceous Cormorant (*Phalacrocorax brasilianus*)

Status: Accidental summer visitant.

Documentation: Photograph: adult in nuptial plumage, 5–6 June 1990, Squaw Creek (BJ, C. Wilson et al.; photos in MRBRC file).

Habitat: Lakes, marshes, and rivers.

Comments: A nuptial plumaged adult was found and photographed at Squaw Creek on the evening of 5 June 1990 (BJ, C. Wilson). The bird was present and photographed again the following day (DAE, D. Mead, LG).

This species will undoubtedly appear again in the near future, as there is a minimum of a dozen records of its occurrence during the summer in Oklahoma (Wood and Schnell 1984) and Kansas (Thompson and Ely 1989). Browning (1989) established that the correct specific epithet for this cormorant is *brasilianus,* and not *olivaceus.*

Family Anhingidae: Anhinga

Anhinga (*Anhinga anhinga*)

Status: Casual transient and summer visitant in Mississippi Lowlands, accidental elsewhere; former summer resident in Mississippi Lowlands.

Documentation: Photograph: female, 23 June 1956, Mingo (R. Coy, D. Reynolds-**BB** 23[8]:1; Fig. 1).

Habitat: Rivers and edge of heavily-wooded swamps.

Records:

Spring Migration: Widmann (1907) mentions that this species was a common summer resident but gives no dates of spring arrivals. The earliest recent-arrival record is of an adult male, 19 Apr 1981, Mingo (SS, T. Weyrauch-**BB** 48[3]:16). There are at least three additional recent records from the Mississippi Lowlands: 1, 21 Apr 1987, Caruthersville, Pemiscot Co. (LG, FL-**BB** 54[3]:12); a female, 21 May 1976, Mingo (BE-**BB** 44[2]:14); 1, 11–26 May 1964, Mingo (J. Rogers-**BB** 31[2]:9).

Summer and Fall: This species was "a fairly common summer resident" in Dunklin and Pemiscot counties—in the bootheel—until as late as 1896 (Widmann 1907). Apparently, it was even more widespread prior to the drainage and deforestation of the southeastern section of the state (Widmann 1907). Recently, apparent postbreeding wanderers have been found at a borrow pit at Caruthersville, Pemiscot Co.: 4 (1 male, 3 female plumaged), 25 Aug 1983 (JW et al.-**BB** 51[1]:39); male, 15 June 1986; female, 26 June–11 July 1986 (CP, JW-**BB** 53[4]:15); 2, adult male and an immature, 5 Aug–16 Sept 1986 (m.ob.-**BB** 54[1]:33). These birds may be from a small breeding colony just across the Mississippi R. at Reelfoot L. in Tennessee.

Comments: Another indication that the Anhinga was more widespread in the state before the turn of the century, not only as a breeder in the southeast, but as a migrant elsewhere, comes from information compiled by Harris (1919b). Specifically, he mentions two records: a small flock was observed on the Missouri R., in what is now Kansas City, in 1882 by B. Bush; and a specimen was procured (but now cannot be located) by J. Bryant at Parkville, Platte Co. on the Missouri R. in 1898.

In addition to the above records, there are a few sightings from other areas of the state that either lack details or the details were unsatisfactory for distinguishing this species from the Double-crested Cormorant.

Family Fregatidae: Frigatebirds

Magnificent Frigatebird (*Fregata magnificans*)

Status: Accidental vagrant.

Documentation: Photograph: adult female, 28 Sept 1988, Longview L., Jackson Co. (CH, RF et al.-**BB** 56[1]:12; Fisher 1989; Fig. 2).

Records:

The above record occurred about three weeks after a storm system, rem-

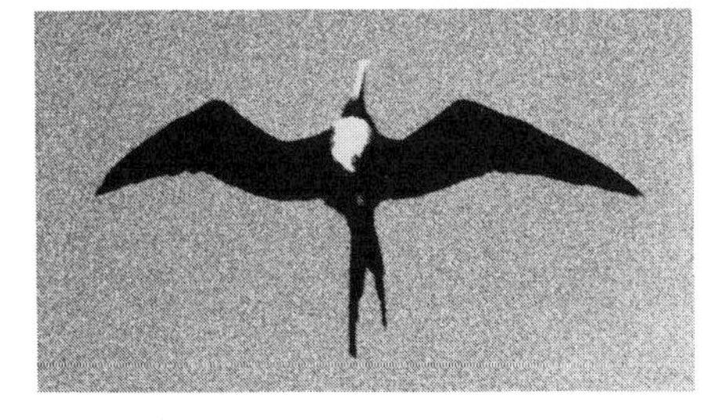

Fig. 2. Adult female Magnificent Frigatebird photographed at Longview Lake, Jackson Co., 28 September 1988. Photograph by Larry Werner. It represents the first confirmed record for the state.

nants of hurricane Gilbert, passed through the state. Several additional observations were made in the interior of the country during this same period (Lehman 1989).

Although the Magnificent Frigatebird is far more likely to appear in Missouri than any other species, there is a specimen of the Great Frigatebird (*F. minor*) from Oklahoma (Wood and Schnell 1984) and a record of the Lesser Frigatebird (*F. ariel*) from Maine (Snyder 1961). In light of these other records and the fact that all of these species are very similar, observers should make note of every detail, especially the ventral surface, of any frigatebird seen. The following records—both of single birds—represent observations where critical details for identification were not noted: immature or adult female, observed during passage of severe storm front, 2 Apr 1982, Marshfield, Webster Co. (JS-**AB** 36:857); 1, appeared as a storm front passed through the area, 27 Apr 1983, near DeSoto, Jefferson Co. (M. Wiese-**BB** 50[3]:10). Two other observations have been published, but in both cases the observer's integrity has been questioned.

Order Ciconiiformes

Family Ardeidae: Bitterns and Herons

American Bittern (*Botaurus lentiginosus*)

Status: Uncommon transient; rare summer resident; accidental winter visitant.

Documentation: Specimen: female, 8 Dec 1926, Fayette (CMC 71).

Habitat: Primarily found in marshes; occasionally seen in wet meadows and at the edge of ponds and lakes.

Records:

Spring Migration: The first individuals arrive at the end of Mar. Numbers increase through Apr and usually peak in early May. Earliest records: 1, 27 Mar 1990, Duck Creek (BL, HF-**BB** 57[3]:135); 1, 28 Mar 1976, near Alton Dam (B. Boesch-**NN** 48:44). High counts: 17, migrants seen leaving the marsh at dusk and headed north, 15 May 1979, Squaw Creek (MBR, DAE, TB-**BB** 46[3]:6); 14, 7 May 1976, Squaw Creek (NJ-**BB** 44[2]:15). Late date of obvious migrant: 1, 28 May 1980, Big Oak Tree SP (DAE).

Summer: Currently this species is a much less common breeder than at the turn of the century, as Widmann (1907) stated it was a "fairly common summer resident." Most breeding records for the past quarter century have been in marshes north of the Missouri R. The best areas are at Squaw Creek (several pairs bred there in 1988 and 1990 [DAE, RBE]), Ted Shanks, and Marais Temps Clair.

Fall Migration: Migrants begin appearing as early as mid-Aug and peak in mid- to late Sept. Occasionally an individual is seen into early Dec (4 early Dec records). Earliest record: 1, 18 Aug 1980, Springfield (W. Holloway-**BB** 48[1]:10). High count: 7, 10 Sept 1973, Squaw Creek (FL-**BB** 41[1]:2).

Winter: An accidental winter visitant with only one CBC record: 1, 27 Dec 1959, Squaw Creek CBC. There are two records during mid-winter: 1, 28 Jan 1971, St. Charles Co. (JEC et al.-**NN** 43:29); 1, early Feb 1966, Squaw Creek (HB-**BB** 33[1]:13).

Least Bittern (*Ixobrychus exilis*)

Status: Locally uncommon summer resident.

Documentation: Specimen: male, 23 May 1965, L. Contrary (NWMSU, DAE 1012).

Habitat: Almost exclusively marshes, especially those with cattails.

Records:

Spring Migration: The earliest record is of an individual on 16 Apr 1977, near Alton Dam (JE et al.-**NN** 49:38); however, this secretive species is normally not encountered until the very end of Apr or early May. High count: 5, 26 May 1968, Squaw Creek (FL-**BB** 36[1]:4).

Summer: Although much of its habitat has been lost, as a result of the draining and filling of marshes across the state over the past eighty years, it still is found in relatively large numbers primarily in the floodplain marshes of the Missouri and Mississippi rivers: 35+ nests, summer 1984, Ted Shanks (F. Reid-**BB** 51[4]:11); 31 birds, 27 June 1990, Squaw Creek (RBE); 20–23 birds, 26 June 1982, Marais Temps Clair (BR, TB, PB-**BB** 49[4]:14); ca. 20 birds, 23 June 1974, Squaw Creek (FL); 19 nests, average 4.1 eggs/nest, 6 June–7 July 1962, Beverly and Cooley lakes (JRI, T. Anderson-**BB** 29[3]:20).

Fall Migration: Relatively large numbers are still encountered at breeding sites until mid-Sept, e.g., 6, 13 Sept 1980, Squaw Creek (FL), with only single individuals reported through early Oct. Latest dates: 1, 11 Nov 1986, L. Contrary (LG-**BB** 54[1]:33); 1, 19 Oct 1941, Squaw Creek (WC-**BB** 8[11]:81).

Great Blue Heron (*Ardea herodias*)

Status: Common transient and postbreeding summer visitant; locally uncommon summer resident; rare to locally uncommon winter resident in south.

Documentation: Specimen: female, 16 Jan 1979, Scott Co. (GL 366).

Habitat: Utilizes virtually every type of shallow water habitat.

Records:

Spring Migration: Migrants begin augmenting the winter residents in

Fig. 3. This Great Blue Heron was drawn at Dry Fork Creek near Salem, Dent Co., where about 20 nests were in a large Sycamore. Pen and ink by David Plank.

the south by early Mar. The first arrivals appear in the north by mid-Mar. Peak is usually in mid-Apr.

Summer: This species is conspicuous and widespread during the early summer, even though most of the large nesting colonies are located in the southern half of the state. The Missouri Department of Conservation (MDC) conducted annual statewide surveys from 1978 to 1989 to determine the status and distribution of the state's herons. Published results are as follows: They found 103 colonies of this species in both 1982 and 1983; only 4 of these were north of the Missouri R. in Chariton, Howard, Livingston, and Putnam counties (**BB** 49[4]:13). The largest colony in 1982, which consisted of 140 nests, was in Greene Co. In 1984, a total of 129 nesting colonies, with an average of 29.3 nests/colony, were located; only 5 were found north of the Missouri R. (**BB** 51[3]:7). Censuses were also made in 1988 and 1989 with the following results: 149 colonies, ranging from 1–298 nests/colony, with an average of 28.2 nests/colony (**BB** 55[4]:125); and 170 colonies, the largest colony with 340 nests, with an average of about 30 nests/colony (**BB** 56[4]:145).

By late July postbreeding birds begin congregating at favorable feeding sites: 250, late July 1971, Squaw Creek (HB-**BB** 39[4]:7).

Fall Migration: The postbreeding congregations continue to increase through Aug and peak generally in late Aug and Sept: 350, first half of Sept 1967, Squaw Creek (HB-**BB** 35[1]:9); 300, 31 Aug 1969, Squaw Creek (FL). When feeding conditions are ideal, these congregations can continue into early Oct: 150, 1 Oct 1963, Squaw Creek (FL-**BB** 30[4]:14). Numbers drop off significantly by the end of Oct, and by late Nov only very small-sized groups or, more commonly, only single birds are encountered.

Winter: Status varies considerably depending upon the severity of the winter, with more individuals present during mild years; but regardless of the weather, very few overwinter in the north. High counts: 100+, at heronry, 10 Feb 1990, Taney Co. (JH, PM-**BB** 57[2]:101); 58, 11 Feb 1987, Rockaway Beach, Taney Co. (PM-**BB** 54[2]:12); 52+, 28 Dec 1989, Montrose and Truman Dam, Benton Co. (MBR).

Great Egret (*Casmerodius albus*)

Status: Uncommon transient and summer visitant; locally rare summer resident; accidental winter visitant.

Documentation: Specimen: female, 11 May 1973, near Maryville (NWMSU, DLD 45).

Habitat: Found utilizing virtually every type of shallow water habitat, especially marshes.

Records:

Spring Migration: Birds begin appearing at the end of Mar and peak in early to mid-May. Earliest records: 4, 4 Mar 1951, Alton Dam (RA); 10+, 22 Mar 1989, Caruthersville, Pemiscot Co. (BRE-**BB** 56[3]:89). Highest count: ca. 200, 17 May 1959, Squaw Creek (K. Krum-**AFN** 13:374).

Summer: Very few breeding colonies are known, and all are found in southeastern Missouri. During the MDC's surveys, totals of 4, 7, and 4 colonies were located in 1982, 1983, and 1984. In 1988, the Miner heronry (formerly the Bertrand colony) contained ca. 15 nests, while the Charleston and Caruthersville sites were estimated to have two to three times that many nests (**BB** 55[4]:125). The 1989 survey revealed 10 nests at the Charleston site and 20 at the Caruthersville heronry (**BB** 56 [4]:145). Like the other larger herons, postbreeding concentrations begin to appear in late July.

Fall Migration: The postbreeding assemblage usually peaks during mid- to late Aug, and numbers remain relatively high through the second week of Oct: 35, 11 Oct 1987, Alton Dam (RA-**BB** 55[1]:9). Numbers drop off rapidly thereafter, and usually only single birds are seen from late Oct until mid-Nov; it is only accidentally seen after the third week of Nov. High counts: ca. 1,000, 10 Aug 1940 and 1942, Creve Coeur L. (JEC-**BL** 44:10; **BB** 9[9]:61); ca. 1,000, in early 1940s, near Sugar L., Platte Co. (E. O'Conner-Rising et al. 1978); 700, 30 Aug 1969, St. Charles Co. (PB-**BB** 38[3]:6); 500, 9 Aug 1986, Caruthersville, Pemiscot Co. (DAE, FL, LG). Latest date: 1, 3 Dec 1972, Busch WA (m.ob.-**NN** 45:14).

Winter: Only one winter record: 1, 26 Dec 1973, Gray Summit CBC.

Comments: This species has been increasing gradually since the decline of the plume trade that flourished during the end of the last century and the beginning of this one. Apparently this species bred by the hundreds in the southeastern corner of the state up until the 1890s, but by the time Widmann finished his treatise on Missouri birds, it was almost extirpated from the state (Widmann 1907). This is also echoed in Harris' work (1919b); he lamented that it had "not been seen in the vicinity [Kansas City area] for twenty years."

Snowy Egret (*Egretta thula*)

Status: Rare and very local summer resident in Mississippi Lowlands only; rare transient and summer visitant elsewhere.

Documentation: Sight records only: see below.

Habitat: Mainly found in marshes and flooded fields; occasionally seen at the edge of ponds and lakes.

Records:

Spring Migration: Single birds begin appearing in early Apr, with most recorded from the end of Apr through May. More numerous at all seasons in the southeast. Earliest date for southeastern corner: 3, 22 Mar 1989, Caruthersville (BRE-**BB** 56[3]:89). Earliest date for northern Missouri: 1, 28 Mar 1962, Kansas City (JRI-**AFN** 16:418). High count outside Mississippi Lowlands: 5, 26 Apr 1983, Horseshoe L., Buchanan Co. (FL-**BB** 50[3]:10).

Breeding commences in May at the two known heronries in the southeastern corner. The heronry just east of Sikeston (variously referred to as the Miner or Bertrand heronry) had 3 or 4 pairs in 1962, 7 nests in 1982, and 16 nests on 31 May 1984 (JW-**BB** 52[3]:11). The more recently discovered heronry at Caruthersville, Pemiscott Co., contained 10 nests in 1983 (JW-**BB** 50[4]:31). A 1988 census yielded the following results: 20 nests at Miner, and an estimated 2 to 3 times that many nests at the Charleston and Caruthersville sites (**BB** 55[4]:125). In 1989, 10 and 20 pairs were noted at Charleston and Caruthersville, respectively (**BB** 56[4]:145–46). Apparently the Miner heronry was abandoned after the 1988 season.

Summer and Fall Migration: Breeding continues into June at the above mentioned sites. Postbreeding wanderers begin augmenting the breeding population in mid- to late July, and numbers continue to build, with the peak recorded statewide during Aug and early Sept: ca. 100, 9 Aug 1986, Caruthersville (DAE, FL, LG); 28, 18 Aug 1957, Bull Shoals L., near Forsyth, Taney Co. (NF-**BB** 24[8–9]:4); ca. 20, 8 Aug 1948, Creve Coeur L. (m.ob.-**NN** 20:47). Latest record: 2, 22 Sept 1984, Aldrich (B. Dyer-**BB** 52[1]:27).

Comments: Like the Great Egret this bird suffered greatly from the millinery trade at the turn of the century and was almost extirpated from the state during that period (Widmann 1907).

Little Blue Heron (*Egretta caerulea*)

Status: Locally common summer resident in Mississippi Lowlands, accidental elsewhere; uncommon transient and locally common summer visitant elsewhere.

Documentation: Specimen: male, 9 Apr 1964, Bertrand, Mississippi Co. (SEMO 224).

Habitat: Primarily seen in marshes and flooded fields.

Records:

Spring Migration: Birds begin to arrive in the Mississippi Lowlands at the end of Mar, with a dramatic increase in numbers by the second week of Apr: 300–500, 8 Apr 1964, Bertrand Heronry, Mississippi Co. (PH). It is locally abundant by mid-Apr: ca. 2,000, 24 Apr 1976, Miner, Scott Co. (BE, PH-**BB** 44[2]:14) but is decidedly less common elsewhere in the state, with usually only singles or small groups of fewer than 20 individuals observed. The earliest date was obtained through the retrieval of a bird banded at Glen Allan, Mississippi, on 6 June 1937; it was found dead on ca. 10 Mar 1944, 2 miles south of Dudley, Stoddard Co. (Cooke 1950). Earliest date for northern Missouri: 1, 3 Apr 1965, L. Contrary (FL-**BB** 32[1–2]:9). High count for early in season: 200–300, 28 Mar 1967, Bertrand Heronry (fide PH).

This species was first recorded breeding in the state in 1922 (Bennitt 1932). Only two breeding sites are currently known; the Miner and the Caruthersville sites had ca. 900 and 1,000 nests, respectively, in late May 1985 (JW-**BB** 52[3]:11). About the same number of nests were noted at the above sites (the Miner heronry birds apparently moved east of Charleston) in 1989 (**BB** 56[4]:146). Until the arrival of the Cattle Egret (see below), this species was the most numerous heron in the above colonies.

Summer and Fall Migration: Although this species is now known to breed only in the southeastern corner, it has been found nesting at least twice in the St. Louis area. Wayne Short found an adult and six young on 12 July 1936, Dardenne Island, Missouri R., St. Charles Co. (**BB** 3[7]:68). "Small numbers bred," at Marais Temps Clair in the summer of 1962 (RA-**BB** 29[3]:20).

Birds begin dispersing from breeding localities in mid-July with large numbers concentrating at locally favorable feeding sites statewide: 250+, 29 July 1984, Camden–Fleming bottoms, Missouri R., Ray Co. (CH-**BB** 51[4]:12). Postbreeding dispersal usually peaks in Aug. Numbers drop off significantly after early Sept; however, a few birds typically linger until the beginning of Oct. Largest concentration: ca. 3,000, 9 Aug 1986, Caruthersville (DAE, FL, LG). Latest date: 2, 5 Oct 1982, Squaw Creek (FL).

Tricolored Heron (*Egretta tricolor*)

Status: Casual transient; accidental summer visitant.

Documentation: Photograph: adult, 15–16 May 1978, L. Springfield, Greene Co. (M. Goodman-**BB** 45[3]:15; VIREO x08/11/001).

Habitat: Primarily seen in marshes, lakes, and flooded fields.

Records:

Spring Migration: The earliest record is also the first for the state—a bird that was collected by E. Currier on 13 Apr 1890, near Sand Ridge, Clark Co. (Widmann 1907; specimen not located). There are three additional spring records (all during May): adult, photographed, 3 May 1975, Schell-Osage (NJ, JG-**BB** 43[1]:18); the above mentioned Springfield record; and an adult, 22 May 1976, Swan L. (JR et al.-**BB** 44[2]:15).

Summer: There is only a single record for this season: 1, 7 June 1963, Beverly L. (DAE, T. Anderson, WG, M. Flieg-**BB** 30[3]:18).

Fall Migration: Five records are as follows: 1, 10–31 Aug 1958, St. Joseph (FL et al.-**BB** 25[9]:4); immature, 12–19 Aug 1971, Squaw Creek (MBR et al.-**BB** 38[4]:7); 1, 27 Aug 1952, Parkville, Platte Co. (JB); 1, first week of Sept 1961, near Portageville, New Madrid Co. (L. Jenkins-**BB** 28[4]:21); and 1, 12–19 Sept 1948, Iatan Marsh, Platte Co. (JB, EC, H. Hedges-**BB** 16[3]:3).

Comments: The Tricolored Heron has been recorded breeding on a few occasions in Mississippi Co., Arkansas (James and Neal 1986), which is adjacent to Dunklin Co., Missouri. Thus, this species may eventually be found breeding in one of the heronries in Missouri's bootheel. We were unable to track down the specifics on the Tricolored Heron record given on the 1966 revision of the Squaw Creek checklist.

[**Reddish Egret** (*Egretta rufescens*)]

Status: Hypothetical; accidental fall visitant.

Documentation: Sight record only: see below.

Habitat: Should be looked for in marshes and flooded fields.

Records:

There is a single sight record: 1, 30 Sept 1951, Marais Temps Clair (JEC, A. Bolinger-**BB** 19[2]:3). This record is very likely correct, since the bird was independently identified and the "unusual feeding behavior was noted"; however, since this represents the only record, and no photograph or specimen was obtained, we treat it as hypothetical.

This species should be looked for in the southeastern corner of the state from late July through early Sept.

Cattle Egret (*Bubulcus ibis*)

Status: Locally common summer resident in Mississippi Lowlands; uncommon transient and summer visitant elsewhere.

Documentation: Specimen: male, 9 June 1965, west of Charleston, Mississippi Co. (NWMSU, DAE 897).

Habitat: Short-grass fields, wet weadows, flooded fields, and marshes.

Records:

Spring Migration: Birds begin appearing in early Apr and are fairly numerous by the end of the month in the southeastern corner. By mid-May, nest construction is in full swing there. Like the Snowy Egret, it is known to breed at only two heronries, both in the Mississippi Lowlands: the Miner and Caruthersville sites had ca. 1,650 and 1,500 nests, respectively, in late May 1985 (JW-**BB** 52[3]:11). In 1989, 1,000 and 1,200 pairs were present at the Charleston and Caruthersville areas, respectively (**BB** 56[4]:146). The species is encountered less frequently and in much smaller numbers in other areas of the state. Earliest dates: 1, 16 Mar 1990, Duck Creek (SD, BL-**BB** 57[3]:135); 5, 22 Mar 1989, Caruthersville (BRE-**BB** 56[3]:89); for northern Missouri: 1, 30 Mar 1979, Swan L. (RB-**BB** 46[3]:6).

Summer and Fall Migration: As with most other herons, postbreeding concentrations begin to appear in the latter half of July, peak in Aug, and gradually dissipate by the end of Sept. Moderate numbers (< 50/day) are encountered in Oct, with only individuals or very small-sized flocks regularly seen until the end of Nov; however, there still is no Dec record. Highest count: ca. 8,000, 9 Aug 1986, Caruthersville (DAE, FL, LG). Highest count in northern Missouri: 357, 27 Sept 1982, Squaw Creek (FL-**BB** 50[1]:20). Latest dates, both in the fall of 1973: 1, 24 Nov, Squaw Creek (FL-**BB** 41[1]:2); 7, 2–30 Nov, near Columbia (J. Kare).

Comments: Prior to 1930, the Cattle Egret was restricted to the Old World (Spain, Africa); however, shortly thereafter it appeared in eastern South America, and by the early 1940s it had spread to Florida in the United States (Crosby 1972). It was first detected and photographed in Missouri on 19 Apr 1955, in St. Charles Co. (V. Balsevic-**BB** 23[2]:1). It was not recorded again until one was seen in the northwestern corner at Squaw

Creek on 14 June 1959 (FL, SR-**BB** 26[6]:2), although undoubtedly the bird was appearing in the southeastern corner during this interim. The first nesting record was established in May 1963 when 4–5 pairs bred at the Bertrand Heronry, Mississippi Co. (PH-**BB** 30[2]:31). Shortly thereafter, the breeding population began to increase dramatically, and birds began appearing statewide. By the late 1970s it had replaced the Little Blue Heron as the most common nesting species in the Bertrand/Miner Heronry.

Green-backed Heron (*Butorides striatus*)

Status: Common summer resident; accidental winter visitant in east.
Documentation: Specimen: male, 7 Sept 1933, Jackson Co. (KU 20217).
Habitat: Marshes, along streams, at the edge of ponds and lakes.
Records:

Spring Migration: The initial migrants appear in mid-Apr, but the bird is not readily encountered until the very end of Apr or early May. Rarely are more than 1 or 2 birds encountered together. Earliest dates in the south: 1, 1 Apr 1988, Duck Creek (BRE-**BB** 55[3]:86). Earliest dates in the north: 1, 11 Apr 1965, Spanish L., St. Louis (RA-**BB** 32[1–2]:9); 1, 11 Apr 1982, Squaw Creek (FL).

Summer: This species is relatively common and normally encountered in isolated pairs throughout the state; however, there are at least two records of large colonies. The first report was of ca. 100 nests on 12 July 1936 near Dardenne Island on the Missouri R. in St. Charles Co. (WS, S. Jenner, T. Baskett-**BB** 3[7]:68). The second report was of ca. 100 nests within the city limits of Puxico, Stoddard Co., during the summer of 1968 (J. Toll-**BB** 36[2]:15).

Fall Migration: It remains relatively common until at least mid-Sept. Large numbers, however, may regularly pass through the state in early Oct, as the following intriguing records indicate: 50+, heard at night, 1–3 Oct 1989, Taney Co. (JH-**BB** 57[1]:45); large movement of undetermined numbers heard migrating at night, 3 Oct 1986, L. of the Ozarks (PM, JH et al.). After early Oct there are less frequent observations that usually only involve single birds. By the end of Oct this species is very rare, with only two Nov and one early Dec records. High count late in season: 9, 30 Oct 1984, Aldrich (CB-**BB** 52[1]:27). Latest dates: 1, 5 Dec 1978, Springfield (B. Dyer, CB-**BB** 46[2]:11); 1, 15 Nov 1979, Busch WA (m.ob.-**NN** 52:4).

Winter: Only one record: 1, CBC count period, Dec 1980, Orchard Farm CBC.

Black-crowned Night-Heron (*Nycticorax nycticorax*)

Status: Uncommon transient statewide; locally uncommon summer resident in Mississippi Lowlands, and rare ? elsewhere; accidental winter visitant.

Documentation: Specimen: female, 28 Apr 1973, Bigelow (NWMSU, DLD 157).

Habitat: Primarily seen in marshes, but regularly seen at edge of lakes and ponds.

Records:

Spring Migration: The first birds appear at the very end of Mar or early Apr and usually peak in late Apr. Nesting commences in May at the two known breeding sites in the southeastern corner. The Caruthersville and Miner sites were estimated to have ca. 100 and 150 nests, respectively, in late May 1985 (JW-**BB** 52[3]:11). Only 50–60 nests were located at each of these sites in 1989 (**BB** 56[4]:146). Earliest dates: 1, 24 Mar 1963, Beverly L. (Rising et al. 1964). High count outside Mississippi Lowlands: 57, 13 Apr 1980, Squaw Creek (FL-**BB** 47[3]:11).

Summer and Fall Migration: Breeding status outside the Mississippi Lowlands is less clear; however, it has been recorded nesting, at least very sparingly, statewide: 500+ nests, summer 1937, Marais Temps Clair (WS-**BB** 4[7]:84). Sixty birds were seen at Squaw Creek on 22 June 1990, and 3 ground nests were located there on 27 June (RBE). Apparently it was much more common and more widespread as a breeder prior to the early 1900s; both Widmann (1907) and Harris (1919b) mention that it was formerly more numerous. Currently, both adults and immatures are seen in small numbers statewide throughout the period; nevertheless, it is uncertain whether these birds represent breeders, nonbreeders, or postbreeding wanderers.

Postbreeding concentrations are composed of significantly fewer birds than that of most other heron species. The largest numbers reported outside the southeastern corner are: 76 (over half were immature), 21 Aug 1988, opposite Chain of Rocks, St. Louis (RA-**NN** 60:58); and 57, late Aug 1968, Squaw Creek (HB-**BB** 36[2]:16). Relatively large numbers are seen through mid-Sept, with small-sized groups observed from late Sept through Oct. Most of the late Oct and a greater majority of the Nov records are of immatures; furthermore, all of the Nov records are of single birds. Latest record: immature, 25 Nov 1972, near Columbia (BG-**BB** 36[2]:16).

Winter: Only four records: 1, all winter, 1972–73, at Southern Hills L., near Springfield, Greene Co. (NF-**BB** 40[2]:7); 1, 21 Dec 1975, Trimble CBC; adult, 31 Dec 1988, Orchard Farm CBC; adult, 20 Feb 1990, Branson, Taney Co. (M. Strange).

Yellow-crowned Night-Heron (*Nyctanassa violaceus*)

Status: Uncommon transient; rare summer resident.

Documentation: Specimen: male, 13 June 1934, Cardwell, Dunklin Co. (MU 297).

Habitat: Marshes, rivers, edge of lakes and ponds.

Records:

Spring Migration: Birds begin arriving in early Apr, but nest construction does not begin until the latter half of May. The largest concentrations are recorded in the latter half of May: 40, 22 May 1971, Squaw Creek (FL). Earliest dates, south: 1, 27 Mar 1966, Aldrich (SH-**BB** 33[2]:6); and north: 1, 2 Apr 1967, Squaw Creek (JHA-**BB** 34[2]:5).

Summer and Fall Migration: This heron nests, usually in isolated pairs, along streams, rivers, and at the edge of wooded swamps and marshes statewide. The largest concentration of breeding birds was reported at Duck Creek, where "several pairs" were located in summer of 1982. Also in 1982, it was reported as "numerous" at the Ted Shanks area (**BB** 49[4]:13).

Postbreeding concentrations are relatively small and less frequently recorded than other herons. Largest numbers appear in late Aug and early Sept: 29 (2 adults, 27 immatures), 7 Sept 1947, near Parkville, Platte Co. (JB-**BB** 15[1]:3). By the end of Sept it is rarely encountered, and with the exception of one record, all Oct sightings are of single birds. There are three Nov observations, with the two latest as follows: 1, 24 Nov 1962, L. Contrary (FL-**BB** 29[4]:22); adult, 13 Nov 1984, Springfield (L. Confer-**BB** 52[1]:27).

Family Threskiornithidae: Ibises and Spoonbills

White Ibis (*Eudocimus albus*)

Status: Accidental spring transient; casual summer visitant.

Documentation: Sight records only: see below.

Habitat: Marshes, swamps, and flooded fields.

Records:

Spring Migration: Only a single observation: 1, 2 May 1968, Bertrand Heronry, Mississippi Co. (PH).

Summer and Fall Migration: All ten records, five of which are from the Mississippi Lowlands, are of postbreeding wanderers from the end of June through late Aug. Most of the records are of first- or second-year birds. The first record was of two adults that were taken from a flock of "about one hundred" on 10 July 1910 at Old Monroe, Lincoln Co. (Williams 1913). The earliest date is of 1 immature, 26 June–20 July 1963 at Trimble (C. Blanchard-**BB** 30[3]:19). The latest date is of 2 first-year birds, 29 Aug 1964, Duck Creek (JHA et al.-**BB** 32[3]:16). Recent high count: 4, 3 Aug 1982, Mingo (SD, BL-**BB** 50[1]:21).

Glossy Ibis (*Plegadis falcinellus*)

Status: Rare transient; accidental summer visitant; accidental summer resident in Mississippi Lowlands.

Documentation: Specimen: male, 30 Apr 1975, Bigelow Marsh (DAE, MBR-**BB** 43[1]:18; NWMSU, DLD 164; Fig. 4).

Habitat: Marshes and flooded fields.

Records:

Spring Migration: The first arrivals usually appear in mid-Apr. The majority of records are in May. Most observations are from the southeast and the St. Louis area; however, it does occur infrequently in other sections of the state. Besides the above mentioned specimen, there are no fewer than five additional late Apr and May records for the northwestern corner (Holt and Buchanan counties). Earliest date: 1, 12 Mar 1983, Mingo (SD-**BB** 50[3]:11). High counts: 7, 19 or 21 May 1968, Bertrand Heronry, Mississippi Co. (PH); 3, 19 Apr 1974, St. Louis (A. Bromet-**BB** 41[3]:3).

Fig. 4. This adult Glossy Ibis was photographed by David Easterla at Bigelow Marsh, Holt Co., on 30 April 1975. The light-colored area in front of the eyes is bare skin in this nuptial-plumaged individual. In the nuptial-plumaged White-faced Ibis the bare facial skin is bordered with a broad band of white feathers that extends behind the eye.

Summer: There is only one nesting record: a pair with 3 young (ca. 2–3 weeks old) were first found on 1 July 1968; they were photographed on 7 July, Bertrand Heronry, Mississippi Co. (PH; VIREO x05/2/007–009; Fig. 5). There are at least three other June records: 1, 4 June 1982, Squaw Creek (FL); 3, 10 June 1982, near Little Bean Marsh, Platte Co. (JW, M. Nelson-**BB** 49[4]:14); 1, 30 June 1968, Bertrand Heronry, Mississippi Co. (PH).

Comments: A few pairs of the Glossy Ibis were also found nesting in Mississippi Co., Arkansas (adjacent to Missouri's bootheel) during the 1960s and 1970s (James and Neal 1986). There are no definite fall records due to the difficulty of distinguishing this species in basic and immature plumage from the very similar White-faced Ibis (*P. chihi*). Also, many early records (prior to the mid-1960s) were referred to as simply "Glossy Ibis" because at that time the two currently recognized species were considered

Fig. 5. This adult and young Glossy Ibis were photographed at Bertrand Heronry, Mississippi Co., on 1 July 1968 by Paul Heye. This is the only instance that this species has been recorded nesting in Missouri.

to be conspecific. Observers should carefully examine the facial pattern on any spring or early summer adult ibis.

White-faced Ibis (*Plegadis chihi*)

Status: Rare transient in west, casual in east; accidental summer visitant in west.

Documentation: Specimen: female, 26 Apr 1975, Bigelow Marsh (DAE; NWMSU, DLD 163).

Habitat: Marshes and flooded fields.

Records:

Spring Migration: Although there are a few records for mid-Apr, the majority of the observations are during the last week in Apr and in May. It is primarily recorded from the western half of the state, but there are a few records for the St. Louis area. What were probably the first records (and specimens) for the state were collected around the turn of the century (no specific dates) in St. Charles Co. (Widmann 1907). Recent St. Charles Co. records include: 1, 20–25 Apr 1978 (JVB et al.); 4, 6–8 May 1978 (PS et al.-**BB** 45[3]:15); 1, 11 May 1984 (D. Becher-**BB** 51[3]:8). There are a couple of additional records for eastern Missouri. Earliest date: 4, 9 Apr 1967, Squaw Creek (HB-**BB** 35[4]:5). High counts: 35, 28 Apr 1990, Squaw Creek (RBE-**BB** 57[3]:135); 16, 26 Apr 1975, Bigelow Marsh (DAE, MBR-**BB** 43[1]:18).

Summer: Only two unquestionable records: 2, 18–24 June 1967, Squaw Creek (HB-**BB** 34[2]:14); adult, 3 July 1990, Squaw Creek (DAE et al.). There are additional summer observations from western Missouri that probably pertain to this species.

Fall Migration: There are only a couple of definite records: 3, one collected (MDC B0082), 18 Sept–2 Oct 1977, Schell-Osage (JR, H. Gregory-**BB** 45[1]:24); 1, 30 Aug 1981, St. Charles Co. (M. Scudder, C. Roberts-**BB** 49[1]:13). There are, however, a number of fall observations that are probably referable to this species. Most of these are from mid-Sept to mid-Oct, the earliest being 2, 11 Aug 1986, at Squaw Creek (LG-**BB** 54[1]:33); and the latest, 1, 22–25 Oct 1980, Browning L. (FL).

Roseate Spoonbill (*Ajaia ajaja*)

Status: Accidental summer visitant.

Documentation: Photograph: immature, 1–19 Aug 1986, Schell-Osage (Wilson 1986; photo in ref and VIREO x05/3/018; x05/3/001).

Habitat: Marshes and flooded fields.

Records:

The above represents the only state record. It was first found by J. Shatford and later verified and photographed by J. Wilson. The bird was not seen after the initial date of discovery until J. Heatley, who was unaware that a spoonbill had been seen on the first, saw what was undoubtedly the same individual on 19 August in the same area (fide B. Heck).

Family Ciconiidae: Storks

Wood Stork (*Mycteria americana*)

Status: Casual summer visitant.
Documentation: Sight records only: see below.
Habitat: Marshes, swamps, rivers, and flooded fields.
Records:

All records are of postbreeding wanderers between mid-July and mid-Sept. There are only two records since the mid–1930s: 1, 16–26 July 1952, near Seckman, Jefferson Co. (B. Tanner-**BB** 19[9]:3); and an adult, 14 Sept 1975, Schell-Osage (G. Seek-**BB** 44[1]:18).

Formerly, this species apparently was recorded nearly annually in the Mississippi Lowlands, and Widmann (1907) states that it "is a regular summer visitant in the Peninsula [bootheel]." He mentions one specific record for central Missouri: 7, 11 Aug–11 Sept 1902, New Haven, Franklin Co. (Dr. Eimbeck). There was a major influx in August 1932: 130 were seen on 24 August, at St. Louis (G. Foster, R. Zahn), and 44 were observed near Cape Girardeau (R. Dede-**BL** 36:374). The last large flock reported was of 35–40 on 14 Aug 1937 at Cape Girardeau (H. Bolen-**BL** 39:476), although a flock of 23 birds was discovered by WG just southeast of St. Louis in Illinois in 1963 (Anderson and Bauer 1968).

The paucity of recent records in Missouri appears to be a reflection of the poor breeding performance of this species in the southern United States and along the eastern coast of Mexico. Kushlan and Frohring (1986) argue that the Wood Stork population was stable in southern Florida (the principal nesting population in the United States) until 1967; they then document a 75% decrease in that population between 1967 and 1981–82. Thus, if this is true, the decline in Missouri observations could not be explained by the breeding dynamics of the southeastern United States population.

Interestingly, postbreeding observations of this species in Texas declined during the same period as Missouri (Oberholser 1974). Since the Texas breeding population was already on the verge of extirpation prior to the postbreeding decline in Missouri and Texas, this would appear to indicate that the postbreeding populations probably originated from breeding colonies along the eastern coast of Mexico.

The purported photographing of 4, on 23 June 1956, at Mingo (**BB** 23[8]:1) is in error, as the photos were of birds in Florida (R. Nolf, pers. comm.; SJM).

Family Cathartidae: American Vultures

Information from a variety of data sets (behavioral, morphological, and biochemical; Ligon 1967; Vanden Berge 1970; Sibley and Ahlquist 1985, respectively) unequivocally indicate that the American Vultures are most closely related to the storks (Ciconiidae). This will soon be standard treatment, so we have placed the family Cathartidae after the Ciconiidae.

Black Vulture (*Coragyps atratus*)

Status: Rare transient in south, accidental in north; rare summer resident in south; rare to locally common winter resident in the extreme south.

Documentation: Photograph: 18 Feb 1990, below Table Rock L. Dam (DAE; VIREO x08/31/001).

Habitat: Primarily found in forest bordering swampy areas, lakes, rivers, and along bluffs.

Records:

Spring Migration: Although data are scant on when the bulk of the birds return to southern Missouri, it is clear that at least some migrants arrive at the end of Feb and in early Mar. Most observations are at the end of Apr and during May; nesting is already in full swing by early May: 1 nest, 2 May 1981, Mingo (JW-**BB** 48[3]:17). Extralimital records: 1, 26 Apr 1970, Missouri Botan. Arboretum (D. Hays-**NN** 42:80); 1, 27 Apr 1986, Stockton L. (JS-**BB** 53[3]:11); 1, early May 1972, north of Warrenton, Warren Co. (B. Massie-**BB** 49[4]:10); 2, 17 May 1981, near Brighton, Polk Co. (JS-**BB** 48[3]:17).

Summer: A local breeder in the southern tier of counties, extending from at least Barry Co. to the Mississippi R. The northernmost site is at Mingo. It formerly was more common and widespread throughout the extreme south, but apparently the clearing of forest has eliminated suitable habitat (Widmann 1907; Neff 1923). High count: 86, early June 1971, along highway 86, Stone Co. (BBR-**BB** 39[2]:6). Extralimital record: 2, 8 June 1957, Hermann, Gasconade Co. (RA-**BB** 24[7]:2).

Fall Migration: Besides the recent observations (PM, JH et al.) in the Taney Co. area, very little information exists for this period. Mahnkey et al. have noted southerly movements of this vulture between mid-Oct and mid-Nov. In 1989 over 40 were noted flying southward during the above period (PM-**BB** 57[1]:47). High counts, both in Taney Co. by PM et al.: 40+, 24 Nov 1988 (**BB** 56[1]:13); 23, 31 Oct 1987 (**BB** 55[1]:10). Extralimital record: 3, 29 Sept 1990, L. of Ozarks SP (PMC, TB).

Winter: The first winter record was not obtained until 30 Dec 1964 when 23 were seen on the Mingo CBC. An extraordinarily large flock, consisting of 112 birds, was reported on this count on 27 Dec 1968; nonetheless, it has been observed on only 10 of the 26 Mingo counts. Birds were first recorded from the Table Rock L. area in Dec 1966 (BBR-**BB** 34[1]:21); however, the species was not reported again from this region until one was seen on 27 Jan 1986 (PM-**BB** 53[2]:8). Pat Mahnkey and J. Hayes' discovery of winter vulture roosts in 1986 in Taney Co., has dramatically changed our perception of this vulture's winter status. It now appears to be a regular winter resident in the Table Rock L. region. High counts: ca. 140, 3 Feb 1990, Table Rock L. (PM); 139, 18 Dec 1988, Taney Co. CBC.

Turkey Vulture (*Cathartes aura*)

Status: Common transient and summer resident; rare and local winter resident in south, accidental in north.

Documentation: Specimen: male, 22 Mar 1973, near Grant City, Worth Co. (NWMSU, DLD 35).

Habitat: Found in a wide range of habitats during migration. Most common along bluffs and at the edge of forest, especially bordering streams and lakes during the breeding season and winter.

Records:

Spring Migration: The first migrants begin appearing in late Feb and early Mar in the south and by mid-Mar in the north. Peak is usually in late Mar and early Apr in the south and during mid-Apr in the north. Relatively large concentrations (30+ birds) are regularly seen into May. High count: 67, 24 Mar 1963, near Ashland, Boone Co. (D. Hatch-**BB** 30[2]:24).

Summer: Although this species breeds statewide, it is most common in the Ozarks (average 2.3 birds/route) and along the forested bluffs of the Missouri and Mississippi rivers. An average of 1.3 birds/route has been recorded in the Glaciated and Osage plains. It has never been recorded on the single BBS in the Mississippi Lowlands, although it may breed there very locally. High count: 66, 20 June 1973, Honey Creek WA, Andrew Co. (FL-**BB** 40[4]:3).

Fall Migration: Migrants begin passing through in early Sept. Relatively large-sized groups (30+) are seen by mid-Sept, with the peak at the end of Sept or early Oct. Most have left the northern half of the state by early Nov; however, relatively large concentrations (100+) are usually seen at favorable roost sites in the south until mid-Nov. During mild falls these large groups may persist until mid-Dec. High counts: 879, 5 roost sites, 29 Oct 1988, Taney/Stone counties (Hayes 1989a); 728, 4 roost sites, 31 Oct–1 Nov 1987, Taney Co. (PM et al.-**BB** 55[1]:10); 234, 4–5 Oct 1986, L. of the Ozarks SP (m.ob.).

Winter: Status of this species during this period varies considerably with the severity of the winter. Most observations are of single birds in the extreme southern section of the state. During mild winters, relatively large concentrations are observed at the Mingo/Duck Creek refuges and the Taney/Stone counties. There is no winter record for the northern quarter of the state. High counts: 308, 3 Feb 1990, Table Rock L. (PM); 280+, 19 Jan 1986, Duck Creek (JW-**BB** 53[2]:8). Northernmost records: 1, 9 Feb 1944, northern Kansas City, Clay Co. (NF-**BB** 11[3]:16); 1, 8 Feb 1990, Mineola Grade, Montgomery Co. (S. Hazelwood-**BB** 57[2]:103).

Order Phoenicopteriformes

Family Phoenicopteridae: Flamingos

[**Greater Flamingo** (*Phoenicopterus ruber*)]

Status: Hypothetical.
Documentation: Photograph: 1, 13 May–11 June 1965, Squaw Creek (HB et al.-**BB** 33[1]:12; photo in ref.).
Records and Comments:
Besides the above sighting there is another observation of a bird on 1 Sept 1959, at LaGrange, Lewis Co. (Harford 1959). Almost certainly these observations represent escaped birds.

Order Anseriformes

Family Anatidae: Swans, Geese, and Ducks

Fulvous Whistling-Duck (*Dendrocygna bicolor*)

Status: Casual transient; accidental summer visitant.
Documentation: Specimen: female, captured in teal trap, 17 Apr 1953, Squaw Creek (W. Boyd; SJM 73.1.36).
Habitat: Marshes, ponds, and lakes.
Records:
Spring Migration: There are two additional records to the above specimen, both from the northeastern corner: 1, shot, 29 Apr 1909, Lewis Co. (Widmann 1907); 1, appeared after a huge dust storm, being first seen on 3 Mar 1935 and lingering to the end of the month, Knox City, Knox Co. (Musselman 1937).
Summer: There are three records: 2, photographed, 2–21 June 1981, Ted Shanks (TB, JW-Barksdale 1983; birds also photographed by DAE); 2 adults, 30 June–1 July 1990, near Steele, Pemiscot Co. (H. Schanda, BRE); 1, eclipse plumage, 7 Aug 1990, Duck Creek (HF).
Fall Migration: An individual was taken (USNM 120308) from a group of

three in the fall of 1890, near New Madrid, New Madrid Co. (Sparks 1891). In the original publication the locality was erroneously given as "New Albany," but it was corrected by Sparks in a later (same vol.) issue of *Forest and Stream;* however, this correction apparently was overlooked by Widmann (1907) and others. We now consider the published sighting of two birds on 27 Sept 1974, Clyde, Nodaway Co., as questionable (**BB** 42[3]:11).

[**Black-bellied Whistling-Duck** (*Dendrocygna autumnalis*)]

Status: Hypothetical; accidental fall transient.
Documentation: Sight records only: see below.
Habitat: Marshes.
Records:
There are three observations, the initial two are considered to be of wild birds: 4, 11 Nov 1939, Squaw Creek (EC, D. Cole, refuge manager-Rising et al. 1978); and 18, 18–25 Oct 1989, west of Portage des Sioux, St. Charles Co. (J. and H. Belz, J. Schneithorst). The other is of a probable escapee, as the bird was very tame and consorting with domestic waterfowl from 5–18 Jan 1985, Cape Girardeau (V. Moss et al.-**BB** 52[2]:19).

Comments: Since the late 1970s this unique duck has increased dramatically in southern Texas (Lasley and Sexton 1988). Reflecting this increase are the number of recent extralimital records away from southern Texas. During the same week that the above flock of 18 whistling-ducks was seen in eastern Missouri, a bird was collected in Nebraska that showed no signs of having been kept in captivity (J. Anderson-**AB** 44:115). Additionally, there are two records during the 1980s for Kansas (Thompson and Ely 1989).

Tundra Swan (*Cygnus columbianus*)

Status: Rare transient and winter resident.
Documentation: Specimen: female, 28 Oct 1968, near Mound City (NWMSU, DLD 119).
Habitat: Marshes, rivers, large lakes, and reservoirs.
Records:
Spring Migration: Migrating individuals begin appearing in late Feb, with the vast majority of observations during the first three weeks of Mar. There are only two definite Apr sightings. High count: 40, 13–19 Mar 1983, Springfield (CB et al.-**BB** 50[3]:11). Latest record: 3 immature, 4–16 Apr 1984, Cooley L. (HB, CH-**BB** 51[3]:8).

Fall Migration: The earliest migrants appear at the very end of Oct, with most records during the last two weeks of Nov. Earliest date: 3, 27–29 Oct 1968, Trimble (C. Blanchard-**BB** 36[2]:8). High counts: 18, 21 Nov 1985, Squaw Creek (B. Heck-**BB** 53[1]:26); 17, 14 Nov 1968, Squaw Creek (HB).

Winter: Most of the observations for this season are of late fall or early spring migrants; nevertheless, there are several observations of birds overwintering in mild years. High count: 7, 10 Feb 1983, Centralia, Boone Co. (J. Smith-**BB** 50[2]:22).

Trumpeter Swan (*Cygnus buccinator*)

Status: Former common transient statewide and summer resident in north.

Documentation: Specimens: none extant, see below.

Habitat: Marshes, lakes, and rivers.

Records:

This species was still a common transient as late as the 1850s and was even breeding across the northern half of the state until then. But, shortly thereafter, it ceased breeding, and by the late 1890s it had become a very rare transient due to extreme hunting pressure (Widmann 1907; McKinley 1962). Apparently the last definite record was of a bird taken in Apr 1900 at Kansas City (Harris 1919b; the specimen is apparently no longer extant).

In the early 1960s, swans were transplanted from the native, remnant population at Red Rocks NWR, Montana, to LaCreek NWR, South Dakota. Young were produced there as early as 1964 (fide JW). On 7 Dec 1978, three birds from a flock of six, were illegally shot at Thomas Hill Res, Mo. One of the three killed, a female, was banded as a cygnet at LaCreek in Aug 1972 (Burgess 1981). A number of sightings of collared Trumpeter Swans in Missouri during the mid-1980s can be attributed to a reintroduction program in Minnesota (see **BB** 52[2]:14 and 52[3]:11). Attempts have also been made in the late 1980s to established this species at Mingo.

Observers should carefully record information (color and numbers on collars) of any tagged individuals, so that the bird(s) origin may be determined. Presently, the Missouri Rare Bird Records Committee does not consider the Mingo or Minnesota populations to be feral.

[**Mute Swan** (*Cygnus olor*)]

Status: Hypothetical; status uncertain.

Documentation: Sight record only: see below.

Habitat: Ponds, lakes, and marshes.

Records:

Unfortunately this aggressive swan has been introduced to North America from the Old World. There is a sight record of a marked bird of known feral origin. An immature was seen with an adult below Winfield Dam on the Mississippi R., Lincoln Co., from 6 Jan to 9 Feb 1985 (RG et al.). The bird was banded as a cygnet the previous July at Eagle, Wisconsin. There are several additional records during the 1970s and 1980s of single birds of unknown origin.

A few pairs bred in the Springfield and Poplar Bluff, Butler Co., areas during the 1970s and 1980s, respectively. These populations are now extirpated.

Greater White-fronted Goose (*Anser albifrons*)

Status: Uncommon transient in west, rare in east; accidental summer visitant; rare winter resident in extreme west, casual elsewhere.

Documentation: Specimen: female, 26 Apr 1968, near Maryville (NWMSU, CLJ 165).

Habitat: Primarily marshes, lakes, and ponds.

Records:

Spring Migration: The species is significantly more common in the western half of Missouri, thus the following generalizations primarily pertain to that area. The earliest arrivals appear as soon as the larger bodies of water open, usually in late Feb or early Mar. Peak (2,000–2,500) is normally during the third week of Mar, with relatively large numbers (in the hundreds) seen through early Apr. Smaller-sized flocks are regularly encountered through late Apr and even rarely into early May: 13, 9 May 1982, Schell-Osage (M. Cooksey, KH, CH-**BB** 49[3]:16). Solitary birds are occasionally seen until the end of May (some of these records are of sick or injured individuals). High counts: 3,000, 25 Mar 1972, Squaw Creek (HB, fide FL); 3,000, 14 Mar 1981, Squaw Creek (FL). High count for eastern Missouri: 19, 21 Mar 1982, Mingo (SD-**BB** 49[3]:16).

Summer: Three records: 1, 12 June 1979, St. Charles Co. (SP, NJ-**BB** 46[4]:7); 1, 1 June 1980, St. Joseph (FL); 1, early May to early July 1990, Riverlands Environmental Demonstration Area, St. Charles Co. (m.ob.-**NN** 62:52). Probably all of these birds were sick or injured.

Fall Migration: The species is considerably less common in the fall than the spring in western Missouri. The following summary concerns western Missouri: The first arrivals begin appearing in late Sept, with peak (mid-hundreds) usually during mid- to late Oct. Relatively large numbers are regularly seen through mid-Nov, even later during mild falls, with a few birds normally lingering into Dec. Earliest record: 12, 13 Sept 1970, Squaw Creek (FL). High count: 2,000, 2 Oct 1965, Squaw Creek (HB-**BB** 33[1]:5). High count in east: 12, 26 Oct 1963, Alton Dam (C. Kniffin-**BB** 30[4]:15).

Winter: A rare winter resident in the extreme western section, especially in the northwestern corner. Naturally, more birds are recorded during mild winters. High count for west: 73, 1 Jan 1983, Squaw Creek (FL). High count (and first winter record for St. Louis) for east: 4, 14 Jan 1971, Busch WA (JEC, RA-**BB** 38[4]:10).

Comments: A hybrid, Greater White-fronted X Canada, was observed at

Busch WA on 12 Sept 1973 (JEC-**BB** 41[1]:2).

A bird originally reported as a Bean Goose (*Anser fabalis*) at Squaw Creek in late Nov 1985 (J. Vance, "Travelin' Birds," *Dateline All Outdoors*, 31 Jan 1986, 4–5), proved to be an immature Greater White-fronted Goose upon examination of the specimen and photos (DAE, MBR).

Snow Goose (*Chen caerulescens*)

Status: Common transient in west, uncommon in east; casual summer visitant in northwest, accidental elsewhere; uncommon winter resident.

Documentation: Specimen: female, 10 Dec 1966, Holt Co. (NWMSU, LCW 603).

Habitat: Marshes, rivers, ponds, lakes, and cultivated fields.

Records:

Spring Migration: Migrants begin augmenting winter populations in mid- to late Feb, with numbers increasing rapidly and dramatically in early Mar. Peak normally occurs in mid-Mar (now in excess of 350,000 at Squaw Creek), with relatively large concentrations remaining through early Apr. A few flocks (numbering in the low hundreds of individuals) are readily encountered through the third week in Apr. Small groups and single birds are seen into May, but by mid-May mostly cripples or sick birds are seen. This species is much less common away from the western tier of counties; for example, at Swan L., only 160 km east of Squaw Creek, peak normally consists of no more than 20,000 birds (KG, pers. comm.). High count: ca. 460,000, mid-Mar 1969, Squaw Creek (HB-**BB** 38[1]:3).

Summer: With the exception of a couple of records, all summer observations are from Squaw Creek. Records span the entire period, but most of them concern sick or injured birds. A solitary bird was seen on 13 July 1981, at Schell-Osage (SS, TB-**BB** 48[3]:23). High count: 10, 13 June 1971, Squaw Creek.

Fall Migration: A few small-sized flocks begin appearing in late Sept, and by late Oct concentrations number in the hundreds of thousands. This species provides one of the most spectacular migration events in the midwest when up to 450,000 birds (numbers fluctuate considerably from year to year) can be seen during the peak at Squaw Creek in early to mid-Nov. Peak numbers at Swan L. at this same time usually consist of ca. 25,000 (KG, pers. comm.). Concentrations in northwestern Missouri often remain high until the water freezes (usually by mid-Dec). High count: 450,000+, Nov 1985, Squaw Creek (fide B. Heck).

Winter: Concentrations vary considerably from year to year, as a result of the weather, with more birds remaining during mild winters. Even during severe winters, a relatively large-sized flock roosts on the open water generated by the Iatan Power Plant, Platte Co. High counts, both on the Squaw Creek CBC: 200,000, 16 Dec 1984; 130,000, 19 Dec 1982.

Comments: This species has been steadily increasing over the past thirty years as a result of improved breeding production. In Missouri, it is more common in the west than in the east at all times of the year. The most impressive concentrations appear in the northwestern corner, where there are suitable feeding and roosting sites.

Formerly the blue morph, a color phase referred to as the "Blue Goose," was considered a separate species. The white morph usually outnumbers the blue phase, often by a 2.5:1 ratio, although this ratio can vary considerably. For example, the blue morph comprised slightly over one-half of the 451,200 geese that were present at Squaw Creek during the week of 12 Mar 1967 (HB-**BB** 34[2]:6).

The Lesser Snow (*C. c. caerulescens*) is the race positively recorded in the state. (See comments under Ross' Goose regarding hybrids.) Five hybrid Snow X Canada geese were captured and photographed at Swan L. in mid-Feb 1981 (M. Lee, D. Herrman, fide DAE; photos in authors' possession).

Ross' Goose (*Chen rossii*)

Status: Locally uncommon transient in northwest, casual elsewhere; locally rare winter resident in northwest, casual elsewhere.

Documentation: Specimen: male, 25 Apr 1974, Bigelow Marsh (NWMSU, DLD 155).

Habitat: Marshes, rivers, ponds, lakes, and cultivated fields.

Records:

Spring Migration: The species accompanies Snow Geese, having virtually the same migration time schedule. For example, the first migrants begin appearing in late Feb: 3, 25–27 Feb 1984, Kansas City (CH et al.-**BB** 51[2]:17). High count: 7, 15 Mar 1983, Squaw Creek (FL). Latest date: 1, 7 May 1989, Squaw Creek (DAE-**BB** 56[3]:89). There are only three records for eastern Missouri: 1, 28 Feb–5 Mar 1988, Creve Coeur L. (R. Laffey et al.-**BB** 55[2]:87); 1, 13 Mar 1986, Ten Mile Pond, Mississippi Co. (JH); and 1, 15 Mar 1986, Duck Creek (SD-**BB** 53[3]:10).

Fall Migration: Birds begin appearing with the return of Snow Geese; the earliest date is of 1, 14 Oct 1975, Maryville SL (DAE-**BB** 44[1]:18). The largest numbers coincide with the peak of the Snow Goose migration. The high count is 8, 17 Nov 1985, Squaw Creek (FL-**BB** 53[1]:26); however, estimates made during the falls of 1968 through 1970 at Squaw Creek have shown that during the peak of the Snow Goose migration the ratio of Ross' to Snow may be as high as 1:1,000 (Prevett and MacInnes 1972). Thus, perhaps as many as 300 Ross' Geese were present at one time during those falls. At present, we believe this ratio may have increased in favor of Ross' (as predicted by Prevett and MacInnes), and this increase coupled with the increase in Snows (ca. 450,000+, fall of 1985) may result in greater than

500 Ross' at peak in the fall. There are two fall records for eastern Missouri: 1, 2 Nov 1985, St. Charles Co. (D. Becher-**BB** 53[1]:26); 1, 11 Nov 1988, Busch WA (HF, BL, BRE et al.-**BB** 56[1]:12).

Winter: It is regularly found in small numbers with sizable flocks of Snow Geese in the northwestern corner. These are the following records for eastern Missouri: 2 adults, 22 Dec 1980, Orchard Farm CBC; 1, 11 Jan 1985, Grafton Ferry Road, St. Charles Co. (CP-**BB** 52[2]:15); 2, 18 Feb 1990, Alton Dam (RG, T. Goetz-**BB** 57[2]:101). High count: 4 (3 adults, 1 immature), 29 Dec 1988, Schell-Osage WA (MBR-**BB** 56[2]:53).

Comments: The first definite record for Missouri was of a bird taken in late Oct 1963 near Mindenmines, Barton Co. (fide RA-**BB** 31[1]:30). Hybrids between Ross' and Snow geese are occasionally seen: one was netted at Squaw Creek on 8 Oct 1963 (HB-**BB** 30[4]:15); another was taken there on 29 Jan 1963 and preserved (USNM 479427); and another was captured at Squaw Creek on 15 Mar 1964 (HB, DAE, EC, B. Massie). The rare blue morph, which may result from hybridization with the Snow Goose, has not yet been reported from Missouri.

Brant (*Branta bernicla*)

Status: Accidental spring and casual fall transient; accidental winter resident in northwest.

Documentation: Specimen: male, 27 Nov 1956, Swan L. (USNM 464454).

Habitat: Marshes and lakes.

Records:

Spring Migration: Only two records: 3, 3 Apr 1966, Little Prairie L. near Rolla, Phelps Co. (F. Frame and Mrs. Ollar-**BB** 33[2]:6); 1, 28 Mar 1970, Squaw Creek (P. and L. Prevett-**BB** 38[3]:8).

Fall Migration: At least a dozen records with all but two concentrated in the north central (Swan L.) and northwestern (Squaw Creek) sections of the state. Most of these records are the result of birds having been brought in by hunters. Records span from early Oct through early Dec. Earliest record: 1, 1 Oct 1958, Swan L. (fide DAE-**BB** 25[10]:3). High count, and the only record involving more than one bird at this season: 20, first week of Nov 1987, Hanna L., Iron Co. (M. Roux, fide BRE).

Winter: Three records, all from the northwestern corner: 1, 20 Dec 1987, Squaw Creek CBC (G. Shurvington, DAE); 1, photographed, 4–19 Jan 1975, Pony Express WA, DeKalb Co. (FL et al.-**BB** 42[3]:12; Fig. 6); and 1, 2 Jan 1983, Pony Express WA (FL-**BB** 50[2]:23).

Comments: This species probably occurs more frequently than the above records indicate, as it is easily overlooked among Canada Geese with which it usually associates. All records are of the nominate race.

Fig. 6. This Brant of the eastern race was present at Pony Express Lake in DeKalb Co. for two weeks in January 1975. This is one of only three winter records for the state. It was photographed by Bill Goodge on 19 January.

[**Barnacle Goose** (*Branta leucopsis*)]

Status: Hypothetical; casual transient.
Documentation: Photograph: 1, 28 May 1984, Reed WA (S. Cooper-**BB** 51[3]:8).
Habitat: Marshes, ponds, and lakes.
Records:
Spring Migration: 1, 26 Mar 1950, Marais Temps Clair (m.ob.-**BB** 18[1]:1); 1, 25–26 Mar 1976, Squaw Creek (LG et al.-**BB** 44[1]:25); and the above photograph record.
Fall Migration: 2, 6 Nov 1977, Squaw Creek (LG et al.-**BB** 45[1]:24); 1 photographed, 20–26 Oct 1980, Swan L. (D. Graber-**BB** 48[1]:11).
Comments: This is a commonly raised captive goose in North America, so all records to date are suspect. The species should remain on the hypothetical list until a bird banded in its normal range is recaptured in Missouri.

Canada Goose (*Branta canadensis*)

Status: Uncommon to locally common summer resident; common transient and winter resident.

Documentation: Specimen: male, 8 Dec 1947, Holt Co. (KU 49025).
Habitat: Primarily ponds, lakes, and marshes.
Records:
Spring Migration: A common migrant that primarily passes through the state during late Feb and early Mar. The largest concentrations are at Swan L., where ca. 50,000–70,000 are recorded during peak in late Feb (KG, pers. comm.). The largest number seen at Squaw Creek included 18,000, of which 11,000 were of the larger races and 7,000 were of the small races (primarily Hutchison's or Richardson's) during the first half of Mar 1963 (HB-**BB** 30[2]:22).
Summer: Prior to and during early settlement, the species was found breeding all along the Missouri and Mississippi rivers, but by the 1890s it had become rare (Widmann 1907). Today it continues to breed under natural conditions only in the rocky bluffs along the Missouri R. between Jefferson City and St. Charles (Johnson 1975). Although all other breeding populations in the state are the result of reintroductions (most use metal tubs as nest sites), the "bluff" birds are very likely of natural occurrence; residents of this area have recollections of the birds at least as far back as the 1920s (Johnson 1975). McKinley (1961) has chronicled the breeding history of this species in Missouri.
Fall Migration: Migrants begin reappearing by mid-Sept and peak usually in early to mid-Nov. As in the spring the largest buildup is in the Swan L. area, where regularly over 100,000 are recorded in early Nov. High counts, both at Swan L.: 150,000, mid-Nov 1978; 150,000, 4 Nov 1988 (KG, pers. comm.).
Winter: Thousands winter annually in the Swan L. area; the milder the winter, the larger the number. It is commonly found wintering in other areas across the state, particularly where there is a combination of open water and favorable feeding sites. High counts, both on the Swan L. CBC: 132,100, 19 Dec 1976; 127,950, 18 Dec 1977.
Comments: Breeding birds have always been of the race *maxima;* however, birds that are currently breeding in the state, except for those breeding in the bluffs along the Missouri R., are the result of reintroduced birds from captive stock. The reintroductions began in the late 1940s and early 1950s. The very small race, *B. c. hutchinsii,* is commonly seen in migration and winter. The intermediate race, *B. c. interior* is readily encountered as well. Five different races were identified at Squaw Creek on 20 Nov 1965 (HB-**BB** 33[1]:5).
See comments under Snow Goose account.

Wood Duck (*Aix sponsa*)

Status: Common transient and summer resident statewide; rare winter resident, primarily in south.

Documentation: Specimen: male, 4 Apr 1967, near Maryville (NWMSU, LCW 550).

Habitat: Wooded swamps, ponds, lakes, and slow moving streams and rivers.

Records:

Spring Migration: A few migrants begin appearing in late Feb, but numbers remain relatively low until the end of Mar and peak in early Apr; thereafter it is quite common. High count: 1,000, 1 Apr 1988, Swan L. (KG, pers. comm.).

Summer: The most common breeding waterfowl in the state. It is especially common at the Duck Creek and Mingo areas and on oxbow lakes along the Missouri and Mississippi rivers. By mid-July birds begin congregating: 300, 5 July 1988, Swan L. (KG, pers. comm.).

Fall Migration: The late summer concentrations, largely composed of immature birds, continue to build through Aug and peak in early to mid-Sept. Nevertheless, relatively large numbers are observed through mid-Oct: 340, mid-Oct 1963, Squaw Creek (HB-**BB** 30[4]:16). Numbers drop off shortly thereafter and continue to decrease until the end of Nov when usually only a few birds are encountered. High count: 1,500, fall (no specific date given) 1963, Duck Creek (J. Davis-**BB** 30[4]:16); 1,000, 3 Oct 1988, Swan L. (KG, pers. comm.).

Winter: Birds are seen in small numbers and in widely scattered localities in winter; most are observed in the southern third of the state. Very rarely, a relatively large concentration may be encountered: 1,000, 27 Dec 1963, Mingo CBC. Other high counts: 500, late Dec 1963, Squaw Creek (HB-**BB** 31[1]:30); 150, 28 Dec 1965, Mingo CBC.

Green-winged Teal (*Anas crecca*)

Status: Common transient; locally common winter resident in south, rare in north; rare summer visitant in northwest, accidental elsewhere.

Documentation: Specimen: female, 10 June 1978, Bigelow Marsh (NWMSU, LWT 89).

Habitat: Marshes, ponds, and lakes.

Records:

Spring Migration: Migrants begin appearing as soon as the water reopens. Peak is normally at the end of Mar or early Apr. By mid-May it is rare but regularly seen in small numbers until the end of the period. High counts: 25,000, 3 Apr 1966, Trimble/Squaw Creek (FL); 22,200, 19–25 Mar 1967, Squaw Creek (HB-**BB** 34[2]:6).

Summer: A rare but regular summer visitant in the northwestern corner, especially in the Squaw Creek area. It may occasionally even breed there, as pairs are infrequently seen throughout the summer, but to date there is no evidence of nesting. It is accidental elsewhere: 2, throughout the summer of 1979, Thomas Hill Res. (JR-**BB** 46[4]:7). High counts: 10, 3 July 1990, Squaw Creek (DAE); two counts of 8 birds at Squaw Creek in June.

Fall Migration: Migrants begin reappearing in Aug, with large concentrations by the end of Aug or early Sept: ca. 700, 1 Sept 1963, Squaw Creek (HB-**BB** 30[4]:15). Numbers continue to increase until they peak, generally in late Oct or early Nov. Relatively large concentrations (in the hundreds) may still be observed even through early Dec. High count: 55,100, early Nov 1963, Squaw Creek (HB-**BB** 30[4]:15).

Winter: Locally common resident in south, much rarer in north. Numbers vary considerably depending upon the severity of the winter. High counts, both on Mingo CBC: 2,000, 27 Dec 1963; 650, 28 Dec 1965.

American Black Duck (*Anas rubripes*)

Status: Uncommon transient and winter resident; accidental summer resident; casual summer visitant.

Documentation: Specimen: male, 9 Dec 1949, Holt Co. (KU 39026).

Habitat: Lakes, reservoirs, ponds, and marshes.

Records:

Spring Migration: It is more common in the eastern half of the state. Peak is usually in late Mar, and it is virtually absent by May.

Summer: There are surprisingly two nesting records, both from western Missouri: 10, including immatures that were banded in early Aug 1962, Squaw Creek (HB-**BB** 29[3]:21); 2 pairs, with broods of 8 and 5, summer 1973, Schell-Osage (JR-**BB** 40[4]:2). There are several additional records of nonbreeding birds.

Fall Migration: Birds begin reappearing in late Aug and gradually increase in numbers until they peak in early Nov. High count: ca. 850, early Nov 1962, Squaw Creek (HB-**BB** 29[4]:23).

Winter: Widespread, but in relatively small numbers throughout the season. The largest concentrations are seen in the Mingo area. High counts, both on Mingo CBC: 500, 28 Dec 1965; 315, 28 Dec 1962.

Comments: There apparently was little contact between the Black Duck and Mallard prior to 1900; but the release of farm-reared game Mallards in the eastern U.S., coupled with a natural expansion of the Mallard eastward, has resulted in extensive introgression between these two forms (Heusmann 1974). Upon close examination, most, if not all, of the birds seen in the state are intergrades with the Mallard. Even birds that are relatively close and appear pure in the field, when examined in the hand have flecks of green on the head. The above nesting records undoubtedly represent birds that had Mallard genes. Ankney et al. (1986) have shown that the genetic distance between the Black Duck and Mallard is slight, and it is comparable to that found between local populations of the same species. This integration coupled with the fact that these ducks are widely sympatric (thus one cannot be treated as a subspecies of the other), argues for the Black Duck to be treated as a melanistic morph of the Mallard.

Mallard (*Anas platyrhynchos*)

Status: Common transient and winter resident; rare summer resident.
Documentation: Specimen: male, 14 Nov 1965, Holt Co. (KU 51002).
Habitat: Found on virtually every type of water, but most common in marshes and ponds.
Records:
Spring Migration: The most common duck during migration. Winter populations are augmented by migrants in the latter half of Feb. Peak is normally in mid-Mar. Numbers drop off significantly by early Apr, and by early May only smaller-sized flocks are seen (usually numbering under one hundred birds/flock). High count: 100,000, 12 Mar 1967, Squaw Creek (HB-**BB** 34[2]:6).
Summer: A locally rare nester in small numbers throughout the state. Possibly the largest concentration of breeding birds is at Squaw Creek.
Fall Migration: Numbers begin building in Aug. Large flocks are seen by late Sept, with an impressive peak occurring in mid- to late Nov. If the fall is mild, huge concentrations may remain until the water freezes. High counts, both at Squaw Creek: 168,000, 18 Nov 1968 (HB-**BB** 36[4]:11); 160,000, 28 Nov 1963 (HB-**BB** 30[4]:15).
Winter: It remains quite common throughout the winter, especially in areas of open water. Even during severe winters, it is readily found using the larger rivers and reservoirs. High counts, both at Squaw Creek: 210,000, 14–20 Dec 1958 (K. Krum; fide FL); 94,000, 7–8 Jan 1963 (HB; fide FL).

Northern Pintail (*Anas acuta*)

Status: Common transient; accidental summer resident in northwest; rare summer visitant in northwest, accidental elsewhere; uncommon winter resident.
Documentation: Specimen: male, 1 Feb 1956, Gravois Mills, (MU 1586).
Habitat: Marshes, ponds, and lakes.
Records:
Spring Migration: Birds begin arriving as soon as the ice begins to melt on lakes and ponds. Peak is normally during the third week in Mar. Most have passed through by mid-Apr, and few are seen in May. High count: 145,000, 25 Mar 1965, Squaw Creek (HB-**BB** 32[1–2]:10).
Summer: The species is regularly seen in small numbers throughout the summer in northwestern Missouri, particularly in the Squaw Creek area; however, there are only three definite nesting records for the state. In 1921, Harris (**BL** 23:254) stated that it "is quite likely that a few pairs of Shovelers

Fig. 7. Nest and eggs of Northern Pintail at Bigelow Marsh, Holt Co., 20 May 1990. This photograph is the third definite nesting record for the state. Photograph by David Easterla.

and Pintails have succeeded in bringing off young in the lake regions of Platte and Buchanan counties, as family parties of these ducks have lately been seen on the Missouri R. between Leavenworth and St. Joseph." The second record is of a family seen in the summer of 1971 at Squaw Creek (HB-**BB** 38[4]:7). Most recently, a hen was flushed from a nest containing nine eggs on 20 May 1990 in Bigelow Marsh (DAE-**BB** 57[3]:135; Fig. 7).

Fall Migration: Small-sized flocks begin appearing in the latter half of Aug, and by early Sept thousands may be seen at some of the refuges: 10,000, 9 Sept 1978, Swan L. (RB-**BB** 46[1]:22). However, peak generally does not occur until the end of Oct, with thousands still remaining until even mid-Dec in mild years. High counts, both at Squaw Creek: 52,000, 2 Oct 1965 (HB); 50,000, 28 Oct 1968 (HB-**BB** 36[4]:11).

Winter: Although relatively large-sized flocks (in the low thousands) may be seen during the early and latter part of this period, it generally is an uncommon resident. In the area of greatest abundance (at Mingo), it is nearly three times as common as the Gadwall (19.8/pa hr vs. 7.0/pa hr). High count: 6,000, 28 Dec 1965, Mingo CBC.

Comments: A hybrid N. Pintail X American Black Duck was seen on 25 Mar 1962 at Squaw Creek, not St. Joseph (FL-**BB** 29[2]:15). A hybrid male N. Pintail X Mallard was taken at Fountain Grove WA, Linn Co. on 4 Nov 1955 (M. Milonski; MU 1602).

The drought of the late 1980s throughout the prairie pothole country of North America, coupled with hunting pressure, drastically reduced N. Pintail numbers in recent times. Waterfowl biologists estimated that the population of this species was over 6 million in the 1970s but had dropped as low as 2.6 million in 1988. To a lesser extent, these same forces have affected many of the other waterfowl populations in the same way.

Blue-winged Teal (*Anas discors*)

Status: Common transient; rare summer resident in north; casual winter resident.

Documentation: Specimen: male, 9 May 1967, Bigelow Marsh (NWMSU, LCW 477).

Habitat: Marshes, ponds, and lakes.

Records:

Spring Migration: Single birds and pairs begin arriving in early Mar, but it remains rare to uncommon until early Apr when large numbers begin to appear. Peak usually does not come until about the third week in Apr, with large numbers seen into early May. By the third week in May, few birds remain. High counts: 50,000, 23 Apr 1967, Squaw Creek/St. Joseph (FL); 12,000, 11 Apr 1982, Squaw Creek (FL).

Summer: The species breeds locally across northern Missouri. There are no definite nesting records south of the Missouri R., although birds are occasionally seen there at this season: 2, 15 June 1978, east of Charleston, Mississippi Co. (BE-**BB** 45[4]:11).

Fall Migration: Migrants begin arriving in mid-Aug, with large concentrations present by early Sept. Peak is from mid- to late Sept, and most birds are gone by mid-Oct. By mid-Nov it is quite rare. High count for early and late in season, respectively: 6,000, 20 Aug 1987, Swan L. (P. Thomsen, fide KG); 1,500, 2 Nov 1987, Swan L. (P. Thomsen, fide KG).

Winter: This species is very rarely encountered during this season, and almost all of the records are of 1–2 individuals during the latter half of Dec. Only two Jan records: 1, 4 Jan 1969, Alton Dam (H. Hill-**NN** 41:28). High count: 6, 29 Jan 1987, Rockaway Beach, Taney Co. (PM-**BB** 54[2]:12).

Comments: Occasionally hybrids between this and the Cinnamon Teal are recorded. There are at least five records of hybrids, all involving single males: 1 Apr 1972, Sugar L. (FL); trapped, 3 Apr 1952, Fountain Grove WA, Livingston Co. (**BB** 19[12]:2); 13 Apr 1982, Fountain Grove WA, Linn Co. (H. Clark; MU 1596); 10 May 1955, Trimble, Clinton Co. (L. Helm; MU 1592); and 21 June 1969, Squaw Creek (MBR, RAR, DT).

Cinnamon Teal (*Anas cyanoptera*)

Status: Rare spring transient in west, casual in east; accidental fall transient.

Documentation: Specimen: male, late Mar 1986, near Big L. SP (D. Mead, B. Snipes; SJM 89.1.646).

Habitat: Marshes and ponds.

Records:

Spring Migration: A rare but annual migrant through the western part of the state; much scarcer in the east (> 6 records for the St. Louis area). Although it is regularly recorded from mid-Mar to the beginning of the second week of May, the bulk of the records are during late Mar and early Apr. Earliest record: pair, 1 Mar 1977, Trimble (B. West-**BB** 44[4]:25). High count: 2 pair, 31 Mar 1976, Squaw Creek (DAE-**BB** 44[1]:25), although Harris (1919b) mentions a record of 5 birds by Tindall at Lake City, Jackson Co., in 1895 (no specific date given). Latest record: 1, 19 May 1967, Squaw Creek (HB-**BB** 34[2]:6), but see below. One record for the southeastern corner: 2, 15–17 Mar 1986, Duck Creek (SD, BRE, K. Adams-**BB** 53[3]:10).

Summer: A single record involving a bird that was first discovered in late May and lingered until 15 June 1982, Clarence Cannon NWR, Pike Co. (G. Wolff et al.-**BB** 49[4]:14).

Fall Migration: There are only four fall records, all of adult males: 2, 13 Aug 1983, Schell-Osage (JS et al.-**BB** 51[1]:40); 1, 5 Sept 1976, Schell-Osage (NJ, KH-**BB** 44[3]:20); 1, 9–15 Oct 1983, Mingo (BE); 2, 27 Oct 1971, Busch WA (JC-**BB** 39[1]:5).

Comments: See Blue-winged Teal account.

Northern Shoveler (*Anas clypeata*)

Status: Common transient; accidental summer resident in northwest; rare summer visitant in northwest, accidental elsewhere; rare winter resident.

Documentation: Specimen: male, 27 Nov 1965, Duck Creek (GL 46).

Habitat: Marshes, ponds, and shallow lakes.

Records:

Spring Migration: A few individuals begin arriving in early Mar, but relatively large numbers (in the hundreds or low thousands) do not appear until the latter half of Mar. Peak is later than most other waterfowl and normally occurs in early to mid-Apr. By the beginning of May numbers have dropped off significantly, although solitary or even small groups usually persist until the end of May. High counts, both at Squaw Creek: 20,000, 14 Apr 1968 (FL); 15,000, early Apr 1967 (HB-**BB** 34[2]:6).

Summer: Although Harris (see N. Pintail account) mentioned probable nesting records for northwestern Missouri, definite proof was not established until 1 female and 8 downy young were seen on 10 May 1986 in a marsh on the western edge of Mound City, Holt Co. (LG). Apparently another female and brood were seen at Squaw Creek in May 1988 (BJ, fide RBE). It is surprising that there are not additional records for this region, as some individuals (including pairs) are seen virtually every summer in the Squaw Creek area. High count: 30, 1 June 1987, Swan L. (P. Thomsen, fide KG).

Fall Migration: A few birds begin reappearing in Sept. Numbers gradually increase until they peak in early to mid-Nov. Concentrations drop off dramatically at the end of Nov. High count: 5,000, 17 Nov 1968, Squaw Creek (HB-**BB** 36[4]:11).

Winter: A rare but regular winter resident in relatively small numbers (usually of groups of 20 individuals or less). High counts, both on the Mingo CBC: 600, 27 Dec 1963; 195, 28 Dec 1962.

Gadwall (*Anas strepera*)

Status: Common transient; accidental summer resident in north; rare summer visitant in northwest, accidental elsewhere; uncommon winter resident in south.

Documentation: Specimen: female, 25 Oct 1963, Duck Creek (MU 1581).

Habitat: Marshes, ponds, and shallow lakes.

Records:

Spring Migration: Less common than the N. Pintail and N. Shoveler. Peak occurs earlier for this species than for the shoveler—usually in late Mar. Few birds remain by the latter half of Apr. It is rarely seen during May. High count: 5,000, 24 Mar 1985, Squaw Creek (FL).

Summer: Prior to 1990 this duck had not been recorded breeding in the state since the 1910s. Around the turn of the century it bred in the northern section of the state (Widmann 1907; Harris 1919b). At Squaw Creek in 1990, two instances of breeding were documented. An adult with 13 young were seen on 21 June, and a nest with 7 eggs was located there on 25 June (RBE). Over the past thirty years one or more birds have been seen most summers in northwestern Missouri, especially in the Squaw Creek area. Current records away from the northwestern corner: no.?, 23 June 1973, Creve Coeur L. (A. Bromet-**NN** 45:90); female, 31 July 1981, near Taberville Prairie (SS-**BB** 48[3]:23). High count of adults: 6, 8 June 1967, Squaw Creek (FL).

Fall Migration: The first migrants arrive in mid-Sept. Peak is in late Oct or early Nov. Large numbers are readily seen throughout Nov but drop off markedly by early Dec. High counts: 6,000, 26 Oct 1985, St. Joseph/Squaw Creek area (FL); 4,000, 21 Oct 1978, Squaw Creek (FL).

Winter: See comments under N. Pintail. High counts, both on the Mingo CBC: 6,000, 28 Dec 1965; 1,000, 28 Dec 1962.

Eurasian Wigeon (*Anas penelope*)

Status: Casual transient.

Documentation: Specimen: male, head only, 10 Nov 1946, St. Charles Marsh, St. Charles Co. (MU 188; see below).

Habitat: Ponds, small lakes, and marshes.

Records:

Spring Migration: There are five records, all of single birds, and all within a three-week period from late Mar to early Apr: 23 Mar 1947, near Old Monroe, St. Charles Co. (K. Wesseling-**BB** 15[1]:1); 25–26 Mar 1950, Sugar L. (J. Bishop, M. Magner-**BB** 17[4]:1); 1 Apr 1959, Squaw Creek (fide FL); shot by a hunter, 10 Apr 1905 (Widmann 1907); and 25–26 Apr 1943, Marais Temps Clair (WC; Wilhelm 1958).

Fall Migration: Only two records: the above specimen record and an adult male, 24 Oct 1985, St. Joseph (FL et al.-**BB** 53[1]:26).

Comments: The above mentioned specimen has disappeared or has been misplaced since the MU collection was computerized in 1986. William Elder (pers. comm.) could not find it in Apr 1988, and we failed to uncover this specimen during a May 1989 visit.

American Wigeon (*Anas americana*)

Status: Common transient; casual summer visitant in northwest; uncommon winter resident.

Documentation: Specimen: male, 6 Dec 1956, Sumner, Chariton Co. (MU 1580).

Habitat: Ponds, small lakes, and marshes.

Records:

Spring Migration: Winter populations are augmented by migrants by the end of Feb or early Mar. Peak is in late Mar or early Apr, and by the end of Apr most birds have left the state. A few strays regularly linger into May, but it is rarely seen during the latter half of the month. High count: 4,000, 15 Mar 1983, Squaw Creek (FL).

Summer: All records (> 10) are from the northwestern corner. High count: 4, 8 June 1967, Squaw Creek (FL).

Fall Migration: Birds appear as early as the end of Aug, with relatively large congregations observed by mid-Sept: 500, 17 Sept 1967, Squaw Creek (FL). Peak is generally at the end of Oct, with large numbers encountered until the end of Nov. High counts, both at Squaw Creek: 1,800, 24 Oct 1982 (FL); 1,500, 29 Oct 1978 (FL). It is unclear whether a single bird seen on 10 Aug 1963, Busch WA, (JC-**BB** 30[3]:19) was an early migrant or a summer visitant.

Winter: Normally seen in small groups at this season. As with virtually all other waterfowl, this wigeon is considerably more common in the south. High counts, both on Mingo CBC: 4,000, 27 Dec 1963; 1,400, 28 Dec 1965.

Canvasback (*Aythya valisineria*)

Status: Common transient and winter resident on the Mississippi R.; uncommon transient and rare winter resident away from the Mississippi R.; accidental summer visitant.

Documentation: Specimen: male, 25 Oct 1972, Maryville SL (NWMSU, DLD 90).

Habitat: Prefers deep water, large rivers, lakes, and reservoirs.

Records:

Spring Migration: The relatively large winter concentrations on the Mississippi R. are enhanced by migrants during the latter half of Feb and peak in early Mar. Most have passed through the state by the end of Mar; mainly small numbers are seen in Apr, and by May it is exceedingly rare. High counts: 5,000–10,000, early Mar 1963, Mississippi R., St. Louis (RA et al.-**BB** 30[2]:23); 7,000, 5 Mar 1967, Mississippi R., St. Louis (JEC-**BB** 34[2]:6). High count for western Missouri: 2,000, 31 Mar 1974, Squaw Creek (FL). Latest date: pair, 29 May 1972, Maryville SL (MBR-**BB** 39[3]:6).

Summer: Only two sightings: 1, 17 June 1989, Alton Dam (JVB-**NN** 61:54); male, 22 June 1981, Squaw Creek (SD-**BB** 48[3]:23).

Fall Migration: The first migrants generally do not arrive until mid-Oct, with large numbers appearing in Dec. Five birds seen on 1 Aug 1962 at Squaw Creek (HB-**BB** 29[3]:21) were presumably exceptionally early migrants. High counts: 250, end of Nov 1977, Alton Dam (JEC-**BB** 45[1]:24); 200, 29 Oct 1978, St. Joseph (FL).

Winter: Impressive concentrations are seen along the Mississippi R., especially above Alton Dam. The species is rare away from this river. High counts: 5,000–6,000, second week of Jan 1975, Alton Dam (RA-**BB** 42[1]:12); 3,000, 12 Feb 1967, above Alton Dam (RA, PB-**BB** 34[1]:21).

Redhead (*Aythya americana*)

Status: Uncommon transient; casual summer visitant in northwest; rare winter resident.

Documentation: Specimen: male, 9 Nov 1975, Maryville SL (NWMSU, DST 54).

Habitat: Lakes, ponds, rivers, and marshes.

Records:

Spring Migration: Migrants begin arriving with the opening of water. Peak is usually in mid-Mar. Medium-sized groups, numbering between 20–50 birds, may be seen into early Apr. It is very rare by the end of Apr and is of casual occurrence in May. High counts: 1,200, 19 Mar 1939, Bean L., Platte Co. (WC-**BB** 6[4]:36); 1,000, 5 Mar 1967, St. Louis (JEC-**BB** 34[2]:6).

Summer: There are at least six records, all between Kansas City and the Iowa border. At least three of these involved pairs, but there still is no evidence of breeding. High count: 3 pairs, 20 June 1990, Squaw Creek (RBE).

Fall Migration: The initial migrants normally arrive in early Oct, with numbers increasing until the peak in early Nov. Small-sized flocks are regularly seen through Nov, but by early Dec few are present. High counts: 4,000, 6 Nov 1978, Browning L. (FL-**BB** 46[1]:22); 1,000, 29 Oct–2 Nov 1987, Swan L. (P. Thomsen, fide KG).

Winter: It is scarce statewide in winter and seen only in small numbers. High counts: 30, 31 Dec 1973, Big Oak Tree SP CBC; 30, 20 Dec 1980, Kansas City Southeast CBC.

Ring-necked Duck (*Aythya collaris*)

Status: Common transient; casual summer visitant; uncommon to locally common winter resident.

Documentation: Specimen: male, 3 Mar 1936, Rocheport, Boone Co. (MU 262).

Habitat: Lakes, ponds, and marshes.

Records:

Spring Migration: Birds begin arriving as soon as the ice melts. Peak is usually during the third week in Mar, but occasional large-sized flocks (in the high hundreds or low thousands) remain until mid-Apr. The species is casually encountered after very early May. High counts: 5,000, mid-Mar 1963, Mississippi R., St. Louis (RA-**BB** 30[2]:23); 5,000, 15 Mar 1983, St. Joseph/Squaw Creek area (FL); 5,000, 21–23 Mar 1980, Squaw Creek (FL). Latest record: 1, 18 May 1982, Eleven Point R., Oregon Co. (JW-**BB** 49 [3]:16).

Summer: There are a minimum of eight summer records, and all but one observation are north of the Missouri R. The sole record in southern Missouri is of a male on 15 June 1982, Big Piney R., Pulaski Co. (JW-**BB** 49[4]:14). All observations are of single birds.

Fall Migration: The first individuals appear in late Sept, gradually building in numbers through Oct, with a major influx in early Nov. Peak is normally in mid-Nov. Although there is a drop off in numbers in early Dec, it is still frequently seen until the end of the period. Earliest record: 2, 18 Sept 1982, St. Joseph (FL). High counts: 5,000, 17 Nov 1985, Squaw Creek (FL); 3,000, 6 Nov 1978, St. Joseph (FL-**BB** 46[1]:22).

Winter: One of the most common diving ducks at this season. Occasionally relatively large rafts are seen along the Mississippi R. and in southern Missouri. High counts, both on the Mingo CBC: 5,878, 2 Jan 1989; 3,088, 28 Dec 1978.

Greater Scaup (*Aythya marila*)

Status: Rare transient and winter resident; accidental summer visitant in northwest.

Documentation: Specimen: female, 10 Nov 1972, Maryville SL (NWMSU, DLD 88).

Habitat: Rivers, reservoirs, and lakes.

Records:

Spring Migration: Records begin as soon as there is open water and continue through early May; most of the observations are concentrated in Mar through the first half of Apr. High counts: 16, 22 Feb 1981, Alton Dam (RK-**BB** 48[2]:13); 15, 5 Apr 1984, Excelsior Springs, Clay Co. (CH-**BB** 51[3]:9). Latest records: female, 11 May 1989, near Squaw Creek (DAE, TE,

MBR-**BB** 56[3]:89); 1, 6 May 1978, Maryville SL (DAE, MBR).

Summer: Only one record: female, 7 June 1976, Bigelow Marsh (DAE, MBR-**BB** 44[2]:20).

Fall Migration: There are considerably more fall than spring observations. The best data on the fall migratory movements come from detailed observations (primarily by DAE) made at the Maryville SL, since its construction in 1970. Usually single birds or small-sized groups of 2–3 birds begin appearing during late Oct, with a dramatic increase (usually associated with a cold front out of the northwest) in numbers in early Nov. Peak is normally during mid-Nov. Relatively large numbers remain through Nov, but by the second week in Dec only small-sized groups or solitary birds are seen. All of the following records are by DAE from the Maryville SL. Earliest record: 1, 18 Oct 1986. High counts: 60, 16 Nov 1977 (**BB** 45[1]:24); 40, 3 Nov 1984.

Winter: Less frequently encountered at this season (usually only single birds or small-sized groups) than during the migration period. It is primarily found along the Mississippi R. and on the larger bodies of water in the south. High count: 27, 17 Dec 1989, Kansas City Southeast CBC.

Comments: This species is frequently confused with the more abundant Lesser Scaup. The only reliable field mark for distinguishing these two species is the extent of white on the upper side of the primaries. Figure 8 illustrates the rather pronounced differences in this feature between the two species.

Lesser Scaup (*Aythya affinis*)

Status: Common transient; rare summer visitant; uncommon to locally common winter resident.

Documentation: Specimen: female, 16 May 1970, Bigelow (NWMSU, LWT 102).

Habitat: Rivers, reservoirs, lakes, and ponds.

Records:

Spring Migration: Much more common than the preceding species at all seasons. The largest concentrations are seen along the Mississippi R. Birds begin arriving with the thawing of ice. Peak is usually during the latter half of Mar, but large numbers (in the thousands) are still apparent during the first half of Apr; thereafter, numbers drop off dramatically. Nevertheless, groups with a few hundred individuals may be seen into early May. Normally by mid-May only small-sized groups or single birds are seen, but a few invariably linger until the end of the period. High counts: 50,000–70,000, during Mar 1963, St. Louis (RA-**BB** 30[2]:6); 20,000, 21 Mar 1965, Alton Dam (RA et al.-**NN** 37:5).

Summer: Although a few birds linger nearly every year, there still is no nesting record. These summer visitants (all but a few observations are of

Fig. 8. The extended wings of female Greater (upper) and Lesser (bottom) Scaup show the differences in the amount of white in the primaries. Note that the white is much more pronounced and extensive in the Greater than the Lesser. This is true for both sexes. Photograph taken at Maryville Sewage Lagoons on 8 November 1972 by David Easterla.

1–2 individuals) may be found almost anywhere across the state, but most records are from the Squaw Creek area. High count: 14, 11 June 1967, Squaw Creek (FL).

Fall Migration: The earliest migrants begin appearing at the end of Sept. The species is rare until about the third week in Oct. Peak does not occur until early to mid-Nov, with relatively large concentrations encountered into early Dec. High counts: 6,000, 14 Nov 1970, Squaw Creek (FL); 6,000, 6 Nov 1978, Browning L. (FL-**BB** 46[1]:22).

Winter: Generally one of the more common and widespread diving ducks at this season. High counts: 978, 28 Dec 1968, Kansas City Southeast CBC; 790, 31 Dec 1973, Big Oak Tree SP CBC.

Harlequin Duck (*Histrionicus histrionicus*)

Status: Casual transient; hypothetical winter visitant.

Documentation: Specimen: o?, female plumage, taken from hunters,

3 Nov 1972, Maryville SL (not the Nodaway Co. L.), (DAE-**BB** 40[1]:7; NWMSU, DAE 2794).

Habitat: Rivers, reservoirs, and lakes.

Records:

Spring Migration: A single record: 1, collected (specimen not located), 21 Mar 1897, Montgomery Co. (E. Parker; Widmann 1907).

Fall Migration: Two additional records besides the above Maryville record: 1, collected (not located), 29 Oct (prior to) 1884, St. Charles Co. (Widmann 1907). The year 1897, given in Anderson and Bauer (1968), is incorrect. Hurter (1884) published a list of birds, which included the above record, that he had collected in the St. Louis area. The other record is of a female, 22 Nov 1970, Little Dixie L. (IA, BG-**BB** 38[4]:9).

Comments: In addition to the above records there are two records without any specific dates: 1, on the Missouri R., at Eaton Tower, Jackson Co. (Harris 1919b); 1, shot in 1887, at Lake City, Jackson Co., by J. Bryant (Harris 1919b).

Previous authors (Widmann 1907; Harris 1919b; Bennitt 1932) list this species as being a very rare winter resident, without any specific records for that season. To date, we find no evidence for this bird having occurred during the winter season. This species is occasionally seen in association with Bufflehead in the midwest.

Oldsquaw (*Clangula hyemalis*)

Status: Rare transient and winter resident.

Documentation: Specimen: male, 24 Feb 1970, Duck Creek (GL 60).

Habitat: Rivers, reservoirs, and larger lakes.

Records:

Spring Migration: With the exception of four records (two of these are in early Apr), all records are during Mar. Most of the records are from the Mississippi R. in the St. Louis area. High count: 6, 27 Mar 1968, Forest L., Adair Co. (R. Luker-**BB** 36[1]:4). Latest records: 1, 24 May 1985, L. Jacomo (CH, RF, MMH-**BB** 52[3]:12); male, 27–29 Apr 1990, Swan L. (J. Eldridge et al.-**BB** 57[3]:136; photos in MRBRC file).

Fall Migration: Birds first appear at the very end of Oct or early Nov. The number of individuals are too low, and records are too uniformly distributed between early Nov and the end of the period to discern a peak. Most records are of solitary birds. Earliest record: 1, 28 Oct 1981, Alton Dam (BR-**BB** 49[1]:13). High count: 4, 21 Nov 1983, Greene Co. (NF-**BB** 51[1]:40).

Winter: There are more records during this period, usually at least one observation/winter than at other seasons. Most observations are of birds on the Mississippi R. and on the larger reservoirs. High count: flock of 12, most of Jan 1950, Mississippi R., Portage Des Sioux, St. Charles Co. (m.ob.-**BB** 16[5]:3).

Black Scoter (*Melanitta nigra*)

Status: Casual spring transient; rare fall transient; casual winter resident along Mississippi R.

Documentation: Specimen: female, 20 Oct, Maryville SL (NWMSU, DST 88).

Habitat: Rivers, reservoirs, and lakes.

Records:

Spring Migration: Only six unquestionable records: 2, 27 Feb 1949, near Alton Dam (JEC, JVB-**BB** 16[4]:3); 1, 25 Mar 1972, Alton Dam (J. Ruschill-**BB** 39[2]:10); 1, 11 Apr 1959, Alton Dam (JEC, E. Hath-**BB** 26[4]:16); adult male, 17 Apr 1989, near Holt, Clay Co. (E. Eldridge-**BB** 56[3]:89); adult male, 5–7 May 1962, Duck Creek (J. Rogers, N. Holler et al.-**BB** 29[2]:16); and pair, 12 May 1981, on pond in Maryville after a severe thunderstorm (TE).

Fall Migration: Recorded virtually every fall in small numbers (normally only 1–2 birds are seen; the vast majority of birds are immature). The first arrivals appear at the end of Oct. Most records are during Nov, with a sharp drop off in observations by the end of Nov. This species is only casually reported in Dec. Earliest date: 3, 16 Oct 1976, Chain of Rocks, Mississippi R., St. Louis (JE-**BB** 44[3]:20). High counts: 27 (3 adult males, 24 female plumaged), 15 Nov 1977, Alton Dam (CS et al.-**BB** 45[1]:25); 10 (all female plumaged), 3 Nov 1984, Maryville SL (DAE).

Winter: All records (n=ca. 5) for this season are from the Mississippi R. in the St. Louis area. These observations are scattered throughout the period. High count: 7, 20 Dec 1950, Alton Dam (JVB-**BB** 18[2]:3).

Surf Scoter (*Melanitta perspicillata*)

Status: Casual spring transient; rare fall transient in west, casual in east; accidental winter visitant.

Documentation: Specimen: first-year male, 26 Apr 1965, L. Jacomo (DAE et al.-**BB** 32[1–2]:10; NWMSU, mount; Fig. 9).

Habitat: Rivers, reservoirs, and lakes.

Records:

Spring Migration: Only four additional records besides the above specimen record: female, 21 Feb 1971, Squaw Creek (MBR-**BB** 38[4]:10); female, 8 Apr 1972, Alton Dam (RA, PB et al.-**BB** 39[3]:6); female, 19 Apr 1969, Maryville Water Plant (DAE-**BB** 38[1]:3; NWMSU, CLJ 150); 1 shot, 3 May 1876, near St. Louis (Widmann 1907).

Fall Migration: Far more numerous at this season than at any other, and most of the records are from the Maryville SL (DAE et al.). Immatures comprise the majority of the birds seen. Birds begin arriving by mid-Oct, with

Fig. 9. This first-year male Surf Scoter was found on 26 April 1965 at Lake Jacomo in Jackson Co. It represents the latest spring record for the state. Photograph by David Easterla.

the largest number of individuals seen during late Oct and early Nov. Virtually all birds have passed through the state by the end of Nov. The following records are all by DAE at the Maryville SL. Earliest record: 1, 12 Oct 1986. High counts: 7, 15 Oct 1975; 7, 3 Nov 1977.

Winter: Only two records: 1, 15 Dec 1978, Alton Dam (PS, BR-**BB** 46[2]:12); 1, 15 Dec 1984, Kansas City Southeast CBC.

White-winged Scoter (*Melanitta fusca*)

Status: Rare transient and winter resident.

Documentation: Specimen: adult male, 30 Nov 1975, Maryville SL (DAE; NWMSU, DST 47).

Habitat: Rivers, reservoirs, and lakes.

Records:

Spring Migration: Rarer in the spring than in the fall or winter. All records but two are during Mar. There is only one Apr record: 1, 4 Apr 1977, St. Louis (F. Hallett-**BB** 44[4]:25). The extraordinarily late bird seen at Alton Dam on 30 May 1955 is the latest observation (RA-**NN** 27:31). High count: 3, 6–7 Mar 1953, Harbor Point, Mississippi R., St. Charles Co. (JEC, RA-**BB** 21[3]:3).

Fall Migration: The first arrivals are usually seen in mid-Oct. Most records are concentrated in late Oct and early Nov, although birds are regularly seen into the winter period. Earliest dates: 2, 15 Oct 1977, Squaw Creek (MBR, TB-**BB** 45[1]:25); 1, 17 Oct 1970, Little Dixie L. (IA, BG-**BB** 38[4]:9). High count: 10 (all female plumaged), 3 Nov 1984, Maryville SL (DAE).

Winter: This scoter is the one expected in winter. Most records that span the entire period are from the Mississippi R. near St. Louis. High counts: 7, 5 Jan 1987, Springfield (m.ob.-**BB** 54[2]:13); 5, 14 Feb 1976, Labadie, Franklin Co. (D. Hays-**NN** 48:33).

Common Goldeneye (*Bucephala clangula*)

Status: Common transient and winter resident; accidental summer visitant.

Documentation: Specimen: female, 3 Feb 1968, Gravois Mills, L. of the Ozarks (NWMSU, LCW 774).

Habitat: Rivers, reservoirs, and lakes.

Records:

Spring Migration: The largest concentrations are encountered on the Mississippi R. during the spring, fall, and winter. The peak of migration is generally earlier than that of other waterfowl. In fact, the peak is usually in late Feb or early Mar. It is regularly seen in fairly large numbers (40–100 individuals/flock) throughout Mar. Only small-sized groups are seen during Apr, and by early May it is only casually encountered. High counts: 200, 5 Mar 1967, St. Louis (JEC-**BB** 34[2]:6); 200, 24 Mar 1990, Table Rock L. (JH-**BB** 57[3]:136). Latest record: 1, 11 May 1973, Squaw Creek (DAE, JHI).

Summer: Only one record: 1, possibly injured, 26 June 1939, St. Charles Co. (L. Ernst, WS et al.-**BB** 6[7]:61).

Fall Migration: Birds begin trickling in by mid-Nov, with relatively large numbers not seen until the end of the period. Earliest date: no.?, 28 Oct 1962, Alton Dam (RA); 3, female plumaged, 29 Oct 1988, Maryville SL (DAE-**BB** 56[1]:13). High count: 160, 30 Nov 1989, Thomas Hill Res. (S. Vasse-**BB** 57[1]:47).

Winter: Common on the Mississippi R. and the larger, deeper reservoirs and lakes. High counts: 2,500, 19 Feb 1989, between mouth of Missouri R. and Elash, Illinois (RG-**BB** 56[2]:53); 1,000, 2 Jan 1963, Alton Dam (SHA-**NN** 35:3).

Comments: On two occasions hybrids have been reported between this species and the Hooded Merganser: 1, was present for ca. 30 days, first found on 4 Mar 1951, on Mirror L., Parkville, Platte Co. (H. Hedges-**BB** 18[4]:1); the other was photographed on 6–9 Feb 1978, Alton Dam (BR, PB-**BB** 45[2]:12).

Barrow's Goldeneye (*Bucephala islandica*)

Status: Casual transient; accidental winter resident.
Documentation: Specimen: none extant (see below).
Habitat: Rivers, reservoirs, and large lakes.
Records:
Spring Migration: Three unequivocal records: male, 29 Feb 1984, near Fayette, Howard Co. (C. Royall-**BB** 51[2]:18); 1, 14 Mar 1962, Fellow's L. (F. Shumate, P. Weber-**BB** 29[2]:15); 1 first-year male, 29–30 Mar 1947, along Highway H, St. Charles Co. (JEC, K. Wesseling, not Wuestling in Anderson and Bauer 1968).
Fall Migration: Two of the three fall records include birds that were shot but apparently not saved. The first is of a bird taken in the fall of 1890 at "New Albany" (=New Madrid, New Madrid Co; see Fulvous Whistling-Duck account; Widmann 1907); the second is of a bird shot and photographed during the third week of Nov 1918, on the Missouri R., above Kansas City (no specific locality; R. Holland-**BL** 21:54). The only recent observation is of a single bird at L. Jacomo, 10 Dec 1972 (MMH et al.; Rising et al. 1978).
Winter: Surprisingly, there are only three observations: male, 20–21 Dec 1975, L. Jacomo, Kansas City Southeast CBC; female, 8 Jan 1977, Chain of Rocks, Mississippi R., St. Louis (Eades et al. 1978); male, 15 Jan 1973, present over two weeks, Alton Dam (D. McClaren et al.-**BB** 40[2]:7).
Comments: Besides the above records, Harris (1919b) mentioned two additional records for the Kansas City area; however, he gave no data for either record, leaving the question open to whether these were from Missouri.

Bufflehead (*Bucephala albeola*)

Status: Uncommon transient; rare winter resident.
Documentation: Specimen: male, 29 Mar 1969, Maryville (NWMSU, CLJ 151).
Habitat: Generally found on the larger and deeper bodies of water, although occasionally encountered on small ponds.
Records:
Spring Migration: Migrants appear with the retreating ice and usually peak at the end of Mar. Relatively large numbers (50–100/day) are seen until mid-Apr, but only small-sized flocks and solitary birds are seen during the latter half of Apr into early May. High count: 3,500, 24 Mar 1985, Squaw Creek (FL). Latest records: 1, 16 May 1965, Squaw Creek (FL); female, 13 May 1990, Maryville SL (TE, DAE-**BB** 57[3]:136); 1+, 13 May 1990, Squaw Creek (TB, RB, PMC).
Fall Migration: A few birds begin to appear at the end of Oct, but the bulk of the migrants do not arrive until mid- to late Nov. Earliest records: 1, 10

Oct 1904, Kansas City (J. Bryant; Widmann 1907); 1, 16 Oct 1975, Squaw Creek (FL). High counts: 125, 18 Nov 1989, Taney Co. (PM, JH-**BB** 57[1]:47); 100, 14–15 Nov 1985, Maryville SL (DAE).

Winter: A rare but regular winter resident in small numbers on the larger bodies of water. High counts: 190, 18 Dec 1976, Kansas City Southeast CBC; 190, 18 Dec 1988, Taney Co. CBC.

Hooded Merganser (*Lophodytes cucullatus*)

Status: Uncommon transient; rare summer and winter resident.

Documentation: Specimen: male, 9 Nov 1921, Jackson Co. (KU 39033).

Habitat: Most commonly seen on wooded lakes, slow moving rivers, and large creeks; but frequently seen on open bodies of water during migration.

Records:

Spring Migration: Migrants begin appearing in late Feb and peak during the final two weeks of Mar. Most migrants have passed through the state by early to mid-Apr. This species breeds locally statewide, and females are seen with broods by the end of May. High count: 150, 31 Mar 1982, Squaw Creek (FL).

Summer: Although this species breeds in small numbers in oxbows, wooded marshes, and swamps throughout the state, the largest concentrations (several pairs at each site) are found at the Ted Shanks and the Duck Creek/Mingo areas. The species was formerly a more common summer resident, especially along the larger rivers and in the swamps of the Mississippi Lowlands when habitat was more abundant (Widmann 1907).

Fall Migration: Nonsummer residents begin appearing in early Oct, but concentrations are not noted until the end of Oct. Peak is normally in mid-Nov; most of the birds have passed through the state by early Dec. High counts: 450+, 19 Nov 1988, Thomas Hill Res. (M. Goodman, TB-**BB** 56 [1]:13); 300, 15 Nov 1988, Swan L. (KG).

Winter: Usually found in small numbers (1–5 birds/day) scattered across the southern half of the state. High counts, both on the Mingo CBC: 160, 2 Jan 1989; 87, 3 Jan 1987.

Comments: See comments under Common Goldeneye.

Common Merganser (*Mergus merganser*)

Status: Common transient; locally common winter resident; accidental summer visitant.

Documentation: Specimen: female, 20 Nov 1948, Saline Co. (KU 39036).

Habitat: Primarily found on the larger and deeper rivers, reservoirs, and lakes.

Records:

Spring Migration: The largest numbers are encountered in late Feb and early Mar when migrants augment the wintering population. Smaller groups are readily seen into mid-Apr, but thereafter most observations are of single birds. Solitary birds are infrequently seen until mid-May. High count: 2,000+, 6 Mar 1978, above Alton Dam (TB-**BB** 45[3]:16). Latest date: female, 14–17 May 1976, Squaw Creek (MBR et al.-**BB** 44[2]:15).

Summer: There are two observations, both in 1981 at nearby localities (same birds involved?): 2, 1 July, Alton Dam (BR); 3, 11 July, Harbor Point, St. Charles Co. (TB, SS et al.-**BB** 48[3]:23).

Fall Migration: A few individuals begin appearing at the end of Oct, but there is no buildup until the latter half of Nov. The bulk of the birds arrive in early Dec. An unusually early record for this period is of a single bird, 3 Aug 1958, Gravois Mills, L. of the Ozarks (DAE-**BB** 25[8]:2).

Winter: This is the period when the largest concentrations are encountered. Rafts of hundreds of birds, and at times a few thousand, may be observed on the larger rivers and reservoirs, mainly in southern Missouri. High counts: 3,000+, 29 Dec 1985, Stockton L. (MBR, FL); 2,000–3,000, 10 Jan 1989, Smithville L. (DAE-**BB** 56[2]:53).

Red-breasted Merganser (*Mergas serrator*)

Status: Uncommon transient; accidental summer visitant; very rare winter resident.

Documentation: Specimen: male, 4 May 1968, Maryville (NWMSU, DLD 167).

Habitat: Larger rivers, reservoirs, and lakes.

Records:

Spring Migration: Although the first migrants begin appearing at the end of Feb or early Mar, the buildup in numbers is significantly smaller and later than that of the preceding species. Relatively large-sized groups (40+) are not seen until the latter half of Mar. Peak is usually in early Apr, but smaller-sized groups are regularly encountered into early May. Very rarely a large-sized flock may be seen in mid-May: 53, 15 May 1975, Alton Dam (JE-**BB** 43[1]:18). High count: 297, 10 Apr 1980, Thomas Hill Res. (TB-**BB** 47[3]:11). Latest date: 1, 21 May 1974, Squaw Creek (FL-**BB** 41[3]:3).

Summer: A single record: 2, 25 June 1972, Trimble (SP-**BB** 39[4]:9).

Fall Migration: The first arrivals are seen at the very end of Oct. Shortly thereafter it increases in numbers and peaks in mid-Nov. Birds are rarely seen in Dec. Earliest date: 14, 27 Oct 1980, St. Joseph (FL). High count: 100, 14 Nov 1985, Sugar L. (FL).

Winter: About twenty records, with most during late Dec and early Jan. High counts: 6, 17 Dec 1977, Springfield CBC; 5, 5 Jan 1990, Mississippi R., St. Charles Co. (BRO, RG-**BB** 57[2]:102–3).

Ruddy Duck (*Oxyura jamaicensis*)

Status: Common transient; casual summer visitant; hypothetical summer resident; rare winter resident.

Documentation: Specimen: female, 8 Nov 1969, Duck Creek (GL 2).

Habitat: Found in both shallow and deep water, but most commonly seen on the deeper bodies of water.

Records:

Spring Migration: Migrants begin appearing in late Feb or early Mar. Large numbers (in the hundreds) do not arrive until the end of Mar, and peak is usually in early to mid-Apr; numbers drop off rapidly thereafter. Small numbers are regularly seen throughout May, and a sizable flock may rarely be seen in mid-May: 56, 14 May 1978, Squaw Creek, MBR, TB-**BB** 45[3]:15. High counts: 2,500, 16 Apr 1980, Big L. SP/Squaw Creek areas (FL); two counts of 1,500 at Squaw Creek.

Summer: There are a minimum of a dozen records, and all but one observation involve one or two individuals. High count: 6, 17 June 1973, Squaw Creek (MBR-**BB** 40[4]:2). Widmann (1907) mentions that H. Nehrling considered this species to be a former "rare breeder in southwestern Missouri." We treat this statement as hypothetical because there are no substantiating details or records from more northern localities, where one would expect the species to have bred, if at all in Missouri.

Fall Migration: The first migrants usually appear at the end of Sept. Relatively small numbers are seen through the first half of Oct, but by the last week of Oct large numbers (in the hundreds) arrive. Peak is from early to mid-Nov, but few birds remain by early Dec. High count: 1,800, 6 Nov 1978, Browning L. (FL-**BB** 46[1]:23). An unusual record is of an adult male, 5–9 Aug 1972, L. Taneycomo, Taney Co. (NF, SP-**BB** 39[4]:9).

Winter: Seen in small numbers at scattered localities; most numerous on the Mississippi R. High counts: 72, 16 Dec 1978, Kansas City Southeast CBC; 61, 30 Dec 1976, Mingo CBC.

Order Falconiformes

Note: We place the American Vultures (Cathartidae) after the storks (Ciconiidae) to reflect their true phylogeny.

Family Accipitridae: Kites, Eagles, Hawks, and Allies

Osprey (*Pandion haliaetus*)

Status: Uncommon transient; casual summer visitant, former summer resident; casual winter resident.

Documentation: Specimen: female, 16 Oct 1938, Bethany, Harrison Co. (MU 464).

Habitat: Usually associated with large lakes, reservoirs, and rivers.

Records:

Spring Migration: A few individuals appear by the end of Mar, but birds are not regularly seen until early Apr. Peak is during the last two weeks in Apr. Earliest records: 1, 28 Feb 1987, Rockaway Beach, Taney Co. (PM-**BB** 54[2]:13); 1, 7 Mar 1981, near Busch WA (EL-**BB** 48[3]:18). High counts, both at Busch WA: 7, 11 Apr 1987 (JZ-**BB** 55[1]:14); 7, 22 Apr 1990 (JZ-**BB** 57[3]:136). Latest record: 1, 26 May 1962, near Missouri R., Kansas City (JRI-**BB** 29[2]:17).

Summer: Just before the turn of the century this species was considered "uncommon during the breeding season in several parts of the southeast" (Widmann 1907). Apparently it was not restricted as a breeder to this region of the state. Widmann mentions that it was found along the Gasconade and Osage rivers as well as the Missouri and Mississippi rivers. Specifically he states that a pair bred at Creve Coeur L. "thirty years ago" (Widmann 1907). Another interesting record given by him is of a bird seen of 26 June 1906, in Atchison Co. Harris (1919b) states that the last definite record of nesting in the Kansas City area was on the bluffs of the Missouri R., just north of Independence in 1884. However, even by the time Widmann was preparing his treatise on the state's birds, this species was apparently extirpated or nearly so as a nester. Since the early 1960s, there have been nearly fifteen summer observations, but none with evidence of breeding.

Fall Migration: Single birds begin reappearing at the end of Aug. Peak is usually at the end of Sept or early Oct. Only single birds/locality are seen by the end of Oct, and by the end of Nov it is typically absent from the state. Earliest record: 1, 4 Aug 1963, L. of the Ozarks, Camden Co. (K. Stewart-**BB** 30[3]:20). High counts: 40, 14–19 Sept 1943, along Current R. (R. Bennitt-**AM** 147:9); 21, 4–5 Oct 1986, L. of the Ozarks SP (m.ob.). Latest record: 1, 1 Dec 1979, Busch WA (G. and T. Barker-**BB** 47[3]:17).

Winter: At least nine records, all since 1975, span the entire period. Most sightings are of one day only, with the majority at the beginning or at the end of the period. There are two records of a single bird (possibly involving the same individual) spending the entire winter at a fish hatchery, at Table Rock Dam during 1987–88 and 1988–89 (m.ob.-**BB** 55[2]:61 and 56[2]:53). Two additional Jan records: 1, 1 Jan 1983, Taney Co. CBC; 1, 7 Jan 1983, Mingo (JW-**BB** 50[2]:24). Northernmost record: 1, 10 Feb 1980, Winfield Dam, Lincoln Co. (CS-**BB** 47[3]:17).

American Swallow-tailed Kite (*Elanoides forficatus*)

Status: Accidental summer visitant; former summer resident.

Documentation: Specimen: male, 7 Aug 1872, St. Louis (Hurter 1884; STSC).

Habitat: Could turn up anywhere, but most likely to be seen in swampy areas bordered by forest in the Mississippi Lowlands.

Records:

Spring Migration: Formerly this species was of regular occurrence throughout the state, but it was apparently most abundant in the cypress swamps in the southeastern corner and along the Missouri and Mississippi rivers. The earliest arrival date was on 22 Mar 1916 at Courtney in Jackson Co. (B. Bush; Harris 1919b).

Summer and Fall Migration: Widmann (1907) stated that it "is a regular, though, not numerous, summer resident nesting in . . . cypress swamps." He also mentioned breeding records for Clark Co. and for the Kansas City area. Bush (in Harris 1919b) apparently had the last nesting record for the state: two pairs bred near Courtney, Jackson Co., in the summer of 1912. The latest record (Harris 1919b): two birds near Courtney were first seen on 8 July 1916 and remained in the area for two weeks. Shortly thereafter, it apparently became extirpated as a breeder, undoubtedly the victim of the draining and clearing of lowland forest.

The largest concentration ever reported in the state was of a group of 40 birds on 7 Aug 1872, which was present for over a week at St. Louis (Hurter 1884; see above specimen record). Latest date: 1, 4 Sept 1906, Courtney (B. Bush; Harris 1919b).

There are only two observations for this species after Harris' work: 1, 26 July 1972, near Bakersfield, Ozark Co. (D. Kee et al.-**BB** 40[1]:6); and 1, 24 Aug 1975, north of Willard, Greene Co. (Collins 1975).

Black-shouldered Kite (*Elanus caeruleus*)

Status: Accidental transient and summer visitant.

Documentation: Sight records only: see below.

Habitat: Prefers open country: pastures and grasslands.

Records:

Spring Migration: A single record: 1, 15 May 1983, Springfield (GD-**BB** 50[3]:11). There are two old, dubious records by the same observer; see Easterla (1976) for pertinent comments on these records.

Summer: Only record: adult, 14 June 1976, near Pumpkin Center, Nodaway Co. (Easterla 1976).

Comments: Over the past two decades this species has been expanding its range northward in the United States. Between 1982 and 1989 breeding was recorded in Oklahoma, North Dakota, and Kansas (**AB** 36:993 and 42:94–95). If this expansion continues there undoubtedly will be additional Missouri records. In particular, it should be looked for in the open country of the western half of the state.

Mississippi Kite (*Ictinia mississippiensis*)

Status: Rare transient and summer resident in southeast; rarer transient and summer visitant elsewhere.

Documentation: Sight records only: see below.

Habitat: Primarily seen in mature, bottomland forest, especially in areas bordering rivers, streams, and marshes.

Records:

Spring Migration: It is usually not seen until the very end of Apr or early May. Most records are concentrated in mid-May. Nest construction begins by mid-May in the southeastern corner. Along the Mississippi R. it is regularly seen as far north as the Ted Shanks WA. The Mississippi Kite is of casual occurrence away from this river and the southeastern section. There are four records for the northwestern corner: 1, on 17 Mar 1957, near Parkville, Platte Co., is also the earliest date for the state (JRI, T. Pucci-**AB** 11:349); the other three records are during May. An additional early record: 1, 21 Mar 1988, near Springfield (L. Kennard, A. Simmerman-**BB** 55[3]:87). High count: 18 (5 adult, 13 immatures), 28 May 1985, near New Madrid, New Madrid Co. (A. Wildmann-**BB** 52[3]:12).

Summer: Except for recent nesting records at St. Louis (see below), it is known to breed only in the southeastern corner as far north as Cape Girardeau and the Mingo area. Nonetheless, birds are regularly seen throughout the summer as far north as Pike Co. (Ted Shanks; Clarence Cannon NWR). To date there is no confirmation of breeding there, although on 8 June 1989, two adults and 5 immatures were noted at the Ted Shanks area (BJ-**BB** 56[4]:147).

Widmann (1907) found a pair breeding in the St. Louis area for several years in the 1880s, but there was a hiatus in definite nesting records until a nest with eggs was located on 22 June 1987 (RK-**BB** 54[4]:26). The 1988 season produced no fewer than two pairs in St. Louis. Breeding status is uncertain in other regions of the state; for example, an adult and an immature were seen at Poague WA, Henry Co., from 25 June–July 1986 and again in 1987 (K. Kritz, T. Finder-**BB** 53[4]:16); and two were seen on 29 June 1980, McDonald Co. (JW-**BB** 47[4]:11). Widmann observed pairs in Webster and Howell counties in May 1906.

Recent extralimital records include: 1, 3 June 1964, St. Joseph (JHA-**BB** 31[2]:10); 1, 3 June 1976, St. Joseph (FL-**BB** 44[2]:20); 1, 20 June 1975, 11 miles east of Trenton, Grundy Co. (DAE-**BB** 43[1]:23).

Fall Migration: Most observations, away from the known breeding sites, are during Sept. There are three observations for the northwestern corner, the latest is 2, 21 Sept 1986, St. Joseph (FL-**BB** 54[1]:33). The latest for the state: 1, 27 Oct 1968, near Springfield (NF-**BB** 36[4]:11). The northernmost record for the state is of an immature seen just south of the Iowa border near Hamburg, in extreme northern Atchison Co., on 14 Sept 1982 (DAE).

Bald Eagle (*Haliaeetus leucocephalus*)

Status: Uncommon transient and winter resident; currently a very rare and local summer resident.

Documentation: Specimen: male, 27 Feb 1941, Taney Co. (MU 2).

Habitat: Primarily seen along the major rivers, larger marshes, and at the larger reservoirs.

Records:

Spring Migration: Most of the wintering population is returning north at the onset of this period. The largest numbers are seen at the end of Feb and early Mar. A few lingerers are regularly seen at some of the refuges and along the major rivers through early Apr. Nonbreeding individuals are infrequently seen through the end of the period. Nest construction of breeding birds is initiated during this period.

There are numerous accounts of this species breeding along several of the state's rivers and in the southeastern swamps prior to 1850. However, by 1900 it was on the verge of being eliminated as a breeder (Widmann 1907). Nevertheless, there have been occasional reports of nesting birds spanning the period from Widmann's time up to the present (Griffin and Elder 1980). The first documentation of an active nest since Widmann was of an incubating adult photographed in Camden Co. on 26 Mar 1962 (Easterla 1962c).

At present, this species appears to be increasing as a breeder, with 5–6 widely scattered pairs along Ozark rivers and large water impoundments in the southern half of the state (Osage R., Niangua R., Table Rock L., and Duck Creek/Mingo). High count: 43 (31 adults, 12 immatures), 12 Mar 1981, Squaw Creek (FL).

Summer: Besides the nesting records mentioned above, apparent nonbreeding individuals are casually seen.

Fall Migration: Nonbreeding birds (mainly immatures) are casually seen in Aug and early Sept. The first migrants (also largely immatures) begin appearing at the end of Sept. Numbers gradually increase from late Oct through Nov until they peak in early Dec. During the average peak over 200 eagles may be concentrated at Squaw Creek, and when conditions are ideal, nearly 300 birds may be present. High count: 250+, 5 Dec 1976, Squaw Creek (fide NJ).

Winter: At the beginning of this period large concentrations (a few hundred birds) may still be present at Squaw Creek and Swan L. However, as soon as the water freezes and the waterfowl leave, numbers drop off rapidly. A relatively large population winters along the Mississippi R.; fewer numbers winter along the Missouri R. and at southern reservoirs. A study of wintering, color-marked immatures in the state revealed that most originated from Minnesota and Saskatchewan (Griffin et al. 1980). High count: 417 (291 adults, 126 immatures), 19 Dec 1978, Squaw Creek CBC. Annual midwinter counts, sponsored by the MDC and the U.S. Fish and Wildlife Service, are conducted during the first week of Jan across the state. Totals for the past 9 years are given in Table 5.

Comments: All high counts were made of ground-level aggregations and are not of birds in migratory flight.

Table 5. Midwinter Bald Eagle Counts in Missouri. The percentage (rounded to the nearest integer) of immature birds is given.

Date	Count	Percentage
1981	957	38
1982	779	32
1983	908	33
1984	975	33
1985	758	34
1986	977	30
1987	745	38
1988	966	36
1989	1,449	33

Northern Harrier (*Circus cyaneus*)

Status: Uncommon transient and winter resident; very local and rare summer resident, primarily in the Osage Plains.

Documentation: Specimen: male, 6 Nov 1948, Saline Co. (KU 39044).

Habitat: Prefers open areas, especially marshes and prairies.

Records:

Spring Migration: Migrants begin augmenting the winter population in late Feb and early Mar. Peak appears to be in mid-Mar, but relatively high numbers of migrants can be seen into late Apr: 18, 19 Apr 1984, Tyson Valley RC (MP-**NN** 56:52). It is rarely encountered away from breeding sites after the first week in May. High count: 75, 29 Mar 1964, Squaw Creek (FL).

Summer: Now a very rare and local nester; primarily in ravines on native prairies in the Osage Plains. Nests are typically in unburned prairie with a few years of accumulated dead vegetation. The highest concentrations (usually no more than 2–3 pairs at each site) are at Prairie State Park, Barton Co., and Taberville Prairie. The 7 pairs in 1984 at the former site is

Fig. 10. Nest and eggs of Northern Harrier at Taberville Prairie, St. Clair Co., 1 June 1966. This species is now on the state's Rare and Endangered Species List because most of its nesting habitat—prairie—has been converted to cropland. Photograph by David Easterla.

the largest number recently recorded (L. Larson, pers. comm.). Thirteen nests (11 in the Osage Plains) were located during the 1990 "Breeding Bird Atlas" work (JW, pers. comm.). Nonbreeding birds are occasionally seen throughout the state. Prior to settlement, this species undoubtedly was more common and more widespread in the prairie region of the state, but by 1900 it already had declined considerably (Widmann 1907).

Fall Migration: Single individuals begin appearing in late Aug. Numbers gradually increase through Sept and peak in mid- to late Oct. Numbers remain high through Nov but taper off in early Dec. High counts: 25, migrating in 1 hr, 16 Oct 1976, St. Charles Co. (JE-**BB** 44[3]:21); 15, 19 Nov 1961, Callaway Co. (WG, DAE).

Winter: Widespread in low numbers, but locally common in prairies and marshes. CBC data show that it is more common in the Mississippi Lowlands than any other region of the state (Map 8). High counts: 67, 19 Dec 1978, Squaw Creek CBC; 64, 16 Dec 1989, Grand River CBC; 63, 21 Dec 1969, Squaw Creek CBC.

Sharp-shinned Hawk (*Accipiter striatus*)

Status: Uncommon transient and winter resident; rare summer resident, primarily in Ozarks.

Documentation: Specimen: male, 13 Apr 1974, Round Springs, Shannon Co. (MU 1519).

Habitat: Prefers relatively large tracts of forest, especially with Shortleaf Pine stands, for breeding. Can be encountered anywhere during migration and winter.

Records:

Spring Migration: This migration is inconspicuous compared to the fall. An influx of birds is seen during Mar through mid-Apr, with a few birds

Map 8.
Early winter relative abundance of the Northern Harrier among the natural divisions. Based on CBC data expressed as birds/10 pa hrs.

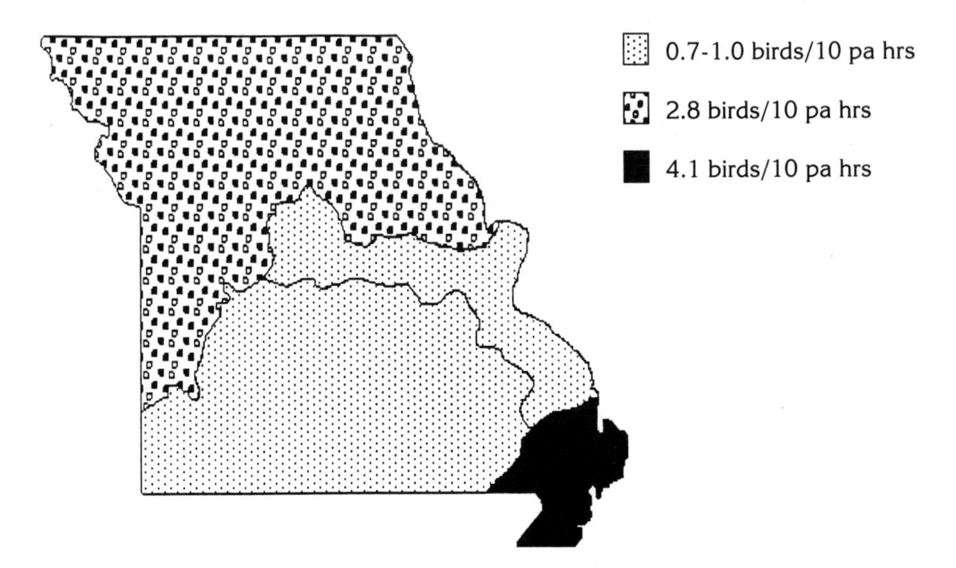

still passing through in early May. High counts: 66, 23 Apr 1989, Busch WA (JZ-**BB** 56[3]:89); 41, 24 Apr 1987, Imperial, Jefferson Co. (MP-**BB** 55[1]:14).

Summer: Until the detailed study of Kritz (1989), the breeding status of this species was very poorly known. Kritz located 17 nests during 1985–86 in the south central Ozarks and Ozark Border (Map 9). All but one of these nests were located in Shortleaf Pine (*Pinus echinata*). This species is, and apparently always has been, a scarcer breeder than the Cooper's Hawk.

Fall Migration: The earliest arrivals begin appearing at the end of Aug. There is a gradual increase until the last ten days of Sept when suddenly, usually following a front, relatively large numbers (> 50 birds/day) may be seen at favorable sites for observing hawk migration. Peak occurs at the end of Sept or early Oct. It is frequently encountered in small numbers through early Nov, but thereafter, only 1–2 individuals/day are sporadically observed. High counts: 125, 4–5 Oct 1986, L. of the Ozarks SP (m.ob.); 122, 4 Oct 1986, Busch WA (JZ, fide MP).

Winter: An uncommon winter resident in small numbers. It is most common in the Ozarks. We have lumped CBC data for both Sharp-shinned and Cooper's hawks, given the difficulty observers have in distinguishing these two species from each other. An average (expressed in birds/100 pa hrs) of 4.0 are recorded in the Ozarks, 3.0 in the Ozark Border, and 1.0 in the other regions.

Map 9.
Counties where Sharp-shinned Hawk nests were located in 1985 and 1986 (Kritz 1989).

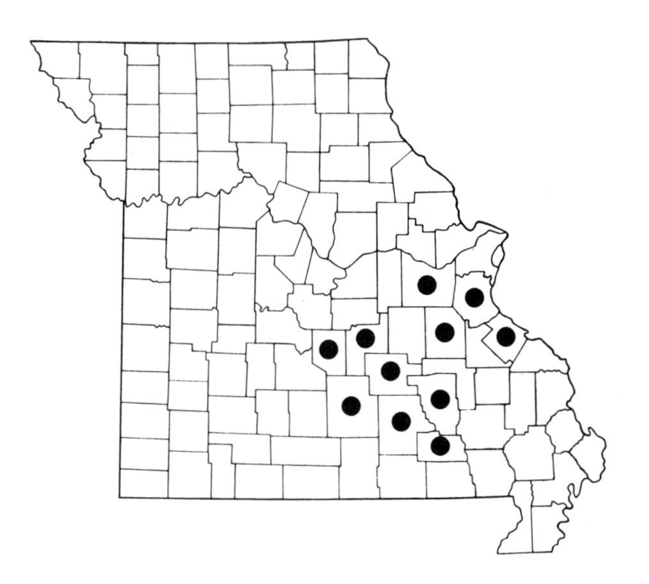

Cooper's Hawk (*Accipiter cooperi*)

Status: Rare transient and winter resident; rare summer resident in Ozarks and Ozark Border, extremely rare elsewhere.

Documentation: Specimen: female, 10 Nov 1946, McBaine, Boone Co. (MU 1081).

Habitat: Prefers mature forest for nesting, especially with Shortleaf Pine; usually encountered in woods during migration and winter but may be seen anywhere.

Records:

Spring Migration: The migration pattern is very similar to that of the Sharp-shinned, but the Cooper's is decidedly less common. High count: 4, 11 Apr 1987, Busch WA (JZ).

Summer: This species has always been more numerous as a breeder than the Sharp-shinned. Widmann (1907) mentioned that it was "a fairly common summer resident in all parts of the state." Undoubtedly, even in Widmann's time, it was least abundant in the less forested regions of the state. Kritz's (1989) study revealed that the Cooper's center of abundance is in the most heavily forested areas of the state, the east central Ozarks and Ozark Border (Map 10). During 1985 and 1986 Kritz located 43 nests, and 67% of them were situated in Shortleaf Pine.

There have been very few nesting reports north of the Missouri R. in dec-

Map 10.
Counties where Cooper's Hawk nesting has been documented since 1985. Most of the information is based on Kritz's study (1989).

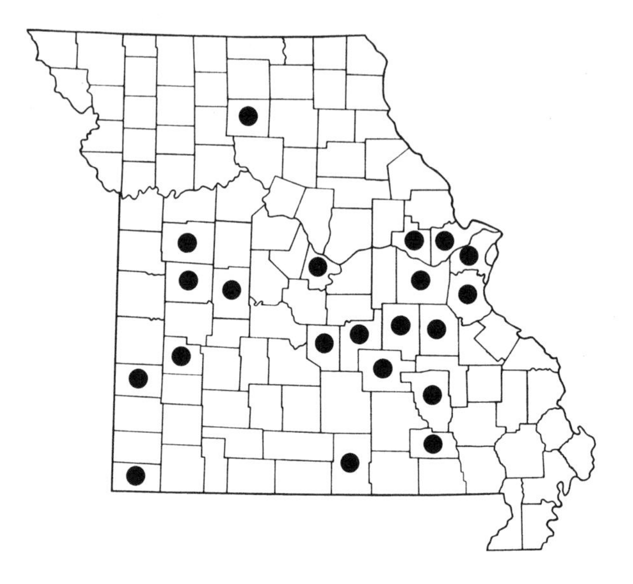

ades (except in St. Charles and Warren counties, which border the river), although undoubtedly there are a few widely scattered pairs going undetected there. One of the few and last definite breeding records for the northwestern corner is of an egg collected from a nest in Maryville on 23 Apr 1878 (C.S. Allen; USNM 17775).

Fall Migration: Its temporal pattern of migration is quite similar to the Sharp-shinned, but it is rarer. Even during the peak, this species is seen only in low numbers (< 10 birds/day). High count: 31, 4–5 Oct 1986, L. of the Ozarks SP (m.ob.).

Winter: As during migration, it is less common than the Sharp-shinned, and like that species, it is most common in the Ozarks. See remarks under winter season of Sharp-shinned Hawk.

Northern Goshawk (*Accipiter gentilis*)

Status: A rare and irregular transient and winter resident.

Documentation: Specimen: female, 25 Mar 1973, near Osceola, St. Clair Co. (CMSU 328).

Habitat: Like the other accipiters, it prefers forest and forest edge.

Records:

Spring Migration: Most records for this period are concentrated in Mar; however, after a major invasion in the previous fall, birds may be encountered until early May. An extraordinary testament to how major the movement was in the fall of 1972 was the 5 birds seen in the Taberville Prairie area on the late date of 21 Apr 1973 (KH et al.-**BB** 40[3]:5). The latest date for the state was also obtained that same spring: 1, 19 May 1973, Swope Park (NJ, JG). There are at least three additional early May records.

Fall Migration: During most falls no more than a total of 1 or 2 birds are seen in the state, and in some years, none are found. However, during years when there is a "crash" in the hare population (the primary food source on the breeding grounds for this species) there is a mass movement of birds to the south, with the first individuals appearing in Missouri at the end of Sept or early Oct. Most of the birds do not appear until the end of Oct through mid-Dec. The falls of 1972, 1982, and 1983 are the recent examples of particularly impressive N. Goshawk incursions in the state. One of the earliest documented invasions in the state was during the fall of 1906 when 5 birds were shot between mid-Nov and mid-Dec (Widmann 1907). Earliest record: adult, 26 Sept 1984, Tyson Valley RC (MP, JZ, SRU-**NN** 56:78). The 26 Sept 1985 observation (**NN** 57:78) should not have been published because there was some question of the bird's identity (MP, pers. comm.).

Winter: As mentioned above, this species is quite rare or absent in the state during most winters; relatively large numbers are seen only during invasion years. High counts: 3, 16 Dec 1972, Kansas City Southeast CBC; 3, 19 Dec 1982, Trimble CBC.

Red-shouldered Hawk (*Buteo lineatus*)

Status: Uncommon permanent resident in the Ozarks and Ozark Border, rare in Osage Plains, and north central and northeastern sections of Glaciated Plains. Extirpated as breeder in the northwestern corner of Glaciated Plains.

Documentation: Specimen: male, 5 Mar 1934, Nevada, Vernon Co. (MU 146).

Habitat: Almost exclusively found along riparian forest, except in migration.

Records:

Spring Migration: Birds return to nesting locales and are seen passing through nonbreeding areas primarily in early to mid-March: 6, 9 March 1985, St. Louis (MP-**NN** 59:61). The nesting cycle is well advanced by mid-Apr.

Summer: It is most common along Ozark streams, rivers, and in the Mingo area: 22 nests, summer 1984, along Shoal Creek, Newton Co. (W. Perkins-**BB** 51[4]:13); 13 nests, summer 1983, Mingo (T. Humphrey-**BB** 50[4]:32). The Red-shouldered is more local and rarer in the Osage Plains and the north central and northeastern sections of the Glaciated Plains. It apparently is extirpated as a breeder in the northwestern corner of the latter area.

Formerly, it was more common throughout the state (Widmann 1907); in particular, numbers are only a small fraction (perhaps < 5%) of what they were in the southeastern corner before this area was largely drained and deforested at the turn of the century. Furthermore, there has been a retraction in the range of this species in the northern half of the state, especially in the northwestern corner (although it apparently was always rare there). The last nesting record in the St. Joseph area was about 1967 (FL) and near Maryville in June 1970 (MBR). Riparian habitat was severely reduced about the time this species disappeared from the latter two sites.

Fall Migration: Little information exists on the migration during this period as a result of the difficulty in distinguishing migrants from permanent residents. Dates of obvious migrants range from early Sept to mid-Nov, with no clear peak. High count at St. Louis (area of greatest hawk watching activity in the state during the 1980s): 3, 16 Nov 1989, St. Louis (MP).

Winter: Although this raptor is less frequently seen and occurs in smaller numbers at this season, it is still uncommon along Ozark streams and rivers. There is a withdrawal from many of the more northern areas. The largest concentrations are found in the Mingo area. Until about 1972, this species was a rare and local winter resident in the northwestern corner; since then there has been only a single record of a bird overwintering there: immature, 1987–88, Squaw Creek (K. Jackson, LG et al.-**BB** 55[2]:61). High counts, both on the Mingo CBC: 29, 28 Dec 1965; 24, 3 Jan 1987.

Broad-winged Hawk (*Buteo platypterus*)

Status: Common transient, except in extreme west; uncommon summer resident in Ozarks and Ozark Border, rare elsewhere; possibly extirpated from the northwestern corner of the Glaciated Plains; hypothetical in winter.

Documentation: Specimen: female, 10 May 1967, near Pickering, Nodaway Co. (NWMSU, LCW 381).

Habitat: Prefers mature, continuous forest for breeding; encountered almost anywhere during migration.

Records:

Spring Migration: Compared to the fall, much smaller numbers are noted at this season. It appears to be more common in the eastern half of the state. The earliest arrivals, usually single birds, are seen in early Apr, with peak at the end of Apr. Migrants are regularly seen into May, with occasional small-sized groups, comprised primarily of first-year birds seen passing through during the latter half of May. High counts: 572, 22 Apr 1990, Busch WA (JZ-**BB** 57[3]:136); 400, 23 Apr 1989, Busch

WA (JZ-**BB** 56[3]:89). Earliest date: 1, 10 Mar 1980, Busch WA (PS, M. Wiese-**BB** 47[3]:11).

Summer: This buteo is most common in the Ozarks and Ozark Border and rare in all other areas. Within the Glaciated Plains region it is more prevalent in the eastern section. Now extremely rare in northwestern Missouri, it should be looked for on the more forested slopes and in the valleys along the Missouri R. It bred in the St. Joseph area until at least 1976 (FL).

Late migrating groups, composed primarily of first-year birds, are occasionally seen: 7, 4 June 1981, near Ashburn, Pike Co. (MBR, DAE).

Fall Migration: Relatively large kettles (hundreds/kettle) have been reported from all regions of the state except the extreme western section (roughly the two western tiers of counties). A few individuals begin to appear by mid-Aug: 3, 9 Aug 1987, St. Louis (MP); 3, 21 Aug 1986, Squaw Creek (FL). Peak is nearly always between 19–28 Sept. Large concentrations are seen through very early Oct, but by midmonth only stragglers are observed. There are no verified Nov records. Latest date: immature female, 30 Oct 1938, St. Charles Co. (CMC 467). High counts in the east: 5,419, 4 Oct 1986, Jefferson Co. (MP; this unusually late high count was the result of weather conditions during that period); 1,823, 24 Sept 1985, St. Louis (MP-**BB** 53[3]:24). High counts in the west: 1,120, 4–5 Oct 1986, L. of the Ozarks (m.ob.); 1,000+, 19 Sept 1989, near Table Rock Dam, Taney Co. (D. and T. Lundberg-**BB** 57[1]:47).

Winter: Hypothetical; all sightings are very suspect. Photograph or specimen documentation required.

Comments: The rare dark morph of the Broad-winged Hawk has been recorded on at least two occasions in the state: 1, 23 Apr 1969, Squaw Creek (FL, DAE, et al.); no date available, St. Louis (MP, pers. comm.).

Swainson's Hawk (*Buteo swainsoni*)

Status: Uncommon transient in west, rare in east; a rare and local summer resident in west central; hypothetical in winter.

Documentation: Specimen: female, 12 Apr 1969, near Skidmore, Nodaway Co. (NWMSU, PER 96).

Habitat: Prefers open country: pastures and prairies. Usually nests in a small grove of trees surrounded by open country.

Records:

Spring Migration: The first birds arrive in early Apr, with the peak during mid- to late Apr. Migrants are seen through the end of the period. Earliest dates: 1, 14 Mar 1987, Kansas City, Clay Co. (T. Schallberg); 1, 18 Mar 1989, N Jefferson Co. (MP-**BB** 56[3]:89). High count: 19, 14 Apr 1984, Taberville Prairie (L. Moore-**BB** 51[3]:10). All records east of Kingdom City, Callaway Co., are of single birds.

Summer: Nesting is apparently restricted to the west central part of the state, from Jackson Co. south to northern Jasper and Greene counties. It is perhaps most common in Barton Co., but even there it is rare and pairs are widely scattered. Nesting undoubtedly occurs in some of the other counties in the Osage Plains. Surprisingly, there are no breeding records from northwestern Missouri, although there are a number of summer sightings of single birds there. There are no confirmed breeding records for north of the Missouri R.; however, there is one observation of a pair exhibiting nesting behavior in extreme southern Howard Co. (Toland 1986).

Late migrants or nonbreeding birds are regularly encountered and at times in relatively large-sized groups: 24, 12 June 1964, near Forbes, Holt Co. (FL); 7, 21 June 1964, near St. Joseph (JHA-**BB** 31[3]:17).

Fall Migration: Migrants begin reappearing in late Aug, with a gradual increase in the frequency of sightings until the peak in late Sept or early Oct. It is only casually seen after the third week of Oct, with only a single Nov record: 1, 6 Nov 1982, near Springfield (A. Banfield-**BB** 50[1]:22). It is seen more frequently at this season in the eastern half of the state, e.g., 5, between 18–24 Sept 1974, in the St. Louis area (RA, JC-**BB** 41[1]:3).

Until the fall of 1988, the largest group reported in a single day was 4 birds. The discovery that this species is best detected by surveying fields being harvested at the peak of its migration in the upper Missouri R. valley resulted in unprecedented numbers being observed in the fall of 1988 (Robbins 1989). Observations by J. Hilsabeck in the fall of 1990 indicate that large numbers of migrants are not restricted to the Missouri R. valley (see below). Hundreds, perhaps a few thousand birds, may pass through the extreme western section of the state each fall. High counts: 159, 25 Sept 1988, Missouri R. valley of Holt and Atchison counties (Robbins 1989); 125+, 5 Oct 1990, Bolckow, Andrew Co. (JHI).

Winter: Hypothetical; highly questionable sightings. Photograph or specimen verification required.

Red-tailed Hawk (*Buteo jamaicensis*)

Status: Common permanent resident.

Documentation: Specimen: female, 12 Nov 1921, Jackson Co. (KU 39043).

Habitat: Found in a wide range of habitats, but prefers forest adjacent to open land.

Records:

Spring Migration: Movement of all races begins in late Feb or early Mar. Peak is in mid- to late Mar. By early Apr most nonbreeding races are gone from the state (see below). Nest construction usually begins in Mar. High count: 82, 23 Mar 1984, Jefferson Co. (MP-**NN** 56:52).

Summer: BBS routes do not adequately reflect the abundance of this and other raptors, since these birds rarely vocalize, and the routes are rou-

Fig. 11. An adult Red-tailed Hawk perched on an Osage Orange fence post. This species reaches its greatest abundance during the summer and winter in the Osage Plains. Pen and ink by David Plank.

tinely completed by midmorning before they become conspicuous by soaring. Nevertheless, these data are shown in Map 11. We suspect that if more reliable data were available, this ubiquitous hawk would be found more common in the Glaciated Plains than in the Ozarks.

Fall Migration: A few are seen migrating as early as late Aug. The non-summer races do not appear until early Oct (see below). Peak is much later

Map 11.
Relative breeding abundance of the Red-tailed Hawk among the natural divisions. Based on BBS data expressed as birds/route.

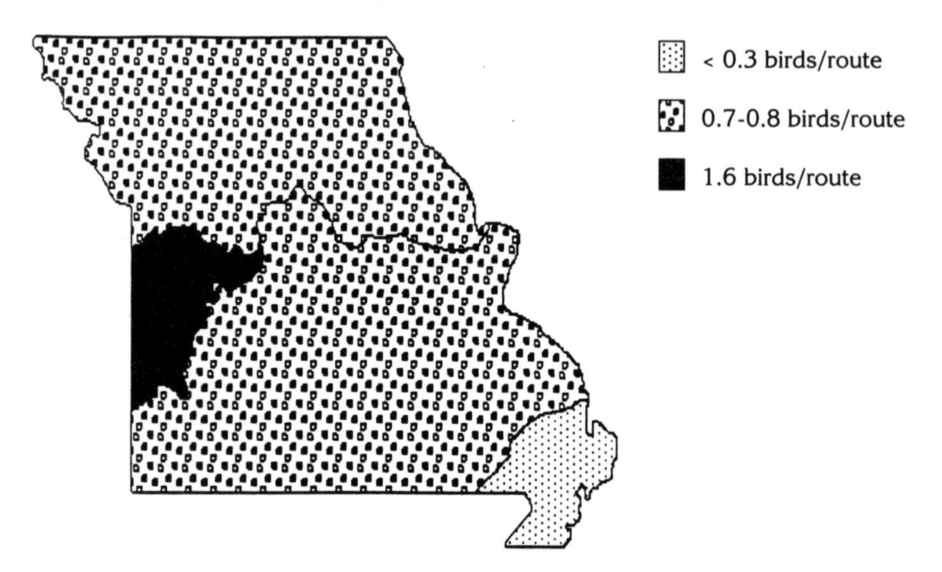

than either Broad-winged or Swainson's hawks and arrives in late Oct and early Nov. High count: 136, 1 Nov 1984, Webster Groves, St. Louis (MP-**BB** 53[3]:27).

Winter: More common at this season than during the summer. It is most prevalent in the Glaciated and Osage plains, where roughly 1.0 birds/pa hr are recorded on CBCs. The Ozark Border and Ozarks have the lowest average, with 0.4 birds/pa hr and 0.6 birds/pa hr, respectively. High counts: 198, 30 Dec 1988, Barton and Vernon counties (145 car miles; MBR-**BB** 56[2]:54); 158, 1 Jan 1990, Springfield CBC.

Comments: The resident race, *B. j. borealis,* is the most common at all seasons. It is the only subspecies present from mid-May through late Sept. The Harlan's Red-tailed Hawk, *B. j. harlani* (23 Oct 1939, Rocheport, Boone Co.; MU 493), which was until recently considered a separate species, is a rare to uncommon migrant and winter resident. It is most frequently encountered in the extreme western section of the state. The earliest date for fall arrival is 1, 25 Sept 1966, Squaw Creek (FL), but most individuals of this race do not appear until late Oct and Nov. High count: 25, 20 Dec 1981, Squaw Creek CBC. The majority have left the state by mid-Mar. Latest spring date: 1, 6 Apr 1984, near Bluffwoods SF (FL).

The Western Red-tailed, *B. j. calurus* (26 Dec 1940, Chariton, Macon Co.; MU 855), is also a migrant and winter resident and is most common in the western half of the state. It is more widespread and more common than *harlani.* The earliest dates for fall arrival are: 1, 3 Oct 1968, Busch WA (KA et al.-**NN** 40:92); 1, 4 Oct 1964, St. Joseph (FL). The latest spring date is 1, 8 May 1989, Schell-Osage (MBR, TE-**BB** 56[3]:89).

All observations of the very pale, Northern Plains race or color morph, *B. j. kriderii* (25 Apr 1968, near Mound City, Holt Co.; NWMSU, CLJ 162), are concentrated from Nov through Mar. Earliest fall date: 1, 17 Oct 1989, near Adrian, Bates Co. (MBR-**BB** 57[1]:47). Latest spring date: 1, 19 May 1963, Squaw Creek (FL).

Ferruginous Hawk (*Buteo regalis*)

Status: Accidental transient in spring; casual fall transient in west, accidental in east; hypothetical winter resident.

Documentation: Sight records only: see below.

Habitat: Prefers open, short-grass pasture and prairie. Also seen in cultivated fields.

Records:

Spring Migration: Based on the observer's reputation (no written details are available), the following observation is likely reliable, even though it is quite late in the season: 1, 24 Apr 1949, Sugar L. (JB-**BB** 16[5]:1).

Fall Migration: We consider only six observations of this commonly misidentified hawk to be reliable. We list them all since many erroneous sightings have been published. There is only one sighting away from the western quarter of the state. With one exception, all of the sightings have involved single, light-phased adults: 1 Oct 1972, Maryville SL (DAE-**BB** 38[4]:9); 2 Oct 1970, Missouri R. bottoms, near Amazonia, Andrew Co. (FL-**BB** 38[4]:9); 11 Oct 1981, Schell-Osage (TB-**BB** 49[1]:13); 19 Oct 1979, El Dorado Springs, Cedar Co. (TB-**BB** 47[1]:15); dark phase, 8 Nov 1953, Missouri R. Valley, near Kansas City (JB-**BB** 21[2]:4); and 24 Nov 1960, St. Louis (RA-**BB** 28[1]:17).

Winter: No acceptable record, although there are several reports. Most, if not all, pertain to "Krider's" Red-tails.

Comments: "Krider's" and leucistic individuals of the Red-tailed Hawk are often misidentified as Ferruginous (there are several published Missouri records of this species that actually pertain to "Krider's" and/or partial albino Red-tails). Care should be taken to note all the details of a bird suspected to be of this species.

The Ferruginous has a very different profile, with proportionally longer and narrower wings and tail than any of the other buteos. The most striking feature on the dorsal surface, when in flight, is the extent of white at the base of the primaries (in both adult and immature).

Rough-legged Hawk (*Buteo lagopus*)

Status: Uncommon transient and winter resident.
Documentation: Specimen: female, 22 Dec 1966, Maryville (NWMSU, DAE 1190).
Habitat: Open country: pastures and prairies.
Records:
Spring Migration: Most have left the state by mid-Mar, but a few are casually seen into Apr (6+ sightings). High count: 20+, 14 Mar 1975, near Cole Camp, Benton Co. (DAE, MBR-**BB** 42[3]:13). Latest date: 1, 29 Apr 1956, Squaw Creek (FL).
Fall Migration: The first arrivals appear about mid-Oct. The largest concentrations are not observed until the very end of this period. Earliest date: 1, 22 Sept 1975, Creve Coeur L. (JC-**BB** 44[1]:19).
Winter: Most common in the Osage and Glaciated plains. High counts: 30, in one field (ca. 60 acres [24.3 hectares]), 20 Dec 1970, 11 miles north of Maryville (MBR-**AFN** 25:586); 25, 24 Dec 1970, Maryville CBC.

Golden Eagle (*Aquila chrysaetos*)

Status: Rare transient and winter resident.
Documentation: Specimen: immature male, 22 Oct 1958, near Centerview, Johnson Co. (MU 1563).
Habitat: Open country; especially in short grass areas.
Records:
Spring Migration: Most have left the state by mid-Mar; however, it is casually seen into Apr (8+ sightings). All records are of single birds. Latest dates: 1, 1 May 1980, Glade Top Trail, Douglas/Ozark Co. (NF-**BB** 47[3]:11); immature, 29 Apr 1969, Cape Girardeau (RAR-**BB** 36[3]:2).
Fall Migration: The first migrants arrive in late Oct, but most observations are in Nov. It is more commonly encountered in the western half of the state. Earliest dates: 1, 16 Aug 1981, Aldrich (NF-**AB** 36:183); adult, 29 Sept 1987, Tyson Valley RC (JZ; Peters 1988). High count: 5, 4 adults and 1 immature, 20 Oct 1974, Swan L. (BG-**BB** 42[2]:5).
Winter: A rare winter resident in open country. There are records from all regions of the state, and most are of single birds. High count: 3, 21 Dec 1968, Squaw Creek CBC.

American Kestrel (*Falco sparverius*)

Status: Common transient and winter resident; uncommon summer resident.
Documentation: Specimen: male, 21 Apr 1976, Maryville (NWMSU, DST 83).

Habitat: Prefers open country; often found nesting in towns, cities, and in abandoned buildings in rural areas.

Records:

Spring Migration: Migrants begin augmenting winter populations in late Feb and peak in late Mar or early Apr. By the end of Apr numbers are significantly reduced. High counts: 44, 17 Apr 1983, St. Joseph/Squaw Creek areas, about 80 car miles (FL); 65, 22 Mar 1971, St. Joseph/Unionville, about 280 car miles (FL).

Summer: Much less common at this season than at any other. BBS data indicate that the kestrel is significantly more common in the Mississippi Lowlands (average 1.3 birds/route) than in any other natural division (average 0.1–0.3 birds/route). High count: 15, 11 July 1986, St. Louis Co. (RG-**NN** 58:67).

Fall Migration: Birds begin to return by late Aug. Peak, however, is not until the very end of Sept or early Oct. Numbers drop off by early Nov. High counts: 58, 4–5 Oct 1986, L. of the Ozarks SP (m.ob.); 15, within 0.5 mile, 30 Sept 1971, near Rosendale, Andrew Co. (JHI-**BB** 39[1]:5).

Winter: The kestrel is also most common at this season in the Mississippi Lowlands and least common in the Ozarks and Ozark Border (Map 12). High counts: 66, 20 Dec 1987, Jackass Bend CBC; 64, 20 Dec 1987, North Cass Co. CBC. At this season, males outnumber females (often a 2:1 ratio) in the western Glaciated and Osage plains (MBR); data are not available for other areas of the state.

Map 12.
Early winter relative abundance of the American Kestrel among the natural divisions. Based on CBC data expressed as birds/10 pa hrs.

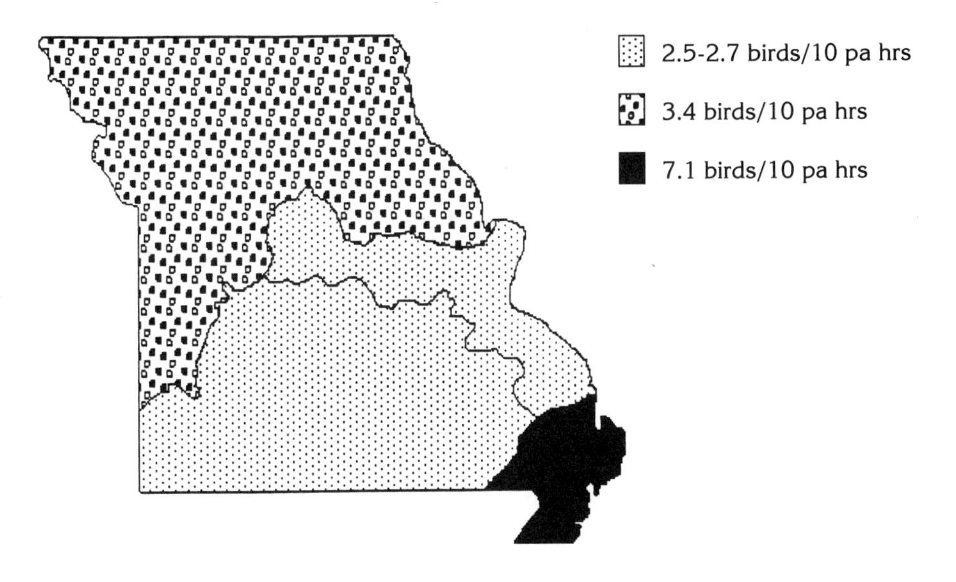

Merlin (*Falco columbarius*)

Status: Rare transient and winter resident.
Documentation: Specimen: male, 19 Sept 1935, Columbia (MU 214).
Habitat: Most commonly seen in open areas near woods.
Records:
Spring Migration: Even though the relative abundance given above for this species is the same for the migration periods and winter, it is more common during the migratory period. All spring reports are of single birds. A few begin to appear by early Mar and peak in early Apr. Birds are regularly seen through the first week in May, but thereafter this species is virtually absent from the state. Latest dates: 1, 14 May 1988, Hamilton, Caldwell Co. (WL, TB et al.); 1, 13 May 1977, Schell-Osage (JR-**BB** 44[4]:25).
Fall Migration: More frequently encountered at this season than at any other. The first individuals are seen in early Sept. Peak is at the very end of Sept and early Oct. It is regularly seen into early Nov, but thereafter it is only sporadically observed. Earliest date: 1, 31 Aug 1968, Creve Coeur L. (RA-**BB** 36[2]:16). High counts, both at L. of the Ozarks SP: 9, 4–5 Oct 1986 (m.ob.); 4, 30 Sept 1978 (RA et al.-**BB** 46[1]:23).
Winter: A rare winter resident, and with the exception of two records (2, 27 Dec 1979, Squaw Creek [MBR]; 2, 2 Jan 1983, Orchard Farm CBC) all observations involve single birds. Records are statewide and span the entire period.
Comments: Two races have been recorded in Missouri. The vast majority of the birds are of the nominate race, but few observers have made racial determinations. There are only four definite records of the pale, western race *richardsoni,* but it undoubtedly is more common than these few records indicate. Widmann (1907) mentions that a bird identified as *richardsoni* was found dead near Billings, Stone Co. (no date given). The other three records are as follows: female, 29 Sept 1988, Maryville SL (MBR, DAE-**BB** 56[1]:13); 1, 3 Oct 1978, Squaw Creek (MBR-**BB** 46[1]:23); and a female, 31 Oct 1989, extreme western Nodaway Co. (DAE-**BB** 57[1]:47).

Peregrine Falcon (*Falco peregrinus*)

Status: Rare transient and winter resident; former summer resident.
Documentation: Specimen: immature o?, 21 Sept 1964, Independence, Jackson Co. (CMSU 172).
Habitat: Can be encountered virtually anywhere during migration; but most commonly seen near shorebird or waterfowl concentrations.
Records:
Spring Migration: The earliest migrants appear as early as the beginning of Mar, but very few are seen until mid-Apr. Peak occurs during the last week of Apr or in early May. It is regularly seen through mid-May, but it is

casual during the last ten days of May. Most records are of single birds, but there are a few records of two birds seen simultaneously at shorebird assemblages. Latest date: 1, 31 May 1971, Creve Coeur L. (PS-**NN** 44:79).

Summer: Up until the 1850s "many" Peregrines nested on the cliffs along a number of our rivers, especially the Missouri and Mississippi rivers (for a review see Widmann 1907). But, by 1880 and the early 1890s, apparently only a few pairs were still known to breed in the state. There is a mounted specimen of an adult male in the MU collection, which was taken about 1885 along the Missouri R. bluffs between Portland and Bluffton, Callaway Co. Possibly the last breeding observation was during July 1911 when an adult was observed along a cliff face on the Meramec R., Crawford Co. (Baldwin 1911).

There are five sightings since 1940 of single birds during 2–27 July. There are no June records.

Fall Migration: Peregrines begin to reappear by the end of Aug. Peak is at the end of Sept and early Oct. Birds are regularly seen through Oct but thereafter are very rarely observed. Earliest date: 1, 10 Aug 1982, St. Charles Co. (K. Boldt-**BB** 50[1]:23). High count: 5, 4–5 Oct 1986, L. of the Ozarks SP (m.ob.).

Winter: Much less common at this period than during migration; however, records span the entire period statewide.

Comments: Peregrines are making a comeback from the population lows of the 1960s and early 1970s. It is conceivable that this species may eventually return as a breeder to some of its former haunts in Missouri.

[**Gyrfalcon** (*Falco rusticolus*)]

Status: Hypothetical; accidental winter visitant.

Documentation: Sight record only; see below.

Comments: Only one reliable sighting for the state: 1, 18 Dec 1977, Squaw Creek (Barksdale and Rowlett 1981). Barksdale and Rowlett (1981) comment on other questionable records. We treat this species as hypothetical since the above record represents the only unquestionable observation. There are nearly twenty records for Illinois (Bohlen and Zimmerman 1989) and several for Kansas (S. Seltman, pers. comm.).

Prairie Falcon (*Falco mexicanus*)

Status: Rare transient and winter resident; former accidental summer resident.

Documentation: Photograph: 28 Nov 1978, south of Mound City, Holt Co. (B. Heck-**BB** 46[1]:23; Fig. 12).

Habitat: Open country: pastures, prairies, and cropland.

Fig. 12. This Prairie Falcon was found with a broken wing beneath a power line near Squaw Creek NWR in Holt Co. The Prairie Falcon is now a regular transient and winter resident in the western half of Missouri. The above individual was photographed on 28 November 1979 by Bill Bennett.

Records:

Spring Migration: Probable winter residents apparently remain until the end of Mar, as the following four observations suggest: 1, 21 Mar 1985, Overton Bottoms, Cooper Co. (TB-**BB** 52[3]:13); 1, 23 Mar 1989, SW corner of Barton Co. (Childers 1989); 1, 27 Mar 1986, Alton Dam (CP-**BB** 53[3]:12); and 1, 28 Mar–3 Apr 1976, Busch WA (MS et al.-**BB** 44[2]:15).

Summer: One old breeding record: a nest (in a tree) with 2 eggs was collected near Maryville on 28 Apr 1880 (Goss 1891; Widmann 1907).

Fall Migration: At least sixteen records, all of single birds, mainly from western Missouri. With the exception of two records, observations range from late Oct through the end of the period. Two-thirds of the records are clustered in late Oct and early Nov. Earliest record: 1, 23 Sept 1983, northwest of Maryville (DAE, TE).

Winter: There are now about 35 records (all involving single birds), and most are from western Missouri and along the Missouri R. valley to St. Louis. Sightings span the entire period. A bird that was banded at Jackson, Wyoming, on 10 June 1939, was hit by an automobile at Carrollton, Carroll Co., on 4 Feb 1940 (Cooke 1941).

Comments: There are over 50 records, and all but five of these have been since the fall of 1972. Although most of this increase in sightings can be directly attributed to an increase in the number of observers, as well as their expertise, there nevertheless has been an increase in the relative abundance of this species over the past fifteen years. Birds have recently appeared in areas where there were no prior records, even though these areas have been thoroughly covered for over thirty years.

Order Galliformes: Gallinaceous Birds

Family Phasianidae: Grouse, Turkeys, and Quail

Gray Partridge (*Perdix perdix*)

Status: Introduced; rare immigrant from southern Iowa populations; permanent resident now breeding in at least northwestern Missouri.

Documentation: Sight records only: see below.

Habitat: Cultivated fields and pastures with fencerows.

Comments: This game bird is a recent arrival from Iowa. The species has been increasing and expanding southward for the past two decades from established populations in northern Iowa (Dinsmore et al. 1984; Dinsmore, *in litt* to MRBRC). The Gray Partridge first appeared in the northeastern corner of Missouri in 1987 or 1988. Subsequently, relatively large numbers appeared in the northwestern corner (between 50 and 75 birds were noted in the fall of 1988). The largest flock was comprised of 20 individuals near the Riverbreaks State Forest, Holt Co. At this same time single birds were noted near Watson, Atchison Co., Elmo, Nodaway Co., and at Pony Express WA, Dekalb Co. (B. Bennett, *St. Joseph News Press Gazette,* 8 Jan 1989). Additional records occurred in western Nodaway Co. in 1990 (fide DAE). The only sighting with written details is of a male seen on 18 June 1990 near Watson (DAE et al.).

Ring-necked Pheasant (*Phasianus colchicus*)

Status: Introduced; uncommon permanent resident primarily in the northern half of the Glaciated Plains; rare and more local in southern half of the Glaciated Plains and the Mississippi Lowlands.

Documentation: Specimen: male, 26 Mar 1977, near Rock Port, Atchison Co. (NWMSU, DST 100).

Habitat: Most common in open, cultivated country with hedgerows.

Comments: Introductions by private owners began as early as the late 1890s (Cary 1984; **NN** 56:38–39). The MDC began releasing birds in 1904 but not with a concerted effort until late 1920–30s. There have been, and continue to be, numerous additional introductions involving several subspecies.

Pheasants are now established in 36 counties in Missouri. Except for several counties in the Mississippi Lowlands, it is principally found north of the Missouri River (Hallet 1989). The species has done well only in the northern half of the Glaciated Plains and in the Marais Temps Clair area in extreme eastern Missouri. Nevertheless, a relatively small but stable population persists across much of the remainder of the Glaciated Plains (north of the Missouri R.) and in the Mississippi Lowlands.

Late-summer roadside surveys (routes in 112 counties) in 1987 recorded a statewide mean of 10.4 pheasants/30 miles. The highest mean, 14.7 birds/30 miles, was recorded in the northeastern section of the Glaciated Plains (Hallet 1988). Some of the highest densities are found at Squaw Creek (300–400) and Marais Temps Clair.

High winter count: 358, 2 Jan 1982, Orchard Farm CBC.

Ruffed Grouse (*Bonasa umbellus*)

Status: Rare permanent resident statewide (based on reintroductions), but most common in the more forested regions of the state. Former common permanent resident.

Documentation: Specimen: male, 12 Jan 1942, Deer Run State Forest, Reynolds Co. (MU 973).

Habitat: Most common at the edge of forest clearings and forest border.

Comments: Virtually all the birds now present in the state are the result of reintroductions. Formerly, prior to the 1880s, this species was "numerous in most wooded parts of Missouri," but by 1900 it was considered rare (Widmann 1907). Small natural populations still persisted in Adair, Clark, Lewis, Montgomery, St. Genevieve, and Warren counties in 1955 (**BB** 22[5]:4). Beginning in 1959, the MDC began extensive reintroductions (primarily with stock from Ohio and Indiana) across the state (Lewis et al. 1968), and as a result, this species has been restored to many of its former haunts in the state.

Greater Prairie-Chicken (*Tympanuchus cupido*)

Status: Local, rare permanent resident in the Osage Plains, on the verge of extirpation in the Glaciated Plains. Accidental transient in these two regions. Formerly an abundant permanent resident in the Osage and Glaciated plains.

Documentation: Specimen: female, 18 Dec 1940, Milan, Sullivan Co. (MU 976).

Habitat: Only found on or near virgin prairie.

Comments: The Missouri presettlement prairie-chicken population may have been as high as one million birds (Wywialowski and Christisen 1989). As early as 1880, this once common to abundant bird had been drastically reduced in numbers and was considered to be rare (Widmann 1907). By 1900 it was no longer of regular occurrence in the Kansas City area, where the last nesting record was 1891 (Harris 1919b). It has continued to decline dramatically and become extirpated from many areas of its former range, primarily as a result of habitat loss.

In 1940, Schwartz (1945) estimated the state's population to be about 12,000 birds (Map 13). By 1983 it was known to exist in only one area north of the Missouri R. (Cannon and Christisen 1984; Christisen 1985; Map 14). The 1988 estimate for the northern Missouri population, primarily surviving in Audrain Co. and the surrounding area, was between 150–200 birds (Wywialowski 1988). Between 1983 and 1988, the male population in the state was estimated to have decreased by 36%, with the state's estimated total chicken population comprising only about 3,000 birds in 1988. Two-thirds of the state's population is now restricted to Barton and Dade counties (Wywialowski 1988).

Presumed migrants are occasionally seen outside areas of permanent residence: 1, 5 May 1973, Bigelow Marsh (DAE-**BB** 40[3]:5); 1, 4 Mar 1983, Squaw Creek (KG-**BB** 50[3]:12); 1, 3 May 1984, Sullivan Co. (R. Thom-**BB** 51[3]:11). Another unusual record (possibly of a migrant as well), and the only record of this species in the southeastern corner of the state, is of a single female bird taken at New Well, Cape Girardeau Co., on 15 Nov 1939

Map 13.
The 1940 range of the Greater Prairie-Chicken in Missouri. Reproduced with permission from Cannon and Christisen (1984).

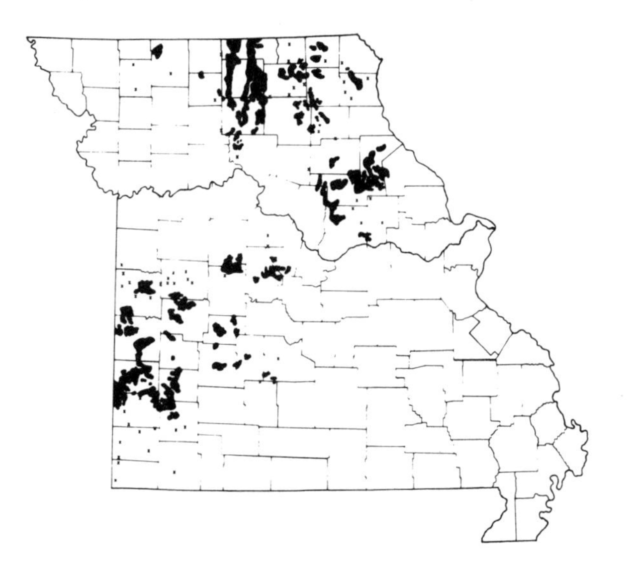

Map 14.
The 1983 breeding range of the Greater Prairie-Chicken in Missouri. Reproduced with permission from Cannon and Christisen (1984).

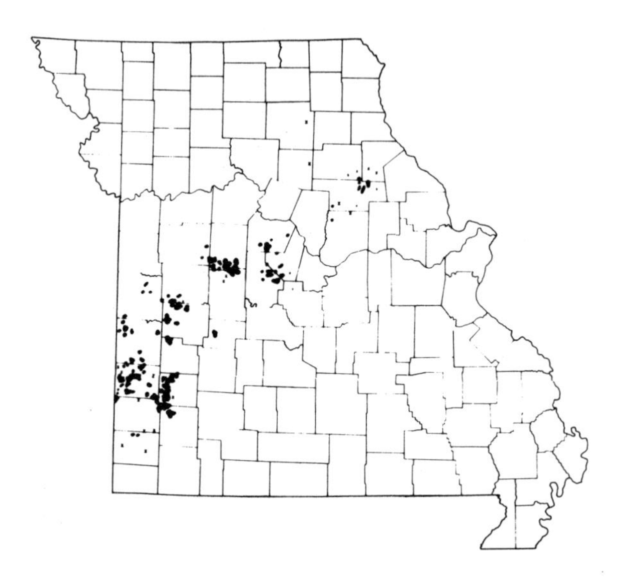

(H. Bolen-**BB** 7[1]:43; SEMO 50). Twenty-five were released at Swan L. in 1960, but this reintroduction failed.

As a result of recent reintroductions in adjacent Iowa, birds have appeared on a relatively large prairie near Eagleville, Harrison Co., Missouri (*St. Joseph News-Press/Gazette,* 29 May 1988). Three booming grounds, consisting of a total of 30–40 birds, were unexpectedly located in the spring of 1990 in extreme western Carroll Co. (J. Hiser). Presumably these birds are of a remnant natural population and not the result of recent reintroductions.

[**Lesser Prairie-Chicken** (*Tympanuchus pallidicinctus*)]

Status: Hypothetical.

Comments: Widmann (1907) treated this species as hypothetical based solely on the presence of about thirty individuals in a New York market that were said to have originated from a dealer in Pierce City, Lawrence Co., Missouri (specimen, a mount, USNM 74007, 18 Jan 1887).

The Lesser Prairie-Chicken was significantly more common in the late 1880s than today, and there are even a few old specimen records of birds occasionally wandering in winter to eastern Kansas (Goss 1891; Thompson and Ely 1989). However, the uncertainty surrounding the original collecting locality, coupled with the unlikelihood of a flock of thirty birds straggling so far east of their normal range, leaves us to believe that the above birds were almost certainly obtained in Kansas.

Wild Turkey (*Meleagris gallopavo*)

Status: Rare permanent resident in Glaciated and Osage plains; uncommon in the Ozarks, Ozark Border, and along the Missouri and Mississippi rivers.

Documentation: Specimen: male, 6 Dec 1871, St. Louis (STSC).

Habitat: Found in forest and woodland border.

Comments: It was formerly a common resident throughout the state, prior to the 1880s (although it probably was always less common on the Glaciated and Osage plains). Widmann stated that it was extirpated from northern Missouri by 1906, and at that time it was present only in small numbers in the Ozarks and swamplands of the southeast. Reintroductions of farm-reared game turkeys failed, and turkey populations remained relatively low (< 2,500 birds in 1952) until the MDC began a management program for the species (J. Lewis, pers. comm.). By the late 1950s this program began to show positive results by boosting numbers through habitat manipulation and the transplanting of native birds (Lewis 1961). The restoration program has been so successful that by 1984 hunting was permitted in all counties. An impressive 36,035 gobblers were harvested in the spring of 1987 (Lewis, pers. comm.).

High winter counts: 195, 18 Dec 1983, Mingo CBC; 180, Grand River CBC.

Northern Bobwhite (*Colinus virginianus*)

Status: Common permanent resident.

Documentation: Specimen: male, 13 Apr 1907, Shannon Co. (AMNH 229446).

Habitat: Woodland edge, hedgerows, and open country with scattered cover.

Comments: This is a relatively common bird throughout the entire state. However, both BBS (see Map 15) and CBC data reveal that it is decidedly less common in the Mississippi Lowlands than in any other region of the state. An average of ca. 1.0 birds/pa hr are recorded on CBCs in the Ozarks, Ozark Border, and the Glaciated and Osage plains, whereas only about 0.1 birds/pa hr are observed in the Mississippi lowlands. High winter count: 255, 28 Dec 1975, Columbia CBC.

The bobwhite is negatively affected by winters when there is prolonged snow and ice cover. Graph 1 depicts the decline of this species on BBS routes following the severe winters of 1976, 1978, and 1983.

Map 15.
Relative breeding abundance of the Northern Bobwhite among the natural divisions. Based on BBS data expressed as birds/route.

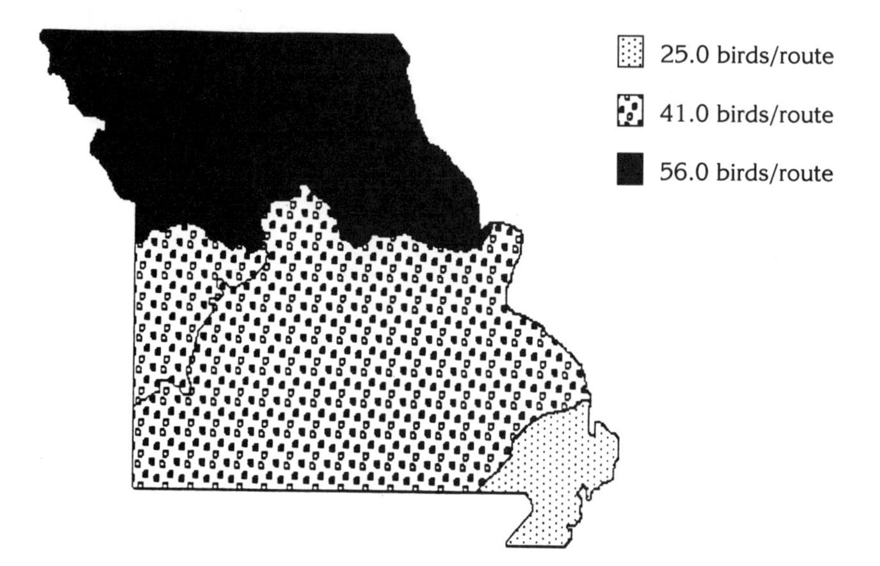

Graph 1.
Missouri BBS data from 1967 through 1989 for the Northern Bobwhite. See p. 23 for data analysis.

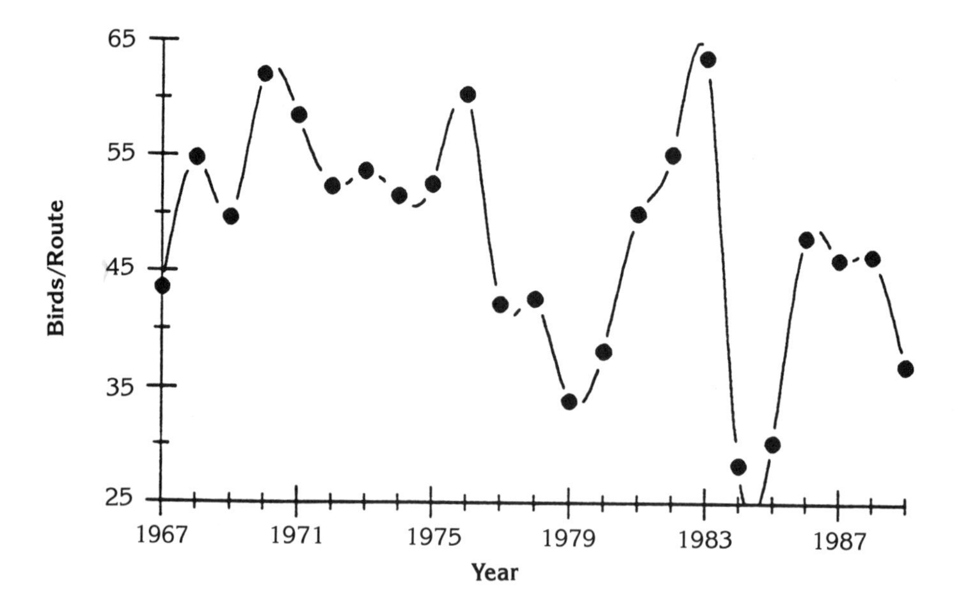

Order Gruiformes: Cranes, Rails, and Allies

Family Rallidae: Rails, Gallinules, and Coots

Yellow Rail (*Coturnicops noveboracensis*)

Status: Rare transient.
Documentation: Specimen: o?, 2 May 1961, Tucker Prairie, Callaway Co. (NWMSU, DAE 324; Easterla 1962b).
Habitat: Wet meadows, prairies, and grassy marshes.
Records:
Spring Migration: A secretive, rarely encountered species that is undoubtedly more common than the relatively few records indicate. Fortuitous events, such as the burning of prairies during mid-Apr (Easterla 1962b), have given a better indication of just how common this species is during migration in Missouri.

The first individuals appear in early Apr and peak during the third week of Apr. By early May most have passed through the state. Earliest date: 1, 27 Mar 1876, St. Louis (Hurter 1884; Widmann 1907). High counts: 7, during burning of wet prairie, 17 Apr 1979, Squaw Creek (TB-**BB** 46[3]:8); 6, 13–14 Apr 1984, Tucker Prairie, Callaway Co. (TB, CH-**BB** 51[3]:11). Latest date: 1, 5 May 1904, Corning, Holt Co. (C. Dankers; Harris 1919b).

Fall Migration: Records range from early Sept through the third week in Oct but are mostly concentrated in late Sept or early Oct. Earliest record: 1, early Sept 1984, Independence (Y. Wasson-**BB** 52[1]:29). High count and latest record: 2, 20–21 Oct 1984, Busch WA (J. Loomis et al.-**NN** 56:84).

Black Rail (*Laterallus jamaicensis*)

Status: Casual transient; accidental summer visitant.
Documentation: Specimen: o?, killed during hail storm, 1 May 1933, Fayette, Howard Co. (Jenner 1934; CMC 46).
Habitat: Marshes, especially in spike-rush (*Spartina*), and wet meadows.
Records:
Spring Migration: Even more secretive than the Yellow Rail, with only eight records (six since 1963) ranging from late Apr through late May. Earliest date: 1, 20 Apr 1963, Beverly L. (T. Pucci-**BB** 30[2]:24). High count and latest record: 2, singing during midday (voices recorded, Cornell Library of Natural Sounds, 37378), 26 May 1986, Bigelow Marsh (MBR-**BB** 53[3]:12).
Summer: A single record: 1, 17 July 1968, Swan L. (L. Kline-**BB** 36 [2]:16). Although there is only one record for this period, there is suitable nesting habitat at several sites across the Glaciated and Osage plains. Moreover, it has been found breeding on several occasions in Kansas (Thompson and Ely 1989) and at least formerly in Illinois (Bohlen and Zimmerman 1989).

Fall Migration: Only two records: 1, 21 Sept 1963, near Ozark, Christian Co. (NF-**BB** 30[4]:17); and o?, tower kill, 23 Sept 1969, Cape Girardeau (RAR-**BB** 38[3]:6; USNM 564977). The bird reported as a Black Rail (Oct 1954, Wright City, Warren Co.) in Anderson and Bauer (1968) was actually a Yellow Rail (R. Conyers-**BB** 21[1]:3).

King Rail (*Rallus elegans*)

Status: Rare transient and summer resident.

Documentation: Specimen: male, 14 May 1967, near Bigelow (NWMSU, LCW 427).

Habitat: Marshes and swamps, especially in cattails.

Records:

Spring Migration: The first migrants appear in early Apr, but peak is not until early May. Earliest date: 1, 31 Mar 1957, Swan L. (DAE, D. Snyder-**BB** 24[7]:4). High count: 4, 4 May 1983, Marais Temps Clair (D. Ulmer-**BB** 50[3]:12).

Summer: A rare and local breeder throughout the state. Most readily found at Squaw Creek, Ted Shanks, Marais Temps Clair, Mingo/Duck Creek, and Schell-Osage.

Fall Migration: Most records for this period are of breeding birds during Aug and early Sept. Definite migrants (birds seen away from breeding sites) are observed in Sept and Oct. High count: 30, includes family groups, 27 Aug 1971, Squaw Creek (HB-**BB** 39[1]:5). Latest records: o?, 25 Oct 1898, Jackson Co. (KU 39053); 1, 14 Oct 1984, Ted Shanks (J. Smith-**BB** 52[1]:29).

Virginia Rail (*Rallus limicola*)

Status: Uncommon transient; casual summer resident.

Documentation: Specimen: male, 9 Oct 1968, Cape Girardeau (SEMO 270).

Habitat: Marshes and swamps, especially in cattails; also wet fields.

Records:

Spring Migration: A few birds are infrequently seen as early as the beginning of Mar, but it is not regularly encountered until late Apr or early May, with peak in mid-May. Earliest date: 2, seen and heard, 25–27 Feb 1986, Binder L., Jefferson City (D. Kurz-**BB** 53[2]:8). High count: 30, 14 May 1967, Bigelow Marsh (DAE, SHA).

Summer: Status poorly known because of its secretive nature. Although there are 15 sightings for this period, no more than 10 could be considered nesting records. The last definite nesting records were in the early 1970s from the St. Louis area (JEC-**BB** 40[4]:3), Squaw Creek (young observed;

HB, pers. comm.), and near Rosendale, Andrew Co. (adult with two young; JHI, pers. comm.). There have been at least five sightings, involving at least one bird, since the mid–1970s. It is perhaps a regular but low density breeder in the marshes of the upper Missouri and Mississippi river valleys.

Fall Migration: Most observations are of single birds between early Aug and mid-Oct. Peak may be in late Sept, but more information is needed. High count: 9, 20 Sept 1969, Squaw Creek (FL). Latest date: 1, 15 Nov 1972, Maryville (DAE-**BB** 40[1]:8).

Sora (*Porzana carolina*)

Status: Common transient; casual summer resident.

Documentation: Specimen: o?, 29 Mar 1977, Silver Mines SP, Madison Co. (MU 1437).

Habitat: Marshes, swamps, and wet meadows.

Records:

Spring Migration: Birds typically first appear by mid-Apr and peak in early to mid-May. Earliest date: 1, 17 Mar 1986, Duck Creek (BRE-**BB** 53[3]:12). High counts: 100+, 9 May 1972, Squaw Creek (MBR); 100, 14 May 1967, Squaw Creek (DAE, SHA).

Summer: At this season, there are fewer sightings for the Sora than for the Virginia Rail, with only four definite breeding records. Three of these are pre–1900 records (Widmann 1907). The only recent record is of an immature at Marais Temps Clair on 26 June 1982 (BR-**BB** 49[4]:16).

Fall Migration: Relatively large numbers begin appearing by late Aug: 50, 29 Aug 1971, Squaw Creek (MBR, FL). Peak is probably in early to mid-Sept. Few are reported after mid-Oct. High count: 100+, 22 Sept 1979, Squaw Creek (FL). Latest date: 1, 6 Nov 1983, Squaw Creek (FL-**AB** 38:208).

Purple Gallinule (*Porphyrula martinica*)

Status: Rare transient and summer visitant in the Mississippi Lowlands, casual elsewhere; hypothetical summer resident; accidental winter visitant.

Documentation: Specimen: male, 12 May 1970, near Rosendale, Andrew Co. (JHI, DAE-**BB** 38[3]:8; NWMSU, DAE 2375).

Habitat: Marshes and swamps.

Records:

Spring Migration: Recorded almost annually in the Mississippi Lowlands (especially the Mingo/Duck Creek area), with all but three records in May. There are at least two records for the St. Louis region and two for the northwestern corner: o?, found dead, 9 May 1963, Beverly L. (DAE-**BB** 30[2]:24; NWMSU, DAE 395); and the above specimen record. Earliest records: 1, 18 Apr 1877, St. Louis (Hurter 1884; Widmann 1907); 1, 18 Apr 1981, Mingo (SS, RB et al.-**BB** 48[3]:18).

Summer: Although there are no definite breeding records, this gallinule almost certainly occasionally breeds in the Mississippi Lowlands. We believe this to be the case since it has bred in adjacent Illinois (Bohlen and Zimmerman 1989), and there are at least seven summer observations for Missouri. One of these involved at least four individuals between 8–22 June 1962 at Duck Creek (E. Hartowicz-**BB** 29[3]:22). A single bird was seen at L. Contrary on 13 July 1963 (FL-**BB** 30[3]:20).

Fall Migration: Single record: 1, found injured in the city of St. Charles, St. Charles Co., Oct 1951 (Comfort 1952).

Winter: Unexpectedly, an immature was found starving in the Missouri R. bottom, Jackson Co., on 30 Dec 1915 (Hoffman 1916).

Common Moorhen (*Gallinula chloropus*)

Status: Rare transient and summer resident.

Documentation: Specimen: male, 8 Oct 1941, Cape Girardeau (SEMO 59).

Habitat: Marshes, swamps, and lakes.

Records:

Spring Migration: Much more numerous than the Purple Gallinule in all areas of the state during all seasons. The first migrants appear at the end of Apr and peak in mid-May. Earliest date: Harris (1919b) lists 13 Apr (with no year) for Kansas City. High count: 16+, 15 May 1985, Duck Creek (MBR-**BB** 52[3]:11).

Summer: A rare, local breeder with most records from the Missouri R. northward. Both Widmann (1907) and Harris (1919b) state that the bird was formerly common. High counts: 7 nests, 6–17 June 1962, Beverly L. (JRI, T. Anderson-**BB** 29[3]:22); 15, 28 June 1989, Duck Creek (F. Reed; fide BJ).

Fall Migration: The majority of records are during Aug and early Sept (most are of summer residents); however, there is a movement of birds through the state during Sept through mid-Oct. Latest date: 1, 15 Oct 1962, Rock Port, Atchison Co. (H. and F. Diggs-**BB** 52[1]:29).

American Coot (*Fulica americana*)

Status: Common transient; rare summer and winter resident.

Documentation: Specimen: male, 3 May 1968, Duck Creek (GL 146).

Habitat: Found on virtually every body of water.

Records:

Spring Migration: The first arrivals begin to trickle in during early Mar. By the last week of Mar, concentrations numbering in the low thousands may be seen. An impressive peak occurs in mid-Apr but shortly thereafter drops off dramatically. Depending on water levels, hundreds may still be seen through the first half of May. Heavy rains in May of 1990 provided optimal nesting habitat for this species at Squaw Creek. At the end of May, *76* nests were found (RBE-**BB** 57[3]:135). High count: 25,000, 16 Apr 1980, Squaw Creek (FL).

Summer: The number of birds that remain to breed is highly variable from year to year. When water levels are high more birds nest. During average years only a few scattered pairs breed across the state, with the largest concentrations at Squaw Creek. High count: 300, includes nonbreeding birds, 18 June 1967, Squaw Creek (FL-**BB** 34[2]:14).

Fall Migration: Migrants begin reappearing in early Sept, and a few thousand may be seen by the end of that month. The huge concentrations, however, do not appear until mid-Oct or peak until late Oct and early Nov. Depending on how mild the fall is, large concentrations (in the hundreds or even low thousands) may remain until late Nov. But in average years, few birds are present in the northern half of the state by early Dec. High count: 20,000, 8 Nov 1983, Squaw Creek (FL).

Winter: Numbers encountered in the state are dependent on the weather, with more birds present during mild winters. During most years, virtually none overwinter in the northern half of the state. High counts, both on the Mingo CBC: 6,000, 28 Dec 1965; 286, 18 Dec 1983.

Comments: Much individual variation exists in the shield of this coot; some individuals possess shields similar to the Caribbean Coot (*Fulica caribaea*). An American Coot with an unusually extensive shield was seen at Squaw Creek on 12 June 1990 (DAE).

Sandhill Crane (*Grus canadensis*)

Status: Rare transient in west, casual in east; casual winter resident.

Documentation: Specimen: first-year male, 15 Apr 1974, near Rosendale, Andrew Co. (JHI, DAE; NWMSU, DLD 111).

Habitat: Most often seen in cultivated fields or shallow marshes.

Records:

Spring Migration: Formerly much more common (Widmann 1907; Harris 1919b), but today it is rare and not even recorded every spring. Most observations are during the last two weeks of Mar, with a few sightings in early Apr. With the exception of fewer than five records, all are from the western half of the state. Earliest date: 7, 27 Feb 1977, Squaw Creek (Turlin-**BB** 44[3]:25). High count in the west: 100, 15 Mar 1979, Reed WA, Jackson Co. (B. Eastman-**BB** 46[3]:7); in the east: 30, 14 Apr 1972, St. Louis (J. Strickling-**BB** 39[3]:6). Latest record: 2, 18 May 1854, between Utica and Lexington (Hoy 1865).

Summer: A single record of 2, 23 June 1981, Ted Shanks (J. Boyles-**BB** 48[3]:24).

Fall Migration: The majority of records are from mid-Oct through mid-Nov. Like the spring records, most are from western Missouri (< 5 records

Fig. 13. The adult Whooping Crane and two Sandhill Cranes were photographed on 15 October 1958 near Bigelow, Holt Co., by Don Reynolds. There has been only one sighting in Missouri since the fall of 1958. Photograph courtesy of the St. Joseph Museum.

for the east). Earliest date: 1, 10–17 Aug 1962, Squaw Creek (HB et al.-**BB** 29[4]:24). High count: 100, 4 Oct 1990, Terre du Lac, St. Francois Co. (HF); 27, 10 Oct 1919, Missouri R., above Kansas City (**BL** 21:54).

Winter: At least seven records: four during the last two weeks of Dec and one in mid-Feb. The two in Jan are: 1, 11–12 Jan 1961, Mingo (A. Hanke et al.-**BB** 28[1]:19); 1, shot, late Jan 1920, Marionville, Lawrence Co. (J. Neff-**BL** 22:169).

Whooping Crane (*Grus americanus*)

Status: Accidental transient.

Documentation: Photograph: adult, filmed, 13–15 Oct 1958, Bigelow Marsh (m.ob.-**BB** 26[1]:1–2; Fig. 13).

Habitat: Cultivated fields and shallow marshes.

Comments: There were a number of observations prior to 1900 (Widmann 1907; Harris 1919b). Harris (1919b) mentions three records for the early 1900s, the last of which is a bird observed by C. Dankers on 27 Mar 1913 in Holt Co. The species was not seen in the state again until the above photographed bird. Another individual, an immature, was seen later that same fall, 30 Nov–15 Dec 1958, at Mingo (m.ob.).

The most recent observation is of two birds seen flying west over Independence on the late date of 29 Apr 1970 (KH, C. Huffman; Rising et al. 1978). Although the date is an unlikely one to see these birds, the circumstances of the observation leave little question. Unbeknownst to each other, the observers independently saw the birds within a few minutes and miles of each other and immediately reported them.

Order Charadriiformes: Shorebirds, Gulls, and Allies

Our arbitrary seasonal criteria are least suited for the Charadriiformes than any other group; however, we follow those criteria here as well for consistency's sake.

Readers should be aware that Charadriiformes seen in early June typically represent late spring migrants, whereas individuals observed from late June through most of July represent adults, most still in nuptial plumage, returning south. Unfortunately, very little data exist on the age of shorebirds migrating through the state. Prater et al. (1977) and Hayman et al. (1986) are excellent references for aging shorebirds.

Family Charadriidae: Plovers

Black-bellied Plover (*Pluvialis squatarola*)

Status: Uncommon transient.

Documentation: Specimen: female, 27 Oct 1925, Jackson Co. (KU 39060).

Habitat: Mudflats, sandbars, flooded fields, edge of lakes.

Records:

Spring Migration: Accidental in Mar, and quite rare until the end of Apr. Peak is in mid-May, but a few are seen through the end of the period. Earliest date: 1, 15 Mar 1942, Platte Co. (WC-**AM** 44:10). High counts: 150, 16 May 1964, Squaw Creek (DAE, HB-**BB** 31[2]:11); 60, 14–20 May 1975, Squaw Creek (HB-**BB** 32[1–2]:11).

Summer: Casual in June and July. High count: 2, 12 June 1983, Overton Bottoms, Cooper Co. (PS, TB-**BB** 50[4]:32).

Fall Migration: Normally the first birds appear in early Aug, with peak at the end of Aug or early Sept. Small numbers, 1–2 birds/day, are regularly seen until the end of Nov. There is only a single Dec record. High counts:

500, 1–15 Sept 1963, Squaw Creek (HB-**BB** 30[4]:17); 42, 3 Oct 1984, Squaw Creek (FL-**BB** 52[1]:30). Latest record: 7, 6 Dec 1980, Squaw Creek (J. Robinson-**AB** 35:305).

Lesser Golden-Plover (*Pluvialis dominica*)

Status: Common transient.

Documentation: Specimen: male, 24 Apr 1966, Bigelow (NWMSU, DAE 1189).

Habitat: Plowed fields, short-grass pastures, recently harvested fields, and mudflats.

Records:

Spring Migration: The first migrants usually appear by the third week of Mar. Peak is normally in early Apr, but flocks numbering in the hundreds may be seen from late Mar through mid-May. Earliest dates: 1, 13 Mar 1964, Squaw Creek (HB-**BB** 31[2]:11); 1, 13 Mar 1977, Squaw Creek (FL). High counts: 2,000, 8 Apr 1944, St. Charles Co. (WS-**NN** 16:8); 1,800+, 1 Apr 1975, Big L., Mississippi Co. (JH).

Summer: Casual in June and July. High count: 12 in nuptial plumage, 23–24 June 1990, southern Stoddard Co. (BRE).

Fall Migration: A few individuals are seen by early Aug, but it remains relatively rare until early Oct when flocks consisting of a few hundred birds may be typically seen foraging in recently harvested fields (especially soybean and corn). It is regularly seen in Nov, but there is only one Dec record. Earliest date: 1, 3 Aug 1974, Squaw Creek (FL). High counts: 600, 12 Oct 1980, Swan L. (BG-**BB** 48[1]:11); 500, 17 Oct 1980, Squaw Creek (FL). Latest dates: 2, 1 Dec 1968, Squaw Creek (FL); 4, 30 Nov 1968, Squaw Creek (JHA-**BB** 36[4]:12).

Snowy Plover (*Charadrius alexandrius*)

Status: Rare transient in northwest, accidental elsewhere.

Documentation: Specimen: male, of a pair, 19 Apr 1968, near Bigelow (Easterla 1969b; NWMSU, DAE 1570).

Habitat: Mudflats and shoreline.

Records:

Spring Migration: The first record for the state was not obtained until 1962 when 1–2 birds were observed on 6 May at Squaw Creek (HB, FL, BB, SR-**BB** 29[2]:17). Since then there have been about twenty additional records, with all but two concentrated in the northwestern corner between late Mar and the third week of May. Earliest date: 1, 25–26 Mar 1967, St. Joseph (JHA, FL-**BB** 34[2]:7). High count: there are at least three observations of two birds. Latest date: 1, 21 May 1981, Big L. SP (MBR-**BB** 48[3]:18) and see below. Only reports outside the NW corner: 1, 27 Apr 1981,

Thomas Hill Res., Macon Co. (TB, IA); 2, 21 May 1985, Schell-Osage (MBR-**BB** 52[3]:13).

Summer: Only one observation: 1, 7–14 July 1968, Squaw Creek (FL, HB-**BB** 36[3]:20).

Fall Migration: A single record: 1, 11 Aug 1963, Squaw Creek (JHA-**BB** 30[3]:20).

Comments: This plover regularly breeds as close as central Kansas (Russell, Barton and Strafford counties; Thompson and Ely 1989).

Semipalmated Plover (*Charadrius semipalmatus*)

Status: Common transient.

Documentation: Specimen: female, 9 May 1967, near Bigelow (NWMSU, LCW 501).

Habitat: Mudflats, bare flooded fields, and shoreline.

Records:

Spring Migration: The first arrivals appear in early Apr. Peak is at the very end of Apr or early May. A few are occasionally seen until the end of May. Earliest date: 1, 21 Mar 1982, Creve Coeur L. (CP-**NN** 54:47). High counts: 375, 9 May 1979, Squaw Creek (TB-**BB** 46[3]:8); 300, 27 Apr 1974, Squaw Creek (DAE-**BB** 41[3]:4).

Summer: There are only two June records: 5, 1 June 1990, Bigelow Marsh (DAE); 1, 16 June 1968, Squaw Creek (FL). There are a number of sightings for late July. High count: 50, 31 July 1988, Thomas Hill Res. (RB-**BB** 56[1]:13).

Fall Migration: Relatively small numbers, compared to the spring numbers, are seen throughout Aug and most of Sept. By the end of Sept it is rare. High count: 15, 31 Aug 1963, Squaw Creek (FL, HB). Latest dates: 1, 20–28 Oct 1989, L. Contrary (DAE, MBR, LG et al.-**BB** 57[1]:49); 1, 24 Oct 1971, Big L. SP (FL).

Piping Plover (*Charadrius melodus*)

Status: Rare transient.

Documentation: Specimen: male, 22 Apr 1967, near Bigelow (NWMSU, DAE 1229).

Habitat: Mudflats, bare flooded fields, and shoreline.

Records:

Spring Migration: Usually not encountered until the second week in Apr, with peak at the end of Apr or early May. There is only a single record during the last ten days of May. Earliest date: 2, 2 Apr 1977, Montrose (NJ-**BB** 44[4]:26). High counts: 10, 28 Apr 1962, Squaw Creek (DAE, WG-**BB** 29[2]:17); 10, 1 May 1979, Squaw Creek (TB-**BB** 46[3]:8). Latest date: 1, 27 May 1966, Squaw Creek (DAE).

Summer: No June records; casually seen in late July (5+ observations). Earliest date: 1, 1 July 1962, Duck Creek (D. Snyder-**BB** 29[3]:22).

Fall Migration: Now usually no more than 1–2 birds/season are seen in Aug through mid-Sept. There are no Oct records besides the one mentioned below. High count: 5, 17 Aug 1969, Squaw Creek (MBR, RAR). Latest record: 1, 27 Oct–5 Nov 1981, St. Joseph (FL-**AB** 36:183).

Comments: As a result of steady declines over the past four decades, the Great Lakes population was listed as endangered, and the remaining U.S. populations were given threatened status by the U.S. Fish and Wildlife Service.

This declining plover nests quite close to Missouri. Most recent breeding records for Nebraska and Iowa are from counties along the Missouri R., not more than 60 km north of the Missouri border (Johnsgard 1980; Dinsmore et al. 1984).

Killdeer (*Charadrius vociferus*)

Status: Common transient and summer resident; uncommon winter resident in south, rare in north.

Documentation: Specimen: female, 17 July 1932, Jackson Co. (KU 19379).

Habitat: Encountered in a wide variety of open, bare habitats; usually associated with water.

Records:

Spring Migration: On average, the first migrants appear in mid- to late Feb in the south and by early Mar in the north. Peak is usually in late Mar. High counts: 275, 25 Mar 1969, St. Joseph (FL); a few counts of 150 birds in late Mar. Young may be seen as early as late April.

Summer: BBS data indicate that this plover is most abundant in the Mississippi Lowlands and least so in the Ozarks (Map 16). As early as late June there is a buildup in numbers at favorable sites: 500, late June through July 1967, Squaw Creek (HB-**BB** 34[2]:14); 500, 31 July 1988, Thomas Hill Res. (RB-**BB** 56[1]:13).

Fall Migration: Large concentrations (hundreds of birds/locality) are regularly seen from Aug through mid-Oct. These impressive gatherings are less frequent during the latter half of Oct but may even occur in Nov: 400–500, 27 Nov 1953, Creve Coeur L. (JEC-**NN** 25:53); 500, 10 Nov 1984, Montrose (RF). High count: 1,040, 15 Sept 1963, Squaw Creek (HB-**BB** 30[4]:17).

Winter: During mild winters it is uncommon in the southern half of the state but rare in the north. Even in those mild winters birds are widely scattered in the state, and they are typically absent in the north during Jan and Feb. In severe winters birds may be virtually absent from the entire state. High counts, both on the Mingo CBC: 108, 2 Jan 1988; 47, 30 Dec 1974.

Map 16.
Relative breeding abundance of the Killdeer among the natural divisions. Based on BBS data expressed as birds/route.

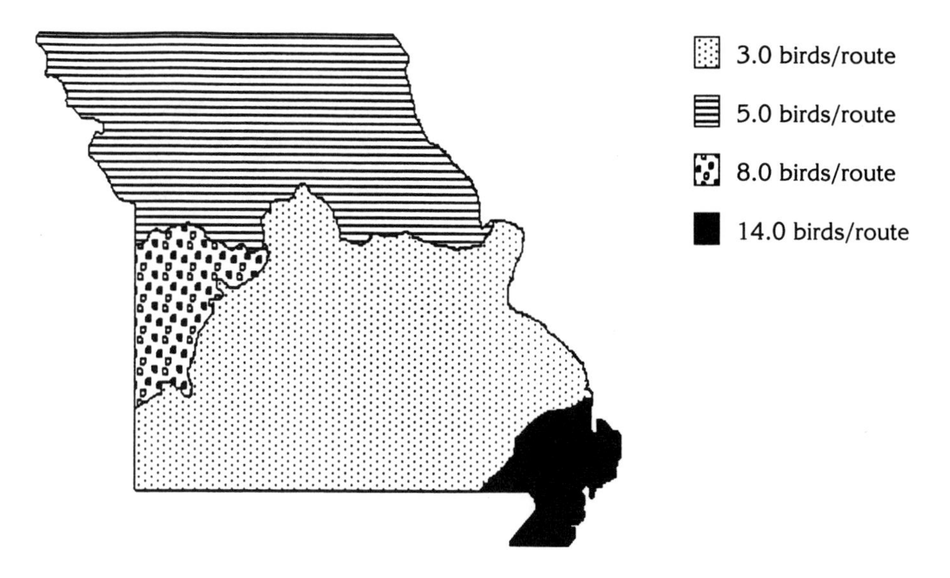

[**Mountain Plover** (*Charadrius montanus*)]

Status: Hypothetical transient and summer resident.
Documentation: Sight records only: see below.
Habitat: Short-grass pastures.
Records:

Spring Migration: The following sighting is the only one for the state that we put any credence in: a single bird was observed on 14 Mar 1953 at Prairie L., Buchanan Co., by reliable observers (BK et al.-**BB** 21[5]:4). However, because this is the only dependable record, we have elected to list this species as hypothetical until more concrete evidence is obtained.

Summer: Harris (1919b) mentions an unlikely observation by Bush, who claimed "that several pairs . . . nested . . . near Courtney [Jackson Co] in 1887." Bush also claimed that they were found in small numbers in other years. We question these observations since no other observer(s) reported this plover during that time, and because the closest it ever was recorded breeding was in western Kansas and Nebraska.

Fall Migration: Documentation of a bird on 10 Nov 1986 (**BB** 54[1]:36) was reviewed by an expert on Mountain Plover identification. We quote in part his summary evaluation: "The description rather conclusively rules out Mountain Plover."

Fig. 14. The Black-necked Stilt, with a Greater Yellowlegs, was present at Schell-Osage WA, Vernon Co. from 1–6 October 1977. "Breeding Bird Atlas" work resulted in the discovery of 3 nests in southeastern Missouri during June 1990. The photograph was taken by Jim Rathert.

Black-necked Stilt (*Himantopus mexicanus*)

Status: Casual transient in west, accidental in east; accidental summer resident in Mississippi Lowlands.

Documentation: Photograph: 1, 1–6 Oct 1977, Schell-Osage (S. Yeskie et al.-**AB** 32:212; Fig. 14).

Habitat: Mudflats and shoreline.

Records:

Spring Migration: There are six records, all of single birds between late Apr and late May. Four of these are from western Missouri. Earliest date: photographed, 22 Apr 1962, Squaw Creek (HB, FL, SR, BB-**BB** 29[2]:17). Latest date: 28 May 1967, Squaw Creek (HB, FL, JHA et al.-**BB** 34[2]:7).

Summer: The discovery of three widely scattered stilt nests on dikes surrounded by rice paddies in southern Stoddard Co., on 28 June 1990 (BJ, photos), was quite unexpected. Two of the nests were believed to have been successful. The adults were initially located on 9 June (BJ). This species has sporadically bred as close as Memphis, Tennessee, since 1982 (L. Coffey-**AB** 36:986–87).

Fall Migration: Three of the four observations, all of single birds, are during Oct in western Missouri. Besides the record above they are: 23–28 Aug 1969, St. Charles Co. (B. Knickmeyer-**BB** 38[3]:6); 15–21 Oct 1939, Iatan, Platte Co. (A. Shirling-**BB** 6[1]:89); photographed, 26 Oct 1974, Muskrat L., Buchanan Co. (FL et al.-**BB** 42[2]:6).

American Avocet (*Recurvirostra americana*)

Status: Rare transient.

Documentation: Specimen: male, 30 Sept 1972, Maryville SL (NWMSU, DST 55).

Habitat: Mudflats and shoreline.

Records:

Spring Migration: This species is more common in the western half of the state. Although there are three late Mar records, normally it is not seen until mid-Apr, with peak in late Apr and early May. It is very rare during the last ten days of May. Earliest date: 1, 24 Mar 1977, Cass Co. (JG-**BB** 44[4]:26). High counts: 100, 8 Apr 1894, near Stotesbury, Vernon Co. (T. Surber; Widmann 1907); 47, 22 Apr 1987, Creve Coeur L. (K. Boldt et al.-**BB** 54[3]:13). Latest date: 1, 31 May 1976, Squaw Creek (MBR-**BB** 44[2]:17).

Summer: No June and only four July observations. The earliest is 1, 8 July 1978, Trimble (FB et al.-**BB** 45[4]:12).

Fall Migration: Migrants are scarce until mid-Aug, and only small numbers (< 10 birds) are normally encountered through Sept. Most of the larger concentrations are observed in Oct. A few are casually seen during the first two weeks of Nov. High counts: 56, 27 Oct 1974, Squaw Creek (FL-**BB** 42[2]:6); 39, 17 Oct 1981, Pony Express WA, DeKalb Co. (FL). Latest date: 2, 14 Nov 1968, Swan L. (L. Kline-**BB** 36[4]:12).

Comments: The American Avocet commonly breeds as close as central Kansas (Barton and Strafford counties; Thompson and Ely 1989).

Greater Yellowlegs (*Tringa melanoleuca*)

Status: Common transient.

Documentation: Specimen: female, 6 Nov 1949, Vernon Co. (KU 39068).

Habitat: Flooded fields, marshes, mudflats, and shoreline.

Records:

Spring Migration: The first arrivals are typically seen by mid-Mar. Peak is during the last two weeks of Apr. It is only casually seen after mid-May. Earliest record: 1, 4 (not 3) Mar 1956, Gravois Mills, L. of the Ozarks (DAE-**BB** 23[4]:3). High count: 1,000+, 12 Apr 1979, Squaw Creek (TB-**BB** 46[3]:8). Latest date: 1, 27 May 1977, Squaw Creek (DAE-**BB** 44[4]:26).

Summer: There are no more than five June sightings, and it is rare in July. Earliest record (probably a late spring migrant): 1, 2 June 1968, Squaw Creek (FL). High count: 20, 31 July 1988, Swan L. (RB-**BB** 56[1]:13).

Fall Migration: By mid-Aug relatively large numbers may be encountered, with peak at the very end of Aug and early Sept. It is regularly seen until mid-Oct but is rare in late Oct and casual in Nov. There is only a single Dec observation. High count: 200, 1–15 Sept 1963, Squaw Creek (HB, FL-**BB** 30[4]:17). Latest records: 1, 1–4 Dec 1987, Swan L. (KM-**BB** 55[2]:61); 1, 30 Nov 1984, Squaw Creek (MBR-**BB** 52[1]:30).

Comments: At all seasons this species is less common than the Lesser Yellowlegs.

Lesser Yellowlegs (*Tringa flavipes*)

Status: Common transient.

Documentation: Specimen: female, 25 Apr 1965, Beverly L. (NWMSU, DAE 877).

Habitat: Same as Greater Yellowlegs.

Records:

Spring Migration: A few birds begin to appear by mid-Mar; however, it is rare until early Apr and peaks in late Apr and early May. By mid-May numbers are greatly reduced, with few individuals rarely seen at the end of the period. Earliest dates: male, 4 Mar 1917, Platte Co. (KU 39069); 1, 5 Mar 1977, Thomas Hill Res. (BG-**BB** 44[4]:26). High counts: 7,000+, 4 May 1979, Squaw Creek (TB-**BB** 46[3]:8); 1,000, 23 Apr 1967, Squaw Creek (FL-**BB** 34[2]:7).

Summer: It is casually seen throughout June and by mid- to late July it is uncommon. High counts: 400, 31 July 1988, Swan L. (RB-**BB** 56[1]:13); 200+, 24 July 1979, Thomas Hill Res. (JR, BG-**BB** 36[4]:7).

Fall Migration: By early Aug relatively large concentrations (in the low hundreds) may be encountered. Peak is usually during late Aug and early Sept. Numbers remain high through mid-Sept, but by late Sept numbers are drastically reduced. It is rarely seen after mid-Oct, and there are only five records for Nov. High count: 1,000, 1 Sept 1963 (FL). Latest dates: 1, 24 Nov 1966, Squaw Creek (FL); 1, 24 Nov 1969, Squaw Creek (FL-**BB** 36[4]:12).

Solitary Sandpiper (*Tringa solitaria*)

Status: Uncommon transient.

Documentation: Specimen: male, 16 July 1932, Fayette, Howard Co. (CMC 150).

Habitat: Marshes, flooded fields, and shoreline.

Records:

Spring Migration: This species arrives later than either of the yellowlegs; the first migrants appear by early Apr in the south and about a week later in the north. Peak is usually in late Apr and early May. Few are encountered after mid-May, and there is only a single record for the last week of May. Earliest date: 1, 29 Mar 1969, Blodgett, Scott Co. (JH). High count: 17+, 15 Apr 1987, Ted Shanks (KM-**BB** 54[3]:13). Latest records: 1, 27 May 1989, Big L. SP (MBR, DAE-**BB** 56[3]:90); two sightings for 23 May.

Summer: This species is often the first scolopacid to be detected returning south. Normally the first individuals are seen by early July. Peak may occur as early as late July. Earliest date: 1, 29 June 1974, Squaw Creek (FL). High count: 10+, 28 July 1932, Fayette (T. Baskett-**BL** 34:342).

Fall Migration: As mentioned above, peak may occur at the very end of

July, but normally it is in early Aug. After early Sept, only single birds are usually encountered. It is accidental in Oct. High counts: 28+, 8 Aug 1987, Ted Shanks (KM-**BB** 55[1]:11); 11, 9 Aug 1973, Mississippi Co. (JH). Latest dates: 1, 25 Oct 1970, Columbia (BG); 1, 12–13 Oct 1966, St. Joseph (FL).

Willet (*Catoptrophorus semipalmatus*)

Status: Uncommon transient in west, rare in east.

Documentation: Specimen: male, 6 May 1914, Jackson Co. (KU 39067).

Habitat: Mudflats, marshes, and shoreline.

Records:

Spring Migration: Normally it is not encountered until the beginning of the last week in Apr, but shortly thereafter relatively large numbers can be seen through mid-May. Peak is usually in early May. It is accidental during the last ten days of May. Earliest date: 1, 4 Apr 1976, Squaw Creek (FL). High counts: 107, 7 May 1967, Bigelow Marsh (DAE-**BB** 34[2]:7); 40, 28 Apr 1990, Swan L. (m.ob.-**BB** 57[3]:137).

Summer: Casual in June and July (more records in July). High count: 6, 4 June 1981, Alton Dam (BR-**BB** 48[3]:24).

Fall Migration: Regularly seen from early Aug through early Sept but never in the relatively large numbers of spring. There are fewer than five records after early Sept. High count: 7, 5 Aug 1980, Springfield (m.ob.-**BB** 48[1]:11). Latest record: 4, 20 Oct 1963, Squaw Creek (FL).

Comments: This species breeds as close as the Sandhill country of Nebraska (Johnsgard 1980).

Spotted Sandpiper (*Actitis macularia*)

Status: Uncommon transient and summer resident.

Documentation: Specimen: male, 10 May 1919, Lexington, Lafayette Co. (CMC 155).

Habitat: Shoreline, flooded fields, and occasionally mudflats.

Records:

Spring Migration: The first arrivals are seen in early Apr in the south and by mid-Apr in the north. Peak is in early May. Earliest dates: 1, 7 Mar 1984, Squaw Creek (FL-**BB** 51[3]:11); 1, 8 Mar 1972, East Prairie, Mississippi Co. (M. Southard-**BB** 39[2]:11). High count: 30, 9 May 1965, Squaw Creek (FL).

Summer: An uncommon summer resident (more local in the Ozarks) statewide.

Fall Migration: Migrants begin augmenting the breeding population in Aug. Few are seen after late Sept (< 5 observations after mid-Oct). Latest dates: 1, 5 Nov 1977, Little Dixie L. (BG); 1, 26–29 Oct 1985, Maryville SL (DAE).

Upland Sandpiper (*Bartramia longicauda*)

Status: Uncommon transient; uncommon summer resident in western half of the Glaciated Plains; rare in eastern Glaciated and Osage plains; very rare to absent elsewhere.

Documentation: Specimen: female, 30 May 1964, Gravois Mills, L. of the Ozarks (NWMSU, DAE 802).

Habitat: Short-grass fields and pasture, and occasionally mudflats.

Records:

Spring Migration: The earliest arrivals appear in the south by the first of Apr, but it is rarely seen in the northern half of the state until mid-Apr. Relatively large numbers (10–15/day) have been observed between early Apr (migrants) and late May (breeders). Earliest record: 1, 26 Mar 1943, St. Louis (WC-**AM** 45:8). High counts: 18, 25 May 1967, Ford City, Gentry Co. (DAE, RAR); 12, 3 Apr 1983, Springfield (CB-**BB** 50[3]:12).

Summer: A rare to uncommon breeder throughout most of the Glaciated and Osage plains. BBS data indicate it is most common in the western half of the Glaciated Plains, where an average of 2.5 birds/route are recorded vs. 0.3 birds/route in the eastern section. The average of 7.7 birds/run on route 32 is the highest mean for any route in the state. A sample of 32 hectares on a prairie in northern Harrison Co. in 1989 produced 27 sandpipers (BJ, C. Wilson, *in litt.*). The estimate for the entire prairie (545 hectares) was over *100* pairs.

An average of 1.0 birds/route has been recorded on the Osage Plains routes. It is more local and rarer in the Ozark Border, especially in the southeastern section. The Upland Sandpiper is even rarer in the Ozarks (recorded only once on BBS routes in the heart of this region), and it is not known to breed in the Mississippi Lowlands.

This species is known to nest as far south as Barton, Dade, and Greene counties, where the Osage Plains interdigitates with the western Ozarks. By the end of July migrating flocks may be seen: 35, 19 July 1964, Amazonia, Andrew Co. (FL-**BB** 31[3]:18).

Fall Migration: Solitary birds and small-sized flocks (5–20 individuals) are seen throughout Aug and Sept. There are no records for Oct, but there is one for Nov. High count: 30, 20 Aug 1979, Kansas City International Airport, Platte Co. (SP-**BB** 47[1]:16). Latest date: 1, 8 Nov 1959, Sugar L. (FL, EC).

Comments: As late as the 1880s this species was described as an abundant migrant and fairly common breeder (Widmann 1907; Harris 1919b); but by the turn of the century it had been so decimated by excessive hunting and the plowing of the prairie that it almost became extirpated. Fortunately, primarily as a result of stricter hunting laws, it has made a modest comeback. The breeding population now appears stable.

Eskimo Curlew (*Numenius borealis*)

Status: Extirpated, on the verge of extinction. Formerly a common transient in spring (especially in the west).

Documentation: Specimen: male, 16 Apr 1894, Vernon Co. (T. Suber; AMNH 752072; see below).

Comments: Apparently this species was still commonly encountered until the turn of the century: flock of 100, 16 Apr 1894, Vernon Co. (T. Suber; Widmann 1907; see above specimen). The last definite record was of a flock of 10, 1 May 1902, Jasper Co. (W. Savage; Widmann 1907). Hurter (1884) mentions a female that was taken on 10 Apr 1871, in St. Louis Co. (live mount; STSC, 4-6-50, J. Hurter 220). Not surprisingly there are no fall records, since this species primarily migrated along and off the east coast in the fall.

During the last twenty years there have been occasional sightings along the Texas coast in the spring and in southeastern Canada and the northeastern U.S. in the fall. This species probably will never be recorded in the state again.

Whimbrel (*Numenius phaeopus*)

Status: Rare spring transient in west, accidental in east; accidental fall transient.

Documentation: Specimen: male, 16 May 1967, near Bigelow (NWMSU, DAE 1489).

Habitat: Mudflats, flooded fields, and wet pastures.

Records:

Spring Migration: Over twenty spring records; all except three are from the western half of the state and occur (85%) during the last two weeks of May. The first record, one of the very few for eastern Missouri, was obtained on 25 May 1936 at St. Charles Co. (WS-**BB** 4[4]:36). Earliest date: 1, 12 Apr 1975, Maryville SL (MBR, DAE-**BB** 43[1]:19). High counts: 11, 19 May 1984, Camden bottoms, Ray Co. (CH, MMH-**BB** 51[3]:12); 9, 20 May 1989, Swan L. (M. Goodman-**BB** 56[3]:90).

Summer: Only three records: 1, 3 June 1954, Missouri R., near St. Charles, St. Charles Co. (B. Dowling et al.-**BB** 22[3]:2); 1, 9 July 1977, Squaw Creek (FL-**BB** 44[4]:30); 1, 9 July 1988, near St. Joseph (K. Jackson, LG-**BB** 56[1]:14).

Fall Migration: There is a report of a single bird in St. Louis Co. on 4 Sept 1949 (JVB-**BB** 17[2]:3).

Long-billed Curlew (*Numenius americanus*)

Status: Casual spring transient in west; formerly more common during spring and fall.

Documentation: Specimen: male, May 1859, no specific locality except "Missouri" (AMNH 738314).

Habitat: Mudflats, flooded fields, and short-grass pasture.

Records:

Spring Migration: Prior to the 1900s this curlew was a common migrant and primarily seen in early Apr (Scott 1879; Hurter 1884; Widmann 1907); but by the turn of the century, it was quite rare. Harris (1919b) stated that there were only six records for the Kansas City region. Since the late 1950s, there have been no more than twelve sightings, with most (75%) in Apr. Earliest date: 1, 31 Mar 1963, Squaw Creek (FL-**BB** 30[2]:25). Recent high count: 6, 4 May 1968, L. Contrary (FL-**BB** 36[1]:5). Latest date: 1, 28 May 1961, Squaw Creek (SR-**AFN** 15:415).

Summer: A single recent record: 1, 8 July 1972, Squaw Creek (FL-**BB** 39[4]:9).

Fall Migration: No recent records. Apparently a former common transient from Aug through Oct (Widmann 1907). Latest date: 4, 2 Nov 1917, on the Missouri R., near Sibley, Jackson Co. (J. Guinotte; Harris 1919b).

Comments: The closest breeding populations are in the Sandhill country in Nebraska (Johnsgard 1980) and in southwestern Kansas (Thompson and Ely 1989).

Hudsonian Godwit (*Limosa haemastica*)

Status: Uncommon to locally common spring transient in west, rare in east; casual in fall.

Documentation: Specimen: female, 25 Apr 1965, Beverly L. (NWMSU, DAE 878).

Habitat: Mudflats and relatively bare flooded fields.

Records:

Spring Migration: Normally the first migrants are observed in mid-Apr, with peak usually during the second week of May (Easterla 1971). Individuals or small groups are regularly seen until the end of May. Earliest record: 1, 3 Apr 1979, Squaw Creek (TB-**BB** 46[3]:9). High counts in the west: 650+, 13 May 1979, Squaw Creek (TB, RB, JR et al.); 400, 29 Apr 1956, Squaw Creek (B. Stollberg-**AFN** 10:337); in the east: all under 10 birds/day. Latest date: 1, 30 May 1960, Squaw Creek (FL).

Summer: Only three records, all at Squaw Creek: 8, 1 June 1969 (FL); 1, 2 June 1981 (MBR); 3, 7 July 1964 (HB-**BB** 31[3]:18).

Fall Migration: Five modern records, all of single birds: 30 Aug 1962, St. Charles airport (C. Kniffen-**BB** 29[4]:25); 21 Sept 1969, Squaw Creek (FL); female, 4 Oct 1988, Big L. SP (MBR-**BB** 56[1]:14; ANSP 180846); 9 Oct 1971, Squaw Creek (FL-**BB** 39[1]:6); and 24 Oct 1971, Big L. SP (FL, MBR).

Marbled Godwit (*Limosa fedoa*)

Status: Rare transient in west, casual in east.
Documentation: Photograph: 2, 21 May 1990, Bigelow Marsh (DAE; VIREO x08/12/002).
Habitat: Mudflats, flooded pastures, and short-grass fields.
Records:
Spring Migration: Prior to the late 1800s this species was considered to be a "fairly common transient," but even by Widmann's time (1907) it was considered rare. Today it is rare in the west, casual in the east, and in years when extensive mudflats are few or nonexistent this species may go unrecorded. It is more frequently encountered in the spring than in the fall.

The first arrivals usually appear by mid-Apr and peak during the last ten days of Apr. Individuals are occasionally seen until the end of May. Earliest date: 1, 12 Apr 1979, Squaw Creek (MBR, DT). High counts in the west: 15, 20 Apr 1970, Squaw Creek (DAE-**BB** 38[3]:8); 12, 23 Apr 1967, Squaw Creek (HB, FL, SR); in the east: no observations of more than one bird. Latest date: 1, 31 May 1981, Squaw Creek (FL).
Summer: Only one June record: 3, 26 June 1966, Squaw Creek (FL-**BB** 33[2]:12). The earliest July date: 1, 21 July 1968, Squaw Creek (FL, HB).
Fall Migration: There are fewer than a dozen modern records (all of single birds) for this period, and all except two are in Aug. Latest date: 1, 20 Sept 1969, Squaw Creek (FL).

Ruddy Turnstone (*Arenaria interpres*)

Status: Rare but regular transient.
Documentation: Specimen: male, 19 May 1967, near Bigelow (NWMSU, DAE 1759).
Habitat: Mudflats and rocky shoreline.
Records:
Spring Migration: This is one of the latest shorebird migrants and is more common in the spring than in the fall. The majority of records are during the last two weeks in May. The earliest and only Apr record is of 7, 30 Apr 1903, Missouri R., Jackson Co. (Harris 1919b). High counts, both at Squaw Creek: 38, 29 May 1960 (FL et al.); 31, 20 May 1976 (MBR, DT-**BB** 44[2]16).
Summer: Only one June record: 1, 2 June 1968, Squaw Creek (FL). Earliest July record: 1, 27 July 1970, Maryville SL (MBR).
Fall Migration: Most observations are of single birds in Aug. Latest and only Oct date: 1, 20 Oct 1917, Missouri R., Jackson Co. (W. Andrews; Harris 1919b).

Red Knot (*Calidris canutus*)

Status: Rare transient.

Documentation: Specimen: male, 7 May 1967, near Bigelow (NWMSU, DAE 1426).

Habitat: Mudflats and shoreline.

Records:

Spring Migration: There are ca. twenty-five records for the state, about evenly divided between spring and fall. Almost all spring records are during mid-May, and all but a couple are from the western half of the state. Earliest and only Apr record: 6, 24 Apr 1970, Squaw Creek (FL). High count: 30, 11 May 1980, Swan L. (BG-**BB** 47[3]:12). Latest date: 1, 25 May 1950, St. Louis (JEC et al.-**BB** 17[7]:1).

Fall Migration: Records range from late Aug through Sept, but most are concentrated in early Sept. The majority of the records are from the eastern half of the state. Earliest date and high count: 3, 23 Aug 1988, Bonne Terre Marsh, St. Francois Co. (HF-**BB** 56[1]:14). Other high count: 3, 12 Sept 1986, opposite Chain of Rocks, St. Louis (RG). Latest date: 1, 6–7 Nov 1981, Alton Dam (BR-**BB** 49[1]:14). There are no Oct records.

Sanderling (*Calidris alba*)

Status: Rare to locally uncommon transient.

Documentation: Specimen: male, 15 May 1965, Mound City (NWMSU, DAE 1010).

Habitat: Mudflats and shoreline.

Records:

Spring Migration: Normally not encountered until the end of Apr, and the largest concentrations and most records are observed in mid-May. Earliest dates: 6, 21 Mar 1902, Kansas City (Harris 1919b); 6, 22 Mar 1940, Squaw Creek (WC, A. Shirling-**BB** 7[4]:28). High counts: 150, 13 May 1979, Squaw Creek (JR, RB, TB-**BB** 46[3]:9); two counts of ca. 50 birds at Squaw Creek in mid-May.

Summer: The latest of two June sightings: 1, 6 June 1965, Squaw Creek FL-**BB** 32[3]:14). Earliest July record: 3, 24 July 1979, Thomas Hill Res. (JR, BG-**BB** 46[4]:7).

Fall Migration: Although records span nearly the entire period, most are concentrated from mid-Aug through Sept. It is only casually observed in Oct, and there are only two Nov observations. High counts: 50, 1–15 Sept 1963, Squaw Creek (HB, FL-**BB** 30[4]:17); 40, 11 Sept 1968, Swan L. (L. Kline-**BB** 36[4]:12). Latest date: 1, 26 Nov 1978, St. Charles Co. (TP-**BB** 46[1]:24).

Semipalmated Sandpiper (*Calidris pusilla*)

Status: Common transient.
Documentation: Specimen: male, 19 May 1967, Holt Co. (NWMSU, NNL 5).
Habitat: Open mudflats and shoreline.
Records:
Spring Migration: One of the most common shorebirds. It is generally seen from early Apr to the end of the period, with peak usually during the third week of May. Earliest record: 2, 21 Mar 1976, Squaw Creek (FL). High counts: 2,000+, 19 May 1963, Squaw Creek (FL); 2,000, 23 May 1965, Squaw Creek (FL).
Summer: There are numerous sightings for early June and the later half of July. High counts: 2,500, 31 July 1988, Swan L. (RB-**BB** 56[1]:14); 1,500+, 1 June 1969, Squaw Creek (FL).
Fall Migration: Quite common in Aug through mid-Sept, especially late Aug and early Sept. Observers should be extremely careful in distinguishing this species from the Western Sandpiper; undoubtedly, many of the purported sightings of the Semipalmated after early Oct pertain to the Western Sandpiper (see next species account). High count: 1,000, 1 Sept 1963, Squaw Creek (HB, FL-**BB** 30[4]:17).
Comments: See Veit and Jonsson (1984) for a thorough treatment on the identification of the smaller *Calidris* species.

Western Sandpiper (*Calidris mauri*)

Status: Rare but regular transient in spring; uncommon in fall.
Documentation: Specimen: female, 18 Aug 1940, Sunshine L., Ray Co. (Newton 1942; MU 835).
Habitat: Prefers open mudflats and shoreline.
Records:
Spring Migration: This species arrives and peaks well before the preceding species. The first birds appear in early Apr and most have passed through by early May. Earliest record: no.?, 31 Mar 1963, near St. Joseph (FL). High count: 12, 2 Apr 1977, Squaw Creek (MBR, DAE-**BB** 44[4]:26). Latest date: 3, 29 May 1976, Schell-Osage (NJ, JG-**BB** 44[2]:16).
Summer: No June records, and the earliest July record is of 1, 4 July 1940, Bean L., Platte Co. (Newton 1942). High count: 50, 31 July 1988, Swan L. (RB-**BB** 56[1]:14).
Fall Migration: Records span almost the entire period, but most are in Aug through early Sept. There are at least five Nov observations, and undoubtedly some—if not all—observations of Semipalmated Sandpipers from mid-Oct through the end of the period pertain to this species. High

count: 55–60, 12 Aug 1988, Swan L. (TB, WL). Latest record: 1, 26 Nov 1975, Busch WA (JC-**BB** 44[1]:20).

Comments: See comments under Semipalmated Sandpiper.

[**Rufous-necked Stint** (*Calidris ruficollis*)]

Status: Hypothetical spring transient.

Documentation: Sight record only: see below.

Habitat: Mudflats.

Comments: A bird in full nuptial plumage was studied for ca. 20 minutes on 20 May 1990, Squaw Creek (DAE-**BB** 57[3]:138). The bird left before it could be photographed.

There is at least one other record for the midwest; a bird in nuptial plumage was seen on 21 July 1962 in Ohio (Ahlquist 1964). It is of casual occurrence on both coasts and is primarily seen during July and Aug.

Least Sandpiper (*Calidris minutilla*)

Status: Common transient; accidental winter visitant.

Documentation: Specimen: female, 7 Sept 1933, Jackson Co. (KU 20207).

Habitat: Less often found in open mudflats preferred by the two previous species. Commonly encountered at the edge of small pools in marshes and along shoreline of ponds and lakes.

Records:

Spring Migration: This species first appears during the second week of Apr. Peak is in early May, but some birds are seen through the end of the month. Earliest date: 1, 27 Mar 1988, near St. Joseph (LG-**BB** 55[3]:88). High count: 500, 7 May 1964, Platte Co. (DAE).

Summer: There are a few records for the first week of June, with a hiatus in sightings until early July. Earliest July record: 4, 1 July 1982, St. Charles Co. (BR-**BB** 49[4]:17). High count: 1,000, 31 July 1988, Swan L. (RB-**BB** 56[1]:14).

Fall Migration: By mid-Aug this species is relatively common. Peak probably occurs in late Aug and early Sept, although observations with detailed numbers are lacking. It is fairly common through the end of Sept and is regularly seen until mid-Oct. There are at least five Nov and three early Dec records. Latest dates: 1, 11 Dec 1987, Swan L. (KM-**BB** 55[2]:61); female, 6 Dec 1975, Maryville SL (DAE-**BB** 44[1]:26; NWMSU, JWG 70).

Winter: There are two winter observations: 1, 27 Dec 1967, Mingo CBC; 3, 2 Jan 1967, Big L., Mississippi Co. (JH-**BB** 34[1]:22).

White-rumped Sandpiper (*Calidris fuscicollis*)

Status: Common spring transient in west, uncommon in east; uncommon fall transient in west, rare in east.

Documentation: Specimen: male, 25 Apr 1968, near Bigelow (NWMSU, NNL 4).

Habitat: Mudflats, flooded fields, and shoreline.

Records:

Spring Migration: A late migrant—not seen before the end of Apr. Peak is normally during the third week of May, but relatively large numbers (in the hundreds) can be seen into the first week of June. Earliest dates: 1, 27 Apr 1974, Squaw Creek (DAE); 2, 27 Apr 1975, Squaw Creek (DAE, MBR). High count: 5,000+, 17 May 1979, Squaw Creek (TB-**BB** 46[3]:8); 1,000, 19 May 1963, Squaw Creek (HB-**BB** 30[2]:25).

Summer: Late migrants are regularly seen during the first half, particularly the first week, of June. Although observations span the period from mid-June to mid-July, there clearly is an increase in sightings and numbers at the end of July (presumably of birds returning from the breeding grounds). High counts: 1,500+, 1 June 1969, Squaw Creek (FL); 45–50, 27 July 1975, Squaw Creek (DAE, MBR-**BB** 43[1]:23).

Fall Migration: Decidedly less common at this season than in the spring. Most observations are of single birds or small groups (< 10 birds) in Aug and early Sept. High count: 50, 31 Aug 1963, Squaw Creek (HB, FL-**BB** 30[4]:17). Latest date: no.?, 13 Sept 1959, Squaw Creek (FL).

Baird's Sandpiper (*Calidris bairdii*)

Status: Uncommon to locally common transient in west; rare spring, uncommon fall transient in east.

Documentation: Specimen: male, 8 Apr 1956, Gravois Mills, L. of the Ozarks (MU 1476).

Habitat: Mudflats, flooded fields, and shoreline, but also seen on completely bare, dry flats.

Records:

Spring Migration: This "peep" is significantly more common in the western half of the state. It is one of the first shorebirds to appear in the spring. The first individuals arrive by mid-Mar, but numbers (> 20 birds/day) are not seen until the second week of Apr. Peak is usually at the end of Apr. By mid- to late May only single birds or small groups (< 5 birds/day) are encountered. Earliest dates: 1, 4 Mar 1988, near St. Joseph (LG-**BB** 55[3]:88); 1, 10 Mar 1985, Squaw Creek (FL); High counts: 205+, 20 Apr 1989, Big L. SP (DAE-**BB** 56[3]:90); 200+, 23–29 Apr 1979, Squaw Creek (TB, JR, RB-**BB** 46[3]:9). Latest date: 3, 31 May 1970, Squaw Creek (DAE).

Summer: Only one June record: 1, 2 June 1981, Big L. SP (MBR). Birds begin to reappear in July; earliest date: 3, 2 July 1961, Squaw Creek (FL).

Fall Migration: Usually only single birds or small groups (< 5 birds) are seen during the first three weeks of Aug. Peak is at the end of Aug and early

Sept. Relatively large-sized flocks may be seen until mid-Oct (25, 20 Oct 1963, Squaw Creek [FL]), but thereafter only single birds are rarely encountered until late Nov. High count: 100, 1–15 Sept 1963, Squaw Creek (HB, FL-**BB** 30[4]:12). Latest date: 1, 30 Nov 1968, Squaw Creek (JHA-**BB** 36[4]:12).

Pectoral Sandpiper (*Calidris melanotos*)

Status: Common transient.

Documentation: Specimen: female, 9 May 1964, Cooley L. (NWMSU, DAE 847).

Habitat: Mudflats, shoreline, and wet short-grass fields.

Records:

Spring Migration: One of the most common shorebirds. The first arrivals appear by mid-Mar, but the relatively large numbers (> 50 birds/day) are not seen until early Apr. Peak is in late Apr and early May. It is readily seen through the end of May. Earliest date: 1, 2 Mar 1981, St. Joseph (FL). High count: 6,500, 8 May 1979, Squaw Creek (TB-**BB** 46[3]:8).

Summer: A few birds regularly linger into early June, but there are very few records from mid-June to early July. However, by mid-July birds begin to reappear. Latest June sighting: 3, 16 June 1968, Squaw Creek (FL). Earliest July date: 2, 2 July 1967, Squaw Creek (FL). High count: 500, 31 July 1988, Swan L. (RB-**BB** 56[1]:14).

Fall Migration: By early Aug relatively large numbers (in the hundreds) are encountered. Peak is usually from mid- to late Aug. It is commonly seen through Sept, and in most years it is seen into early Nov. High count: 750+, 31 Aug 1963, Squaw Creek (HB, FL). Latest dates: 1, 7–9 Dec 1987, Swan L. (KM-**BB** 55[2]:61); 2, 1 Dec 1969, Squaw Creek (FL-**AFN** 23:487).

Comments: The Sharp-tailed Sandpiper (*Calidris acuminata*) should be looked for in fall (especially from mid-Sept through mid-Oct), since this species has been casually recorded in the interior of North America. It is of regular occurrence during the fall along the Pacific Coast south to southern California. Most of the records for North America are of juveniles that retain their distinctive plumage until at least late Nov (Hayman et al. 1986). The juveniles' upper breasts are washed with an orangish-buff that is only finely streaked. Steaking is often restricted to just the lower throat and the sides of the neck. The cap is reddish and the margins of the dorsal feathers are reddish-brown. Pectoral Sandpipers lack the orangish-buff breast and invariably have heavily streaked breasts that are sharply demarcated from the belly. Hayman et al. (1986) provide a thorough treatment of Sharp-tailed Sandpiper identification.

[**Purple Sandpiper** (*Calidris maritima*)]

Status: Hypothetical.
Documentation: Sight record: see below.
Comments:
Although both Widmann (1907) and Harris (1919b) refer to a specimen collected in the spring of 1854 in western Missouri (Hoy 1865), neither actually examined it. Moreover, none of the specimens from that journey can now be located. This lack of evidence coupled with the fact that there were a few other dubious reports from the trip (see Red-necked Grebe species account) makes this record questionable.

There is, however, another record of a bird seen on 4 May 1938, near Orchard Farm, St. Charles Co. (L. Ernest and Miss Elliott-**BB** 5[6]:65). In view of the observers' reputations this is likely a valid record, although the date seems late.

Dunlin (*Calidris alpina*)

Status: Common transient in west; uncommon spring, common fall transient in east; accidental winter visitant.
Documentation: Specimen: female, 9 May 1964, Cooley L. (NWMSU, DAE 568).
Habitat: Mudflats, shoreline, and flooded fields.
Records:
Spring Migration: There are very few records prior to the last week of Apr. Peak is normally during the last two weeks of May. Earliest date: 1, 8 Apr 1977, Big L., Mississippi Co. (JH). High counts: 500+, 19 May 1963, Squaw Creek (FL); 300, 26 May 1968, Squaw Creek (DAE).
Summer: Although there are a number of early June observations, none exist for the remainder of the period. Latest record: 1, 8 June 1969, Squaw Creek (FL).
Fall Migration: This species is only casually seen in Aug and early Sept. Peak is in Oct, and small-sized groups (5–10 birds/day) are seen through mid-Nov. Earliest date: 1, 3 Aug 1974, Squaw Creek (FL). High count: 750, 25 Oct 1986, near Rich Hill, Bates Co. (FL, LG). Latest records: 1, 2 Dec 1972, Squaw Creek (FL-**AFN** 27:623); 8, 2 Dec 1989, near Windfield Dam, Lincoln Co. (m.ob.-**BB** 57[2]:104).
Winter: Only one record: 3, 18 Dec 1970, Big L., Mississippi Co. (JH-**AFN** 25:586).

Stilt Sandpiper (*Calidris himantopus*)

Status: Uncommon to locally common spring transient in west, rare in east; common fall transient statewide.

Documentation: Specimen: female, 25 Aug 1940, Ray Co. (MU 836).
Habitat: Mudflats, flooded fields, and shoreline.
Records:
Spring Migration: Much more common in the western half of the state. Normally the first individuals appear at the beginning of the last week of Apr and peak in mid-May. Birds are seen through the end of May. Earliest dates: male, 9 Apr 1967, near Bigelow (DAE; NWMSU, LCW 790); 6, 16 Apr 1914, mouth of Big Blue R., Jackson Co. (Harris 1919b). High counts in the west: 625, 20 May 1990, Bigelow Marsh (DAE-**BB** 57[3]:138); two counts of 500+ for Squaw Creek. High count in the east: 8, 20 May 1989, Alton Dam area (RG-**BB** 57[1]:53).
Summer: No June records. Birds begin to reappear by early July (all July observations are of groups composed of < 50 birds). Earliest date: 5, 2 July 1961, Squaw Creek (FL).
Fall Migration: Readily seen throughout Aug. Peak is at the very end of Aug or early Sept. Birds are consistently seen through Sept, but are only rarely seen after the first few days of Oct; and it is casual during the first half of Nov. High count: 2,000, 1–15 Sept 1963, Squaw Creek (FL, HB-**BB** 30[4]:17). Latest date: 4, 16 Nov 1975, Squaw Creek (JEC et al.-**BB** 44 [1]:20).

Buff-breasted Sandpiper (*Tryngites subruficollis*)

Status: Rare transient.
Documentation: Specimen: female, 19 May 1967, near Bigelow (NWMSU, LCW 433).
Habitat: Short-grass fields, freshly plowed fields, and mudflats.
Records:
Spring Migration: Less commonly encountered in the spring than in the fall. With the exception of two records, all observations are in May (particularly in mid-May). Earliest dates: 1, 9 Apr 1967, St. Louis (JEC-**BB** 34[2]:7); 1, 29 Apr 1956, Squaw Creek (FL, SR). High counts: 40, 19 May 1984, Camden bottoms, Ray Co. (CH, MMH-**BB** 51[3]:12); 30, 20 May 1967, near Bigelow Marsh (DAE-**BB** 35[4]:5). Latest date: 5, 26 May 1970, Squaw Creek (DAE).
Summer: Only one June record: 1, 1 June 1969, Squaw Creek (FL). Birds begin to return at the very end of July. Earliest July dates: 2, 26 July 1964, Squaw Creek (JHA-**BB** 31[3]:18); 1, 29 July 1987, Mound City, Holt Co. (DAE et al.). High count: 22, 30 July 1970, Maryville SL (MBR-**AFN** 24:690).
Fall Migration: There are more observations and larger numbers seen during Aug and early Sept than during the spring. Peak (largely of immatures) is in late Aug or early Sept, but only small-sized flocks and individuals are seen thereafter. There are only two Oct sightings. High counts:

Fig. 15. This Reeve (female of the Ruff) was present at Big Lake State Park, Holt Co., 6–7 May 1981. The species is of casual occurrence during spring and fall in Missouri. Photograph by David Easterla.

50+, 5–7 Sept 1970, St. Charles Co. (m.ob.-**AFN** 25:65); 50, 2 Sept 1971, Squaw Creek (HB-**BB** 39[1]:5). Latest date: 1, 11 Oct 1947, Marais Temps Clair (K. Wesseling-**BB** 15[1]:1).

Ruff (*Philomachus pugnax*)

Status: Casual transient in west; accidental in east.

Documentation: Photograph: female, 6–7 May 1981, Big L. SP (DAE; VIREO x08/12/001; Fig. 15).

Habitat: Mudflats and shoreline.

Records:

Spring Migration: Five of six observations are of single females. All records are from the western half of the state. Only report of nuptial plumaged male: 1 May 1988, Schell-Osage (SD, BRE et al.-**BB** 55[3]:8). Earliest record: 17 Apr 1982, Schell-Osage (KH et al.-**BB** 49[3]:19). Latest date: 13 May 1990, Schell-Osage (PMC, TB, RB).

Fall Migration: Four records: 1, 12 Aug 1984, Aldrich (JS-**BB** 52[1]:30); 1, 26 Aug 1972, Squaw Creek (SP-**BB** 40[1]:8); 1, 28 Aug 1974, near Alton Dam (MS, H. Wuestenfeld et al.-**BB** 42[2]:6); and 2, 7–10 Oct 1962, Busch WA (R. Laffey et al.-**BB** 29[4]:25).

Comments: The Ruff is an Old World shorebird that regularly migrates through North America. It has bred in western Alaska, and there is speculation that it may nest elsewhere on the continent (*A.O.U. Check-list 1983*). This species was first recorded in Missouri at Squaw Creek on 28 Apr 1962 (WG, DAE; Easterla 1963).

Although eight of the ten records for Missouri are from the western half of the state, we believe this is an artifact of the more consistent prevalence of high quality shorebird habitat in that part of the state. An average of 2–3 birds/year has recently been reported in Illinois (RG, pers. comm.).

Short-billed Dowitcher (*Limnodromus griseus*)

Status: Uncommon to locally common transient.

Documentation: Specimen: female, 17 May 1970, near Bigelow (NWMSU, DST 102).

Habitat: Mudflats, flooded fields, and shoreline.

Records:

Spring Migration: The first migrants appear at the beginning of May. Peak is during mid-May, usually about a week later than the Long-billed peak; some birds are seen through the end of the period. Earliest date: 5, 27 Apr 1974, Squaw Creek (DAE). High count: 300+, 11 May 1979, Squaw Creek (DAE, MBR, TB-**BB** 46[3]:9).

Summer: Except for a single late June record (4, 26 June 1966, Squaw Creek, FL and JHA) of presumably early fall migrants, there are no June observations. The first fall migrants usually appear by mid-July.

Fall Migration: Birds are recorded throughout Aug and peak in mid-Aug. More data are needed to better define the termination of migration throughout the state. High count: 35+, 14 Aug 1975, Squaw Creek/Swan L. (MBR). Latest date: 3, 14 Sept 1969, Squaw Creek (MBR).

Comments: Relatively little detailed data exist on the migration periods for the two species because Missouri observers did not begin distinguishing these two species until the mid–1960s. Data gathered by a few observers over the past twenty-five years indicate that the Long-billed is more common, and its migration period is more prolonged in the spring (Easterla 1970c) and in the fall than that of the Short-billed. However, data are scarce on the termination of the Short-billed's migration in the fall and on the commencement of the Long-billed's fall migration.

Observers should exercise caution in making identifications based solely on plumage characteristics. Hayman et al. (1986) and Kaufman (1990) discuss how to distinguish these species based on plumage.

Specimens collected in spring of the Short-billed are all attributed to *L. g. hendersoni.* An effort should be made to obtain specimens in the fall for subspecific identification, as it is conceivable that the other two races may occur in the state.

Long-billed Dowitcher (*Limnodromus scolopaceus*)

Status: Common spring transient in west, rare to uncommon in east; common fall transient.

Documentation: Specimen: male, 23 Oct 1955, Gravois Mills, L. of the Ozarks (MU 1469).

Habitat: Mudflats, flooded fields, and shoreline.

Records:

Spring Migration: The first arrivals appear in mid-Mar well before the previous species. Numbers gradually build through Apr and peak at the beginning of May. Virtually all are gone by the end of the third week of May. Earliest date: 3, 10 Mar 1985, Squaw Creek (FL). High count: 500+, 10 May 1979, Squaw Creek (TB-**BB** 46[3]:9).

Summer: There are no unquestionable June or July records.

Fall Migration: This species does not begin to outnumber Short-billeds until the end of Aug. The largest numbers are recorded in Oct when flocks of a few hundred birds may be encountered. It is casual in Nov. High count: 250, 20 Oct 1963, Squaw Creek (HB, FL-**BB** 30[4]:17). Latest date: 1, 25 Nov 1984, St. Joseph (FL, MBR-**BB** 52[1]:30).

Comments: See Short-billed Dowitcher account.

Common Snipe (*Gallinago gallinago*)

Status: Common transient; rare winter resident.

Documentation: Specimen: male, 18 Apr 1964, Beverly L. (NWMSU, DAE 565).

Habitat: Marshes, swamps, wet fields, and streams.

Records:

Spring Migration: Birds begin returning with the thawing of ice on streams and marshes, usually at the end of Feb and early Mar. Numbers increase until early Apr when peak is reached. Relatively large concentrations (50–100 birds/day) may be found until the latter part of Apr, but by the beginning of the second week in May few birds remain. It is accidental during the last ten days of May. High count: 400, 7 Apr 1980, Marais Temps Clair (J. Jackson-**NN** 52:43). Latest date: 1, 30 May 1987, Ted Shanks (KM-**BB** 54[3]:14).

Summer: Although there are at least five sightings of this species in June, only one observation involves more than one bird, and there is no evidence of breeding in any of these cases. However, breeding has been documented recently in northern Iowa (Dinsmore et al. 1984), and June courtship display has been noted as far south as southern Illinois (Bohlen and Zimmerman 1989). Therefore, breeding behavior should be looked for in any "summering" birds. Migrants begin to appear at the end of July. There is at least one June record involving two or more birds: 2, 10 June 1978, Bigelow Marsh (DAE, TB).

Fall Migration: By early Aug relatively large numbers may arrive in the northern section of the state: 25–30, 1 Aug 1978, Squaw Creek (MBR, R. Matthews-**BB** 46[1]:24). Numbers continue to increase and peak in late Sept or early Oct. By early Nov relatively few birds remain in the state. Small numbers (1–2 birds/day), however, may be found in favorable areas through the end of the period. High count: 150, 27 Sept 1973, St. Joseph (FL-**BB** 41[1]:3).

Winter: This species occurs in low densities at this season. It is more common in the southern half of the state, especially during milder winters. High counts, both on the Springfield CBC: 19, 20 Dec 1975; 15, 19 Dec 1987.

American Woodcock (*Scolopax minor*)

Status: Uncommon summer resident; casual winter resident in south.

Documentation: Specimen: male, 17 Mar 1977, Nodaway Co. L., Nodaway Co. (NWMSU, JWG 225).

Habitat: Streams bordered by forest, woodland edge, and brushy hillsides adjacent to water.

Records:

Spring Migration and Summer: If the latter part of the winter is mild, birds may be observed displaying as early as the beginning of Feb (see winter section). Even during average years, males are displaying by the beginning of Mar. Most migrants have passed through the state by early Apr. High count: 25 displaying birds, late Mar 1977, Nodaway Co. L. (DAE-**BB** 44[4]:26). Presumably birds are most common in the more heavily wooded areas of the state, such as the Ozarks and Ozark Border, but census data are not available.

Fall Migration: There is clearly a movement of birds into nonbreeding areas at the end of Sept, with a peak in the latter half of Oct. Birds are readily encountered through Nov, but few are seen after the beginning of Dec.

Winter: There are now many records for this period, but most of these involve birds that linger into mid-Dec or are early spring arrivals. There are only a few records of true wintering birds, e.g., 1, Dec–Jan 1963–64, Branson, Taney Co. (D. Schreiber-**BB** 31[1]:32). Undoubtedly, a few birds do overwinter in the southern section of the state during mild winters. During the later half of the 1980s, this species was regularly found displaying at numerous localities (primarily in the southern half of the state) by the first week of Feb. Earliest dates of arrival: 1, 30 Jan 1990, northern Washington Co. (SD-**BB** 57[2]:104); 1, 2 Feb 1989, Eureka, St. Louis Co. (P. Hoell-**BB** 56[2]:54); 4, 2 Feb 1990, near DeSoto, Jefferson Co. (B. Boesch); 1, 12 Feb 1987, St. Joseph (LG-**BB** 54[2]:13). High count for CBCs: 3, 19 Dec 1987, Springfield CBC.

Wilson's Phalarope (*Phalaropus tricolor*)

Status: Common spring transient in west, uncommon in east; uncommon fall transient statewide; accidental summer resident.

Documentation: Specimen: female, 23 Apr 1914, Jackson Co. (KU 39073).

Habitat: Shallow pools in marshes, flooded fields, and on ponds and lakes.

Records:

Spring Migration: The first individuals are seen by mid-Apr and peak in early May. A few birds linger until the end of May. Earliest dates: 1, 25–26 Mar 1988, St. Joseph (C. Fisher, LG-**BB** 55[3]:88); 1, 2 Apr 1966, Squaw Creek (FL). High counts: 1,000, 7 May 1967, Squaw Creek (FL); 600, 2 May 1965, Squaw Creek (FL).

Summer: Apparently this species formerly bred (accidentally) in the state. Widmann (1907) mentions a single breeding record: "H. Nehrling found it with young in July 1884, in Lawrence Co." There are a few recent observations of pairs lingering into June, but with no evidence of breeding: 2 pair, 20 June 1976, Maryville SL, with one male present until at least 23 June (DAE, MBR-**BB** 44[2]:20). Apparently it is an uncommon breeder in central Kansas (Barton and Strafford counties; Thompson and Ely 1989). Migrants begin to reappear by early July.

Fall Migration: It is much less common in the fall than in the spring, both in frequency of observations and number of birds. Small numbers (< 10 birds/day) are seen from early Aug through mid-Sept. Peak is in late Aug or early Sept. There is one late Oct and two Nov records. High count: 30, 1 Sept 1963, Squaw Creek (HB, FL-**BB** 30[4]:17). Latest dates: 1, 17 Nov 1968, Squaw Creek (FL-**BB** 36[4]:12); 2, 12 Nov 1984, St. Joseph (FL-**AB** 39:61).

Red-necked Phalarope (*Phalaropus lobatus*)

Status: Rare transient; accidental winter visitant.

Documentation: Specimen: female, 15 May 1967, near Bigelow (NWMSU, DAE 1544).

Habitat: Shallow pools in marshes, flooded fields, and on ponds and lakes.

Records:

Spring Migration: This species usually does not appear until mid-May, and even then only relatively small numbers (1–6 birds/day) are seen. Peak is typically during the third week of May, but a few are readily observed until the end of the period. Earliest date: 1, 7 May 1972, Squaw Creek (FL). High count: 82, 19 May 1968, Squaw Creek (DAE, FL, JHA-**BB** 36[1]:26).

Summer: There are only two records for this period, both from Squaw Creek: 1, 1 June 1969 (FL); 1, 27 June 1965 (FL-**BB** 32[4]:14).

Fall Migration: There are virtually no records for this phalarope until the end of the third week in Aug. However, the largest concentrations are noted at the very end of Aug and early Sept. Smaller numbers are regularly observed throughout Sept. There are at least three Oct and one Nov sightings. Earliest date: 6, 20 Aug 1970, Squaw Creek (MBR). High count: 14, 4 Sept 1974, Maryville SL (MBR, DAE-**BB** 42[2]:6). Latest date: 1, 3 Nov 1968, Squaw Creek (JHA-**BB** 36[4]:12).

Fig. 16. This basic plumaged Red Phalarope was photographed on 13 May 1972 at Squaw Creek NWR, Holt Co., by Mark Robbins. This phalarope is most often seen during the fall in Missouri.

Winter: A single, extraordinarily late individual was seen by a careful observer on 1 Jan 1938, Iatan Marsh, Platte Co. (WC-Rising et al. 1978).

Comments: Formerly known as the Northern Phalarope.

Red Phalarope (*Phalaropus fulicaria*)

Status: Accidental spring transient; casual fall transient.

Documentation: Specimen: female, 25 Oct 1977, Maryville SL (DAE-**BB** 45[1]:26; NWMSU, DAE 2988).

Habitat: Shallow pools in marshes, sewage lagoons, and lakes.

Records:

Spring Migration: Four records, all between 13 and 26 May, in the northwestern corner: 1, photos, 13 May 1972, Squaw Creek (MBR et al.-**AFN** 26:767; VIREO r08/10/002; Fig. 16); 3, 16 May 1954, Trimble (BK, B. Tordoff et al.-**AFN** 8:315); 2 different birds, one on 24 May (VIREO r08/10/005), the other on 25–26 May 1976, Squaw Creek (MBR, DT et al.-**BB** 44[2]:17).

Summer: One record: 1, nuptial plumage, 28–29 July 1974, St. Louis (MS et al.-**BB** 41[4]:7).

Fall Migration: At least eleven records, four in Sept and seven in Oct, all of single birds. Earliest record: 1, 7 Sept 1986, Thomas Hill Res. (TB et al.-**BB** 54[1]:33). Latest dates: 1, 15–26 Oct 1978, Thomas Hill Res. (JR et al.-**BB** 46[1]:24); 1, 26 Oct 1986, L. of the Ozarks SP (WL-**BB** 54[1]:33).

Family Laridae: Skuas, Gulls, and Terns

Pomarine Jaeger (*Stercorarius pomarinus*)

Status: Casual fall migrant; accidental summer visitant.

Documentation: Specimen: o?, 28 Nov 1915, Missouri R., Jackson Co. (Hoffman 1916; KU 47672).

Habitat: Rivers and large bodies of water.

Fig. 17. This juvenile Parasitic Jaeger was photographed between 24–29 September 1980 at Lake Contrary, Buchanan Co., by Ival Lawhon, Jr. There are only a few definite records of this species for Missouri.

Records:

Summer: A third-year bird was present 9–12 July 1981, Alton Dam (Y. Balsinger, M. Scudder, BR et al.-**AB** 35:945).

Fall Migration: Four additional records besides the one listed above: 1, 3 Oct 1979, Thomas Hill Res. (TB-**BB** 47[1]:16); 1, 7–9 Nov 1979, Alton Dam (Rudden 1980b; photos on file in the Illinois State Museum, RG, pers. comm.); 1, 14–25 Nov 1985, Smithville L. (CH, MMH, FL et al.); and an immature, dark phase, 27 Nov 1982, Alton Dam (SD-**BB** 50[1]:23).

Parasitic Jaeger (*Stercorarius parasiticus*)

Status: Accidental transient.

Documentation: Photograph: juvenile, 24–29 Sept 1980, L. Contrary (FL, DAE et al.-**BB** 48[1]:12; Fig. 17).

Habitat: Usually seen at the larger bodies of water.

Records:

Spring Migration: Harris (1919b) gave details for an unusual sighting of a flock of 5 birds on 23 Apr 1916 at the mouth of the Big Blue R. in Jackson Co. He stated that the birds were observed "at close range," and that "there can be no question of the identification in this case, as the birds were close enough to show their distinctive characteristic, the *sharp* middle tail feathers a few inches longer than the others." Harris also makes the following intriguing comment, "The older river men state that this bird was not uncommon in the days when the river was filled with refuse from the packing houses."

Fall Migration: Besides the above mentioned record, there is only one additional record with specifics. Widmann (1907) states that J. Kastendieck collected a bird (not located) from a millpond near Billings, Christian Co., in Aug 1905. Although these are the only two definite fall records, the following sightings were believed to pertain to this species: 1, 6 Sept 1971, Squaw Creek (FL, MBR, LG, R. Rowlett-**BB** 39[1]:6); 1, late Oct, Smithville L. (T. Schallberg-**BB** 55[1]:11); 1, 3 Nov 1968, Swan L. (JHA et al.-**BB** 36[4]:12); 1, 3 Nov 1972, Browning L. (FL-**BB** 40[1]:8).

Long-tailed Jaeger (*Stercorarius longicaudus*)

Status: Accidental transient.

Documentation: Specimen: none extant (see below).

Habitat: Rivers, lakes, and sewage lagoons.

Records:

Spring Migration: One record: specimen (no longer extant), spring 1910, Bean L., Platte Co. (Holland; Harris 1919b).

Fall Migration: Two records: adult, 12 Sept 1974, Maryville SL (MBR, DAE; Robbins 1975); 2, 3 Oct 1916, Missouri R., Jackson Co. (B. Bush; Harris 1919b).

[**Skua** sp. (*Catharacta* sp.)]

Status: Hypothetical.

Comments: Harris (**BL** 22:170) identified and reported that a skua was collected by a fisherman on 3 Apr 1920, along the Missouri R., near Sibley, Jackson Co. "The head, wings and feet" were saved; however, apparently the specimen is no longer extant. In the absence of a specimen and precedent set by the misidentification of other purported interior specimens of skua, which upon reexamination proved to be jaegers (e.g., Johnsgard 1980), we treat the Missouri record as hypothetical. There are only two verified skua reports for the interior of North America, in New York (*A.O.U. Check-list 1983*) and North Dakota (R. Kreil, C. Grondahl-**AB** 43:1333).

Laughing Gull (*Larus atricilla*)

Status: Very rare transient along Mississippi R.; casual elsewhere; accidental summer and winter visitant.

Documentation: Specimen: first-year male, 28 May 1977, Thomas Hill Res. (JR-**BB** 44[4]:26; MDC B0128).

Fig. 18. This adult Laughing Gull was found at Schell-Osage WA, Vernon Co., on 18 May 1981. Primarily a coastal-inhabiting species, it is a very rare transient in Missouri. Photograph by David Easterla.

Habitat: Rivers, lakes, reservoirs, and mudflats.

Records:

Spring Migration: At least a dozen records, and all except one in May. Six of the records are from the Alton Dam area; the others are scattered throughout the state. Earliest date: 1, 29 Mar–1 Apr 1969, Alton Dam (A. Bromet-**NN** 41:49). High count: 4 adults, 30 May 1983, Alton Dam (PS, TB-**BB** 50[3]:13).

Summer: Two observations: adult, 2 June 1990, Riverlands Environmental Demonstration area, St. Charles Co. (JZ, B. Estill-**NN** 62:53); subadult, 6 June 1981, L. Viking, Gallatin, Daviess Co. (DAE).

Fall Migration: During the past decade, 1980–90, there have been at least twelve records involving at least sixteen individuals from the Mississippi R. at St. Louis (RG, pers. comm.). Fourteen of the birds were observed in Aug and Sept, with the other two in Nov. Some of these records may pertain to only the Illinois side of the river. High count: 2 (adult and immature), 12 Aug 1990, Riverlands Environmental Demonstration Area (M. and J. Holsen). Latest dates: adult, 15–16 Nov 1985, Alton Dam (BR, PS, RA-**NN** 57:90); immature, 16 Nov 1986, Riverlands (RA).

Winter: Only one record: immature, 19 Dec 1976, Fountain Grove WA, Livingston Co. (Goodge 1977; photo in authors' possession).

Franklin's Gull (*Larus pipixcan*)

Status: Common transient in west, uncommon in east; accidental winter visitant.

Documentation: Specimen: male, 3 Nov 1925, Jackson Co. (KU 39075).

Habitat: Marshes, lakes, rivers, and recently plowed fields.

Records:

Spring Migration: Although commonly seen in the spring, it is not seen in the huge numbers as in the fall. It is considerably more common during the spring and the fall in the western half of the state. A few birds begin to appear by mid-Mar, and by the end the month flocks composed of a few hundred birds may be seen. However, relatively large-sized groups (in the high hundreds or low thousands) do not appear until late Apr through the first half of May. Smaller groups or individuals are regularly seen into June. Earliest date: no.?, 6 Mar 1956, St. Joseph (FL). High count: 2,500, 15 May 1966, near St. Joseph (JHA-**BB** 33[2]:7).

Summer: There are a number of records (ca. 20) that span the entire month of June; however, there are only five July observations. High count: 100, 1 June 1969, Squaw Creek (FL).

Fall Migration: A few birds begin to trickle in by the end of Aug, but relatively large-sized flocks (in the hundreds or low thousands) are not seen until the end of Sept. The largest flocks are encountered during the last two weeks of Oct. Few birds are seen after mid-Nov. Earliest date: 1, 8 Aug 1970, Squaw Creek (MBR). High counts: 150,000, 23 Oct 1968, Swan L. (L. Kline-**BB** 36[4]:12); 10,000, 21 Oct 1950, Squaw Creek (K. Krum-**BB** 17[1]):1). Latest date: 1, 1 Dec 1968, Gravois Mills, L. of the Ozarks (DAE-**BB** 36[4]:12).

Winter: Four winter records as follows: 3 adults in nuptial plumage, 23–27 Dec 1988, Big L. SP (MBR et al.-**BB** 56[2]:54); 1, 24 Jan 1948, Marais Temps Clair (JEC et al.-**NN** 20:1); 1, 9 Feb 1957, Alton Dam (DJ-**NN** 29:4); and 1, early Feb 1971, Alton Dam (MS-**BB** 38[4]:10).

Little Gull (*Larus minutus*)

Status: Accidental transient.

Documentation: Sight records only: see below.

Records:

Spring: One sighting: adult in basic plumage, 5–6 Apr 1987, north of Gravois Mills, L. of the Ozarks (WL, TB et al.).

Fall Migration: There are now three observations, all of adults in Nov: 1 Nov 1987, Thomas Hill Res. (BG, IA-**BB** 55[1]:11); 5 Nov 1988, Smithville L. (D. Bryan-**BB** 56[2]:56); 30 Nov 1975, Maryville SL (Easterla 1976).

Common Black-headed Gull (*Larus ridibundus*)

Status: Accidental fall transient.

Documentation: Specimen: adult male, 30 Oct 1976, Maryville SL (DAE-**BB** 44[3]:22; NWMSU, DST 110).

Records: Only one additional record to the one above: adult, 10–25 Sept 1988, Alton Dam (D. Becher, RG et al.).

Bonaparte's Gull (*Larus philadelphia*)

Status: Uncommon transient; rare winter resident.

Documentation: Specimen: male, 11 May 1975, Maryville SL (NWMSU, DLD 166)

Habitat: Larger rivers, lakes, reservoirs, and mudflats.

Records:

Spring Migration: Besides wintering birds, this gull is very rarely seen before the end of Mar, but in early Apr there is an influx that peaks by midmonth. It is readily observed through the first week in May but is only casually encountered thereafter. Earliest date: 1, 4 Mar 1979, L. Contrary (FL). High count: 100+, 10 Apr 1979, Squaw Creek (TB-**BB** 46[3]:9); 100, 25 Feb 1990, Stockton L., Cedar Co. (RF, G. Griffith-**BB** 57[2]:104). Latest date: 3, 25 May 1940, Alton L. (Alton Dam) (L. Ernst-**BL** 42:389).

Summer: A single record: 1 in basic plumage, 31 July 1971, Squaw Creek (MBR, FL-**AFN** 25:389).

Fall Migration: More frequent and observed in larger numbers during this period than in the spring. It first appears in early Oct. and peaks usually in Nov, with smaller numbers lingering until the end of the period. Earliest dates: adult in basic plumage, 13 Aug 1983, Mississippi R., St. Charles Co. (RG); immature, 1 Sept 1974, Swan L. (BG-**BB** 42[2]:6). High counts: 750+, 10 Nov 1990, Thomas Hill Res. (PMC, TB et al.); 600+, 25 Oct 1990, Truman Res., Benton Co. (PMC, J. Hazelman).

Winter: Increasingly seen lingering at the larger reservoirs in the southern half of the state. Now, at least during mild to normal winters, it is recorded annually during Dec and early Jan at Table Rock L. and Stockton L. There are at least a half dozen records for the St. Louis area as well. High counts: 34, 27 Dec 1987, Columbia CBC; 25, 17 Dec–30 Jan 1988–89, Montrose WA (Montrose CBC; LG-**BB** 56[2]:54).

Ring-billed Gull (*Larus delawarensis*)

Status: Common transient; locally common winter resident in the south, rare in the north away from the Mississippi R.

Documentation: Specimen: female, 29 Oct 1925, Jackson Co. (KU 39074).

Habitat: Virtually any body of water as well as recently plowed fields.

Records:

Spring Migration: Large numbers (in the hundreds or low thousands) begin to augment the winter population at the end of Feb. Peak is in early

Mar along the Mississippi R. (where this species is most common at all seasons). In the west, peak is usually in early Apr. Few birds are seen after early May, but occasionally some linger until the end of the period. High counts, along Mississippi R.: 5,000, 5 Mar 1967, Alton Dam (RA, PB-**NN** 39:31); in the west: 2,000+, 10 Apr 1979, Squaw Creek (MBR).

Summer: It is rarely seen throughout this period. High count: 21, 6 June 1979, St. Charles Co. (TP-**BB** 46[4]:7).

Fall Migration: Birds begin to trickle in through Aug, but larger flocks are not seen until the latter half of Sept. But by early Oct, relatively large-sized groups (in the hundreds) are encountered. Peak is normally during mid-Nov. Concentrations remain high, especially along the Mississippi R., until extremely cold weather arrives (often after the end of this period). High counts: 10,000, 17 Nov 1951, along Mississippi R., St. Charles and St. Louis counties (JEC et al.-**BB** 18[12]:3); 3,000, 16 Nov 1986, Alton Dam (RA-**NN** 59:12).

Winter: Large concentrations may remain along the Mississippi R. and at the larger reservoirs in the south until severe, cold weather arrives. Thus, even in early Jan huge concentrations may be encountered along the Mississippi R.: 10,000, early Jan 1975, Alton Dam area (RA-**BB** 42[3]:13).

Comments: At all seasons, away from the Mississippi R., this species is more common than the Herring Gull. Like most larids, it has increased dramatically in numbers over the past twenty-five years, primarily as a result of an increase in refuse dumps and the construction of reservoirs, both of which provide food and open water throughout the winter.

California Gull (*Larus californicus*)

Status: Casual transient in east; accidental in west.

Documentation: Photograph: second year, 6–13 Nov 1981, Alton Dam/ Madison Co., Illinois (BR, RK; photos TB, VIREO x08/7/001).

Habitat: Rivers and large reservoirs.

Records:

Spring Migration: There are five records, four from the Alton Dam area: adult, 7 Mar 1976 (TB-**BB** 44[1]:26); adult, 29 Mar 1987 (RA-**BB** 54[3]:14); adult, 5 Apr 1975 (RA-**BB** 43[1]:20); adult, 7 May 1989, Smithville L. (MBR, DAE, TE-**BB** 56[3]:90); and a third-year bird, 18 May 1986 (RG et al.-**AB** 40:479).

Fall Migration: Four records: the above photographed bird; 1, 4 Oct 1986, Alton Dam (RA-**BB** 54[1]:33); second year, 17–30 Oct 1981, Alton Dam (BR et al.-**BB** 49[1]:14); and an adult, 15 Nov 1986, L. of the Ozarks, below Bagnell Dam (WL-**BB** 54[1]:33).

Comments: This species is likely to occur more frequently in the western half of the state than the above two records indicate.

Herring Gull (*Larus argentatus*)

Status: Common transient and winter resident along Mississippi R.; rare to locally uncommon elsewhere.

Documentation: Specimen: male, 16 Dec 1977, L. Wappapello, Wayne Co. (GL 346).

Habitat: Rivers and reservoirs.

Records:

Spring Migration: By the beginning of this period the large wintering numbers along the Mississippi R. have already headed north. It remains relatively common along the river through early Apr, but few are seen at the end of the month. Occasionally birds linger until the end of May. Away from the Mississippi R., especially in the western half of the state, it is rarely seen, and then only in small numbers (< 5 birds/day).

Summer: It is casually seen during this period (< 8 records and all involving single birds); observations span the entire period, from both sides of the state.

Fall Migration: Occasionally birds are seen at the end of Sept, but it is not regular until early Oct. Numbers begin to build along the Mississippi R. at the end of Oct, but the huge concentrations often do not appear until the end of this period or the beginning of the winter season. Earliest date and the only Aug sighting: 1, 7 Aug 1976, Squaw Creek (FL).

Winter: This species is outnumbered by the Ring-billed Gull statewide, until very cold weather arrives (usually in Jan). At that time, as Ring-bills move south and more Herrings move in, it becomes the most common larid along the Mississippi R. High count: 1,200, 27 Dec 1976, Alton Dam (DJ-**NN** 49:12).

Comments: See comments under Thayer's Gull.

Thayer's Gull (*Larus thayeri*)

Status: Rare winter resident along Mississippi R.; accidental elsewhere.

Documentation: Specimen: adult female, 8 Jan 1978, Chain of Rocks, St. Louis Co. (DAE; NWMSU, MF 88).

Habitat: Larger rivers and reservoirs.

Records:

Fall Migration: Now recorded nearly annually in Nov along the Mississippi R. at St. Louis (RG, pers. comm.). Earliest date: 1, photographed, 11 Oct 1975, Alton Dam (TB-**BB** 29[3]:22). Only record away from Mississippi R.: 2 immatures, 29 Nov 1986, Truman Res., Benton Co. (WL-**BB** 54[1]:33).

Winter: A rare, but regular resident, along the Mississippi R., with all but one observation from the St. Louis area. High count: 11, 19 Feb 1985, Alton Dam (CP-**BB** 52[2]:16).

Comments: There has been considerable taxonomic confusion over the Herring/Thayer's/Iceland gull complex. As a result of Smith's work (1963, 1966), the A.O.U. (1983) treats the western, darker plumaged North American populations as a separate species, Thayer's Gull (Larus *thayeri*). However, recent evidence establishes that Smith's work is suspect (Gaston and Decker 1985; Godfrey 1986; in particular, see Snell 1989). In contrast to Smith's account, Snell and others have shown that *thayeri* freely interbreeds with *kumlieni* (currently considered a race of the Iceland Gull).

These findings would seem to beg for *thayeri* to be considered just a dark race of *kumlieni;* however, the situation is far more complex. *Thayeri* may be more closely related to the Herring Gull (*L. argentatus*), and, in fact, it has been treated as a race of *argentatus.* Moreover, *kumlieni* was once treated as a hybrid between *thayeri* and the nominate Iceland Gull. It is conceivable that *thayeri, kumlieni,* and *argentatus* will ultimately prove to represent a single species.

In view of Snell's findings, which indicates that there is hybridization between these taxa, we should reemphasize that not all individuals may be assigned to one taxon or the other. It is reasonable to assume that hybrids do occur in the state.

Iceland Gull (*Larus glaucoides*)

Status: Casual winter resident along Mississippi R.; accidental elsewhere.

Documentation: Photograph: first year, 28 (originally found on the 19th) Nov 1980, Alton Dam (TB-**NN** 53:4; VIREO x08/7/002).

Habitat: Larger rivers and reservoirs.

Records:

Although there are over twenty-five reports of "Iceland Gull," several later turned out to be Glaucous Gulls, while others pertained to the so-called Thayer's Gull (see above). Probably all *bona fide* sightings of Iceland Gull (at least as it is now defined) in Missouri can be referred to the *kumlieni* race. As Tingley (1983) states, "Definite specimens of *glaucoides* [nominate race] are extremely rare or nonexistent from North America."

There are only about seven unquestionable observations for the state, all since 1980. The only record away from the Mississippi R. at St. Louis is of a single bird seen on 15 Dec 1984 (Kansas City Southeast CBC). There are at least six records (including the above documented record) for the Alton Dam area at St. Louis: first winter, 28 Nov 1983 (RG); 1, 12 Dec 1980 (TB, BR, DJ, PS-**BB** 48[2]:13); adult, 21–29 Jan 1984 (BR, RG et al.-**BB** 51[3]:19); first winter, 5–9 Feb 1983 (RG, JE, SRU et al.-**BB** 50[2]:25); 2, first winter and probable second-winter, 16 Feb 1986 (BRO, RG, PS et al.-**BB** 53[2]:9).

Comments: See comments under Thayer's Gull.

Lesser Black-backed Gull (*Larus fuscus*)

Status: Accidental fall transient and winter resident along Mississippi R.

Documentation: Photograph: adult, 4–23 Dec 1988, Alton Dam (BR, RG et al.-**BB** 56[2]:54; photos by RG in MRBRC file). A bird seen at this same locality on 4 and 18 Feb 1989 was presumed to be the same individual as the one photographed in Dec.

Habitat: Rivers and reservoirs.

Records: Only three additional records besides the one above: second year, 30 Dec 1980, Chain of Rocks, St. Louis (BR, PS, DJ, RG-**BB** 48[2]:13); adult, 11–12 Feb 1984, Alton Dam (BR, RG, m.ob.-**BB** 51[3]:13; photo by B. Rose, VIREO x08/5/002); adult, 16 Feb 1981, Alton Dam (BR et al.-**BB** 48[2]:13).

Slaty-backed Gull (*Larus schistisagus*)

Status: Accidental winter resident.

Documentation: Photograph: adult, 20 Dec–29 Jan 1983–84, Mississippi R., St. Louis (photos by D. Ulmer; Goetz et al. 1986).

Comments: The above extraordinary record was meticulously documented in the Goetz et al. 1986 paper. This northeast Asian species is a regular but rare to uncommon transient to coastal Alaska (Goetz et al. 1986). Prior to the above record this gull had been recorded only once south of Alaska (Victoria, British Columbia; *A.O.U. Check-list 1983*).

Another adult was present on the Mississippi R. in the Davenport, Iowa/Moline, Illinois area from 12–21 Feb 1989 (A. Barker et al.-**AB** 43:322).

Glaucous Gull (*Larus hyperboreus*)

Status: Rare but regular transient and winter resident along Mississippi R.; casual transient and accidental winter resident elsewhere.

Documentation: Specimen: male, 30 Jan 1978, near Trimble (NWMSU, MF 89).

Habitat: Larger rivers and reservoirs.

Records:

Spring Migration: Birds are regularly observed until mid-Mar, and an occasional individual can be seen until mid-Apr. There are two May records, both in the St. Louis area. High count: 4, 9 Mar 1963, Alton Dam (JEC-**NN** 35:2). Latest date: 1, 27 May 1978, St. Charles Co. (RA, PB-**NN** 50:2).

Fall Migration: It is accidental during this period. Earliest dates: 1, first year, 28 Oct 1982, Schell-Osage (JW-**BB** 50[1]:24); 1, 13–18 Nov 1980, Alton Dam (BR, CP, CS-**BB** 48[1]:12); and 1, 15 Nov 1982, Browning L. (FL).

Winter: With the exception of three observations, all records are from the Mississippi R. near St. Louis. High count: 8, Jan 1984, Alton Dam area (m.ob.-**NN** 56:25).

Great Black-backed Gull (*Larus marinus*)

Status: Casual transient and winter resident along Mississippi R.

Documentation: Photograph: adult, 12 Feb 1984, W. Alton, Mississippi R., St. Charles Co. (B. Rose; VIREO x08/5/001).

Habitat: Large rivers and reservoirs.

Records:

Spring Migration: Only two records for this period, both from Alton Dam: 1, 27 Feb 1980 (TP-**NN** 52:28); adult, 17 Mar 1978 (BRO, MS-**BB** 45[3]:17).

Fall Migration: Only one record: immature, 7–8 Nov 1985, Mississippi R. at St. Louis (CP, BR-**BB** 53[1]:27).

Winter: About a dozen records, all in Jan and Feb on the Mississippi R. The only record away from the St. Louis area (and the first for the state) was of an adult at Cape Girardeau on 1 Jan 1945 (Mrs. Findley, Mrs. Harris-**BB** 11[3]:13). High count: 1–3, Jan 1986, Mississippi R., St. Louis (m.ob.-**NN** 58:20).

Black-legged Kittiwake (*Rissa tridactyla*)

Status: Rare winter resident and casual transient along the Mississippi R.; casual transient and accidental winter visitant elsewhere.

Documentation: Sight records only: see below.

Habitat: Larger rivers and reservoirs.

Records:

Spring Migration: Six records, three at Alton Dam. The others as follows: 1, 15 Apr 1984, Aldrich (C. Tyndall et al.-**BB** 51[3]:12); first year, 17 Mar 1984, Springfield (D. Thurman, CB); and J. Bryant apparently took a specimen (no longer extant) in the spring of 1897, near Kansas City (Harris 1919b). Other late date: first year, 15 Apr 1989, Alton Dam (BRO et al.-**BB** 56[3]:90).

Fall Migration: At least eight records for this period, with the earliest in mid-Nov. Four of these are away from the St. Louis area: 1, 15–16 Nov 1980, L. Springfield (M. Thornburg, D. Jones et al.-**BB** 48[1]:12); 1, 15 Nov 1980, Squaw Creek (M. Newlon); 1, 19 Nov 1983, Stockton L. (TB, JD, M. Goodman et al.- **BB** 51[1]:42); 1, 7 Dec 1983, Jackson Co. (CH-**BB** 51[3]:19).

Winter: Since 1967 this species has been recorded on about ten occasions along the Mississippi R. at St. Louis. All except the following record are of single birds: 2 first year, 7–8 Jan 1977, Chain of Rocks, St. Louis (DAE, MBR, LG-**BB** 44[3]:25). Only record away from Mississippi R.: 1 first year, 28 Dec 1989, Stockton L. (MBR-**BB** 57[2]:104).

Comments: With the exception of one, a second-winter bird, all records are of first-year birds.

Sabine's Gull (*Xema sabini*)

Status: Casual fall transient.

Documentation: Photograph: immature, 3–4 Oct 1978, Squaw Creek (MBR et al.-**BB** 46[1]:25; VIREO r08/10/003).

Habitat: Larger rivers, reservoirs, and marshes.

Records:

Fall Migration: At least six records, all involving single birds between mid-Sept and early Oct. All aged individuals (n=3) have been immatures. Earliest date: 12 Sept 1965, Alton Dam (RA-**BB** 33[1]:6).

Widmann (1907) mentions that three birds were collected by C. Worthen in Sept 1900 along the Mississippi R., in the area of Clark Co. Fleming (1912) gave details on two of Worthen's records: an immature male was collected on 15 Sept, "near Fox Island, Missouri," and another male was taken on the same date on the Illinois side of the river at Warsaw. Specifics on the third specimen remain a mystery. None of these specimens appear to be extant.

Caspian Tern (*Sterna caspia*)

Status: Uncommon transient.

Documentation: Specimen: male, 23 May 1972, near Maryville (NWMSU, DLD 120).

Habitat: Larger rivers, reservoirs, and marshes.

Records:

Spring Migration: Rarely seen before early May, with peak at mid-month. Birds are regularly seen through the last week of May. Earliest date: 2, 24 Apr 1982, Duck Creek (SD, BL, K. Adams-**BB** 49[3]:19). High counts: apparently a huge wave of birds was present in the western half of Missouri on 14 May 1978, as over 3,000 were seen in the St. Joseph and Squaw Creek areas (FL), and 247 were counted at Columbia on the same date (JR-**AB** 32:1015). Less unusual high count: 42, 14 May 1977, Montrose (JR-**BB** 44[4]:26).

Summer: It is a rare nonbreeder, with records spanning the entire period. High count: 17, 26 June 1966, Squaw Creek (FL).

Fall Migration: It continues to be rarely encountered in Aug, but by early Sept it becomes more common. Peak is in mid-Sept, but birds are regularly seen until mid-Oct. Thereafter, it is only casually observed; only a single Nov record exists. High count: 250, 14 Sept 1977, Missouri R., Jefferson City (fide BG-**BB** 45[1]:27). Latest date: 1, 8 Nov 1981, Alton Dam (BR, JE-**AB** 36:184).

Common Tern (*Sterna hirundo*)

Status: Uncommon transient.

Documentation: Specimen: female, 17 May 1974, Maryville SL (NWMSU; DLD 153).

Habitat: Marshes, reservoirs, and larger rivers.

Records:

Spring Migration: There are no records for Apr, and normally this tern is not seen until the second week in May. Peak is during the last two weeks of May. Earliest dates: adult, 3 May 1975, Squaw Creek (DAE, MBR); adult, 4 May 1990, Smithville L. (DAE, TE-**BB** 57[3]:138). High count: 50, 23 May 1970, Squaw Creek (DAE).

Summer: Late spring migrants are still passing through by early to mid-June. Presumed nonbreeders or early fall migrants are seen at the end of June and July. High count: 10, 13 July 1983, Alton Dam (BR-**BB** 50[4]:33).

Fall Migration: Birds are readily observed throughout Aug, with peak in early to mid-Sept. There are no definite records after early Oct. High count: 60, 15 Sept 1977, Alton Dam (JE et al.-**BB** 45[1]:26). Latest date: 2, 7 Oct 1962, Browning L. (FL-**BB** 29[4]:26).

Forster's Tern (*Sterna forsteri*)

Status: Common transient.

Documentation: Specimen: male, 19 Apr 1965, Gravois Mills, L. of the Ozarks (NWMSU, DAE 944).

Habitat: Marshes, reservoirs, and larger rivers.

Records:

Spring Migration: This species is more abundant and arrives much earlier than the Common Tern. Birds begin to appear by the beginning of the second week of Apr. Peak is usually in early May. A few birds typically linger through the end of the period. Earliest dates: 1, 31 Mar 1967, Browning L. (FL); 1, 5 Apr 1986, Alton Dam (RG). High count: 60, 12 May 1979, Squaw Creek (FL).

Summer: Like the Common Tern, birds are occasionally seen throughout this period across the state. High count: 35, 20 June 1965, Squaw Creek (FL).

Fall Migration: Regularly seen throughout Aug, becomes more common in Sept, and peaks in mid- to late Sept. Birds are readily observed through the third week of Oct; however, there are no Nov observations. High counts: 30, 3 Oct 1965, Squaw Creek (FL); 25, 15 Sept 1966, Springfield (NF-**BB** 34[1]:9). Latest date: 1, 25 Oct 1972, Maryville SL (DAE-**BB** 40[1]:8).

Comments: The Forster's Tern breeds as close as northern Iowa and central Kansas (Dinsmore et al. 1984; Thompson and Ely 1989).

Least Tern (*Sterna antillarum*)

Status: Rare transient statewide; very local, rare summer resident on the lower Mississippi R.; formerly bred along the Missouri R.

Documentation: Specimen: female, 26 May 1968, Bigelow Marsh (NWMSU, DAE 1776).

Habitat: Rivers, lakes, ponds, and marshes. Breeds on river sandbars.

Records:

Spring Migration: The first birds appear at the beginning of the second week of May. It is most common along the lower Mississippi R. where there are breeding colonies. Now only a few individuals/season are seen away from the lower Mississippi R. region. Presumably these are migrants headed to more northern and western breeding localities in Nebraska and the Dakotas.

Summer: Prior to 1900, it was an uncommon summer resident on the sandbars of the Mississippi and Missouri rivers (Widmann 1907). Birds bred on sandbars of shallow pools adjacent to or on the Missouri R. through the mid–1950s: 23, 4 young being fed by adults, 17 Aug 1941, Big L. SP (WC-**BB** 8[9]:65); 15–20 pairs, 4 nests with 2 eggs each, 23 July 1952, Mud L., Buchanan Co. (BK-**AFN** 7:20); 6 nests, 9–23 July 1955, Missouri R. near Dundee, Franklin Co. (Jackson 1955). In the early 1980s, at least fifteen sandbars with potential nesting habitat existed between Kansas City and St. Louis on the Missouri R. (Smith 1985).

In 1985, the MDC initiated a long-term study of the status of this species along the Mississippi R. (Smith 1985; 1987). A summary of the 1985–87 findings are given below. All colonies were concentrated on the Mississippi R. from Cape Girardeau south to the Arkansas border. In 1985 a total of fourteen colonies, five of which were in Missouri, were located. Fifteen colonies were studied in 1987, and a total of 480 nests were located (Smith 1987). For the 1987 season, a weighted average estimate of clutch size was 2.3 eggs/clutch, with a fledgling success rate of 0.5 fledglings/pair. A record 500+ nests were recorded in this area in May 1989 (J. Smith-**BB** 56[4]:148).

Fall Migration: Birds are seen across the state throughout Aug into early Sept, but the largest numbers are concentrated in late Aug and early Sept. It is only casually seen after mid-Sept. High counts: 36, 5 Sept 1986, Stockton L. (PM et al.-**AB** 41:96); 35, 19 Aug 1965, L. Springfield, Greene Co. (NF-**BB** 33[1]:6). Latest date: 1, 27 Sept 1984, Alton Dam (CP-**BB** 52[1]:30).

Comments: The U.S. Fish and Wildlife Service now considers the interior population to be endangered (U.S.F.W.S. 1985). Sidle et al. (1988) present recent estimates and distribution of the interior populations.

Black Tern (*Chlidonias niger*)

Status: Common transient; former summer resident.
Documentation: Specimen: male, 10 May 1926, Fayette, Howard Co. (CMC 229).
Habitat: Primarily marshes, rivers, and lakes.
Records:
Spring Migration: The first few migrants appear at the very end of Apr or early May. Numbers gradually build until mid-May when huge concentrations may be encountered. Large flocks may still be observed into early to mid-June. Earliest date: 1, 28 Apr 1985, Squaw Creek (FL). High counts: 3,000, 14 May 1978, Squaw Creek/L. Contrary (FL-**BB** 45[3]:17); 1,000, 21 May 1967, Squaw Creek area (FL); 1,000, 13 May 1973, St. Joseph/Squaw Creek area (FL).
Summer: Although records span the entire period, there are two distinct periods when birds are most common. Northbound migrants, occasionally in relatively large numbers (50–100 birds/locality), are still seen passing through the state during early June. Nonbreeding or early southbound migrants are seen during the latter half of July.
Formerly this was a "fairly common" breeder in marshes, but by the turn of the century it apparently was already rare (Widmann 1907). There are very few accounts of it breeding in the state after 1900. A nest with three eggs was located atop a muskrat house on 16 June 1940, Marais Temps Clair (JEC et al.-**BB** 7[7]:46). Apparently the last definite nesting record was of a pair at this same marsh in 1950 (**AFN** 4:278). High count: 125, 18 June 1966, Squaw Creek (FL).
Fall Migration: It is readily seen throughout Aug and early Sept. Peak is normally during the last two weeks of Aug. Numbers drop off dramatically by mid-Sept. High count: 500, 31 Aug 1963, Squaw Creek (FL). Latest dates: 1, 1 Oct 1988, Alton Dam (J. Chain et al.-**NN** 61:3); 1, 28 Sept 1967, Squaw Creek (HB-**BB** 35[1]:11). Widmann (1907) gives a record for 21 Oct (without locality [presumably St. Louis] or year).
Comments: This tern is still a locally common breeder across northern Iowa, Nebraska (primarily the Sandhill country), and central Kansas (Dinsmore et al. 1984; Johnsgard 1980; Thompson and Ely 1989).

Fig. 19. This Band-tailed Pigeon visited a feeder in Marshall, Saline Co., from late November through late February 1983–84. It represents one of only two records for the state. Photograph by Harold Hoey.

Order Columbiformes: Sandgrouse, Pigeons, and Doves

Family Columbidae: Pigeons and Doves

Rock Dove (*Columba livia*)

Status: Introduced; common permanent resident.
Documentation: Specimen: 3 Jan 1963, Cape Girardeau (SEMO 16).
Habitat: Cities, towns, rural buildings and rock quarries.
Comments: No information exists on when it was introduced to Missouri, but presumably it came with the early settlers.

Band-tailed Pigeon (*Columba fasciata*)

Status: Accidental transient and winter resident.
Documentation: Photograph: 1, late Nov–20 Feb 1983–84, Marshall, Saline Co. (H. Hoey et al.-**BB** 51[1]:6; Fig. 19).
Habitat: Both records are of birds at feeders; the above bird was extremely wary.
Comments: In addition to the above record, a bird was also present at a feeder in Kansas City from 29 Mar through 20 Apr 1985 (J. Seaman et al.-**BB** 52[3]:14). There are two records of this pigeon from Kansas (Thompson and Ely 1989).

[**White-winged Dove** (*Zenaida asiatica*)]

Status: Hypothetical; accidental summer visitant.
Documentation: Sight record only: see below.

Habitat:

Comments: A single bird was seen at Schell-Osage, 9 July 1973, by several observers (JR et al.-**BB** 40[4]:3). Unfortunately, no photograph was secured. Given the propensity of this species for postbreeding dispersal, it undoubtedly will reappear in the state. There are at least two sight records for Oklahoma for this same time of year (Wood and Schnell 1984), and four between June and Oct for Kansas (Thompson and Ely 1989).

Mourning Dove (*Zenaida macroura*)

Status: Common summer resident; common winter resident in Mississippi Lowlands, uncommon elsewhere.

Documentation: Specimen: female, road kill, 25 Dec 1965, Gravois Mills (NWMSU, DAE 1051).

Habitat: Most common at edge of woods, and cultivated areas with scattered trees and bushes.

Records:

Spring Migration: Migrants begin augmenting the winter populations at the end of Feb and early Mar. Numbers gradually increase during Mar and peak in mid-Apr.

Summer: A common breeder statewide. By late July large concentrations may be encountered feeding in cultivated fields. Not surprisingly, BBS data indicate that it is most common in the relatively open areas of the state (Map 17). Graph 2 depicts BBS data for the past 23 years in Missouri.

Fall Migration: Very common from Aug through mid-Sept when huge flocks (in the hundreds or low thousands) are observed in cultivated areas. Most have departed by mid-Sept, but relatively large numbers (in the low hundreds) may be seen until mid-Oct.

Winter: CBC data indicate that it is most common in the Mississippi Lowlands (Map 18). High counts: 477, 28 Dec 1975, Columbia CBC; 456, 2 Jan 1983, Orchard Farm CBC.

Passenger Pigeon (*Ectopistes migratorius*)

Status: EXTINCT; formerly an abundant migrant; apparently a local breeder and winter resident.

Documentation: Specimens: male and female, 17 Dec 1896, Attie, Oregon Co. (CAS 15460; 15462; McKinley 1960b clarified the correct spelling and location).

Habitat: Primarily in mature forest, but apparently it was found in a variety of habitats.

Comments: McKinley (1960b) gives a detailed summary of the occurrence of this species in the state based on an extensive literature review. We summarize his findings below.

Map 17.
Relative breeding abundance of the Mourning Dove among the natural divisions. Based on BBS data expressed as birds/route.

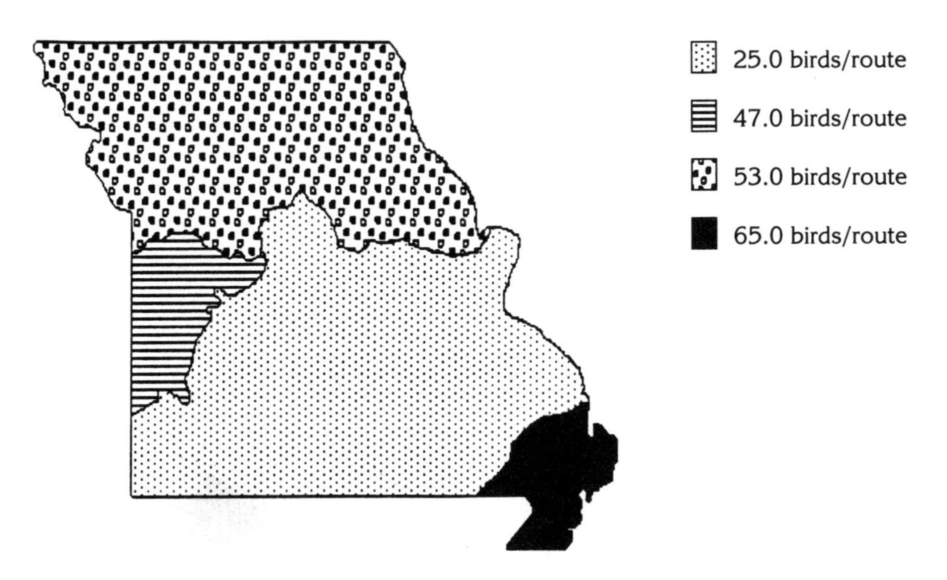

Graph 2.
Missouri BBS data from 1967 through 1989 for the Mourning Dove. See p. 23 for data analysis.

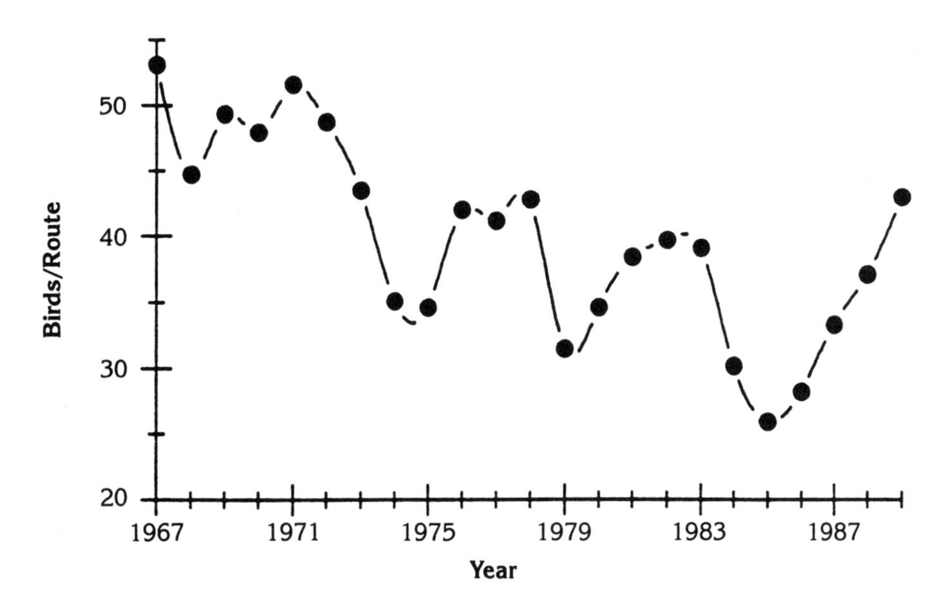

Map 18.
Early winter relative abundance of the Mourning Dove among the natural divisions. Based on CBC data expressed as birds/pa hr.

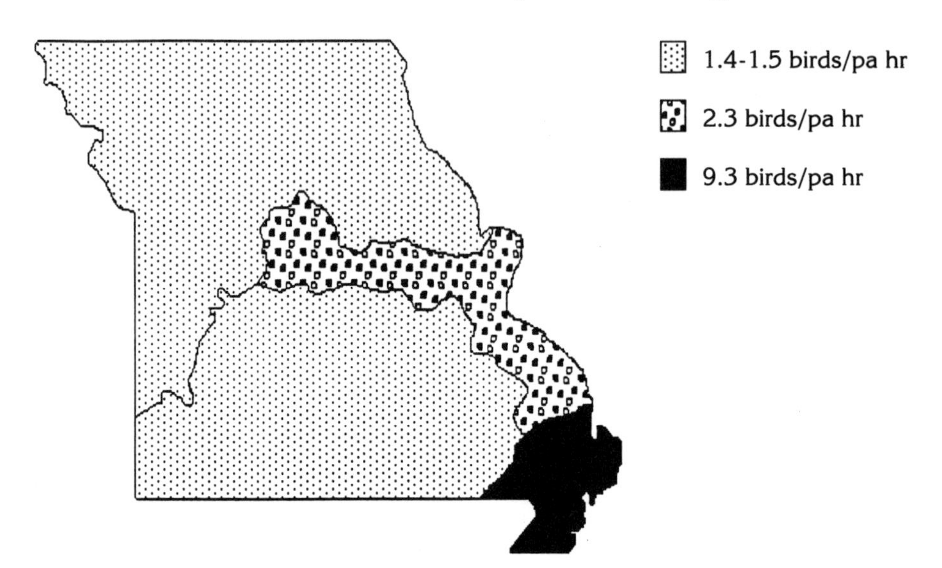

Apparently it was an abundant transient through Missouri, reported on several occasions to "darken" the sky. Spring migration extended from Feb through May, with the largest numbers reported in late Mar and Apr. In the fall, immense flocks were reported as early as Sept, and their numbers increased at roost sites during Oct and Nov. Accounts of nesting and wintering are fewer and more fragmented; nevertheless, it was reported to have bred, at least locally, in huge colonies in a few areas of the state, and there are reports of birds breeding in scattered pairs. Accounts of large winter roosts are mainly from southern Missouri.

This species was still reported in enormous numbers until the 1880s, with one huge winter roost still persisting in Ripley Co. in 1883–84 (McKinley 1960b). However, after that account, there were only occasional reports of pigeons in the state, with none mentioning any large concentrations. Apparently the last report in Missouri (number of birds not indicated) was at New Haven, Franklin Co., on 26 Sept 1902 (Dr. Eimbeck; Widmann 1907).

Inca Dove (*Columbina inca*)

Status: Accidental winter resident.
Documentation: Photograph: 1 at feeder, from mid-Dec to 6 Feb 1987–

Fig. 20. This Inca Dove took up residence at a feeder in Mound City, Holt Co., from mid-December through early February in 1987–88. The species is well-known for its propensity to wander northward following the breeding season. Photograph by Jim Rathert. Courtesy of Missouri Department of Conservation.

88, Mound City, Holt Co. (G. Rockwell, JW et al.-**BB** 55[2]:62; Fig. 20).

Comments: This record fits the postbreeding dispersal pattern of the Inca Dove, as there are about ten and five records for Oklahoma and Arkansas, respectively (Wood and Schnell 1984; James and Neal 1986), and about a dozen for Kansas (Thompson and Ely 1989). Single observations of this species were also obtained during the fall and winter of 1987–88 in Kansas and Nebraska (Williams 1988).

Common Ground-Dove (*Columbina passerina*)

Status: Accidental fall transient and winter visitant.

Documentation: Specimen: male, found injured by C. Bauman on 5 or 6 Oct 1981, near Rosendale, Andrew Co. (JHI; NWMSU, DAE 3099).

Habitat: Open country, especially along hedgerows bordering cultivated fields.

Records:

Fall Migration: Besides the above record there is one additional observation for this period: 1, 29 Oct 1972, ca. 16–17 miles east of Columbia, Callaway Co. (IA).

Winter: Two sightings: 1, late Dec 1956, Squaw Creek (B. Stolberg-**BB** 24[2]:2); and the other possibly involving as many as three birds (minimum of two), 17 Dec–13 Jan 1979–80, Pleasant Hill, Cass Co. (JG et al.-**BB** 47[3]:17).

Comments: The Missouri records coincide with the general postbreeding pattern of dispersal, where most birds in the midwest are seen between late Sept and mid-Dec.

Order Psittaciformes: Parrots

Family Psittacidae: Parrots

Carolina Parakeet (*Conuropsis carolinensis*)

Status: EXTINCT; formerly common permanent resident.

Documentation: Specimen: o?, 4 May 1843, near St. Joseph (J. Audubon and E. Harris; ANSP 136786); see below.

Habitat: Apparently found in a wide variety of habitats, but most commonly reported from river bottom forest.

Comments: A detailed synopsis of the status of this species is presented by McKinley (1960a). We briefly summarize the highlights from his paper below.

The majority of the information for this species comes from people who navigated the state's two largest rivers in the 1800s. Most of these accounts mention that the bird was quite common in forest bordering the rivers. Although there are only a handful of accounts away from watercourses, it is apparent that the parakeet was once widespread and common throughout most of the state.

Apparently it was still common in Missouri through the 1850s (Widmann 1907), but there are very few records thereafter. The last specimen taken in the state may have been the bird collected by J. Bryant in 1894 in Kansas City (Harris 1919b). This specimen apparently had no specific data and has since been stolen (McKinley 1964). The following represent the last reports of this species in the state: 1, 18 July 1905, Notch, Stone Co. (T. Powell; Widmann 1907); 1, present for several weeks in 1912 along the Missouri R., near Courtney, Jackson Co. (B. Bush; Harris 1919b).

Although the original tag on the above mentioned Audubon and Harris specimen states that the bird was collected "above Fort Leavenworth," it is likely that the bird was taken just south of St. Joseph, Missouri (as were the other Harris' parakeet specimens). Harris indicates in his notes that they "shot a number of Parroquets today" along the "Missouri R. below blacksnake Hills" on 4 May 1843 (Street 1948; also see Cooke 1910). On this same date, Audubon and Harris also collected the first specimen of the Bell's Vireo for science. McKinley (1964) also mentions another extant specimen (deposited at St. Benedict's College, Atchison, Kansas) that was purported to have been taken in Platte Co. around the turn of the century. Finally, there is a specimen in the Central Methodist College (CMC 689), which was taken in Franklin Co. in 1875.

Order Cuculiformes: Cuckoos and Allies

Family Cuculidae: Cuckoos, Roadrunners, and Anis

Black-billed Cuckoo (*Coccyzus erythropthalmus*)

Status: Uncommon transient; uncommon summer resident in Glaciated and Osage plains, rare elsewhere.

Documentation: Specimen: male, 9 June 1926, Jackson Co. (KU 39081).

Habitat: Most readily found in migration and during the breeding season in willows that border water areas.

Records:

Spring Migration: The first migrants appear at the very end of Apr. Peak usually occurs during the third or fourth week of May; however, there are years when this and the Yellow-billed Cuckoo do not appear until the end of May. Earliest records: 1, 25 Apr 1990, St. Louis Co. (D. Becher-**BB** 57[3]:138); 1, 29 Apr 1989, Forest Park (MP). High count: 7, 15 May 1977, Honey Creek WA, Andrew Co. (MBR, DAE).

Summer: As a result of its secretive nature and the fact that many observers are unfamiliar with its song, it is often overlooked. Numbers fluctuate considerably from year to year. It is primarily found nesting along rivers and at the edge of marshes and ponds in willows. The Black-billed appears to be most common in the Glaciated and Osage plains and along the Missouri R. drainage (probably also the Mississippi R., but data are lacking). It is apparently absent from the Mississippi Lowlands (Widmann 1907; BBS data); however, a concerted effort has not been made to determine if this species breeds in riparian areas along the Mississippi R. in the southeast.

Fall Migration: Migrants are seen throughout Sept into early Oct, with no discernible peak. Latest date: 1, 21 Oct 1986, Howard Co. (C. Royall-**BB** 54[1]:34).

Comments: During all seasons it is less common than the following species.

Yellow-billed Cuckoo (*Coccyzus americanus*)

Status: Common summer resident; accidental winter visitant.

Documentation: Specimen: male, 1 Nov 1966, 10 miles west of Maryville (NWMSU, DAE 1103).

Habitat: During migration found virtually anywhere; most common during the breeding season in riparian areas; however, it does breed in upland areas.

Records:

Spring Migration: In most years birds begin to appear in early May and peak during the third week of May. However, there are years when birds are not seen until the end of May. Earliest date: 1, 26 Apr 1981, Jefferson City (JW-**BB** 48[3]:19).

Summer: This cuckoo is widespread and common across all natural divisions of the state. BBS data indicate it is most common in the Ozarks and Ozark Border (Map 19). Although it is common, numbers vary considerably from year to year. Graph 3 depicts the dramatic increase in the abundance of this species during the summers of 1977 through 1979.

Fall Migration: Commonly seen through the third week in Sept, but numbers rapidly drop off thereafter. Rare but regularly encountered through mid-Oct. It is casual during the last two weeks of Oct and the first week in Nov. Latest dates: 1, injured, 9 Nov 1947, Kansas City (HH-**BB** 15[3]:3); 2, 2 Nov 1969, one at Maryville, and the other ca. 10 miles north of Maryville (MBR).

Winter: One record: 1, 15 Dec 1987, 4.5 miles west of Parnell, Nodaway Co. (G. Shurvington-**BB** 55[2]:62).

Greater Roadrunner (*Geococcyx californianus*)

Status: Rare permanent resident in southwest.

Documentation: Specimen: male, 24 Jan 1979, Pulaski Co. (MU 1245).

Habitat: Most prevalent in dry, rocky cedar glades.

Records:

The first record was obtained on 19 June 1956 just north of Branson in Taney Co. (Brown 1963). The roadrunner slowly spread northward and eastward out of the cedar glade region of the state until the late 1970s, when the expansion was truncated by three consecutive severe winters (Norris and Elder 1982b; Map 20). At present, it is again reestablishing itself in areas it occupied prior to the late 1970s. It remains most common in the glade area of the southern tier of counties in the southwest. Details that are available for the northern and easternmost records in Map 20 are as follows: 1, summer 1978, T43N, R9W, S29, Osage Co. (R. Walker; W. Elder, pers. comm.); 1, spring 1981, Peck Ranch WA, Carter Co. (K. Kriewitz-**BB** 49[3]:19).

Groove-billed Ani (*Crotophaga sulcirostris*)

Status: Casual fall transient.

Documentation: Specimen: o?, found dead, 13 Nov 1950 (seen alive for ca. 1 week prior to death), Centertown, Cole Co. (D. Campbell; Tulenko 1950; MU 1247).

Habitat: Generally found in open areas with hedgerows, scattered trees, and thickets.

Map 19.
Relative breeding abundance of the Yellow-billed Cuckoo among the natural divisions. Based on BBS data expressed as birds/route.

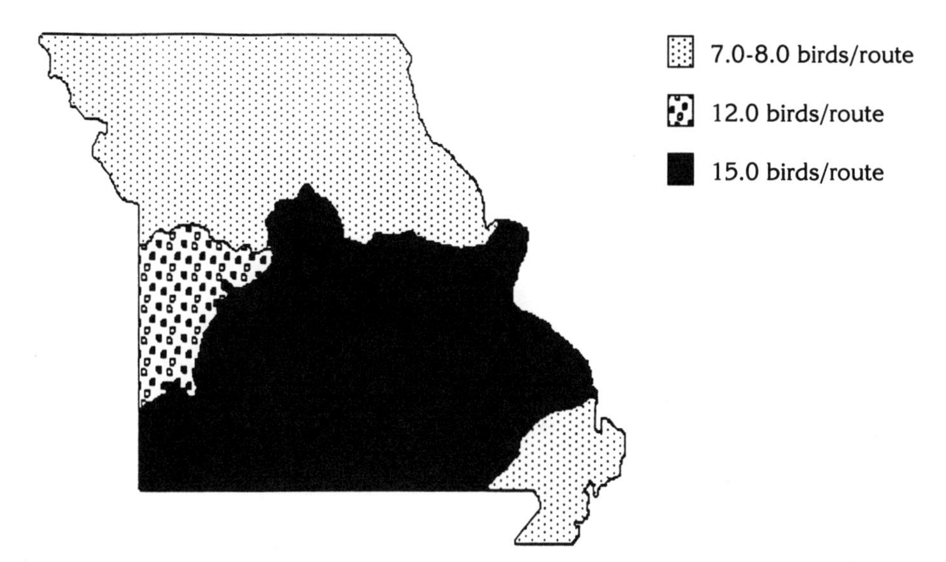

Graph 3.
Missouri BBS data from 1967 through 1989 for the Yellow-billed Cuckoo. See p. 23 for data analysis.

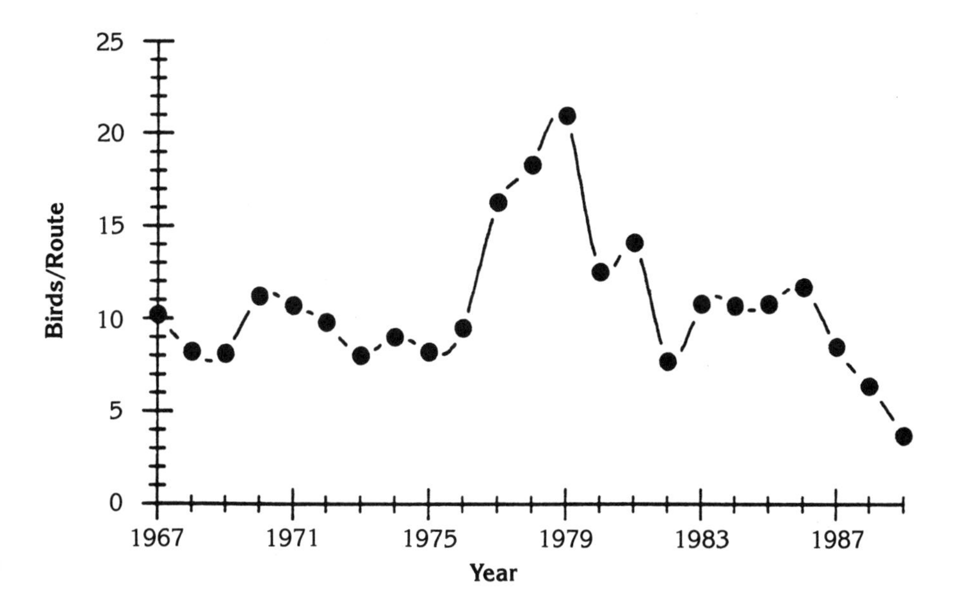

Map 20.
Distribution of the Greater Roadrunner showing the counties occupied in 1976 (diagonal lines) and 1978 (dotted areas). Courtesy of Norris and Elder (1982b) and the Wilson Ornithological Society.

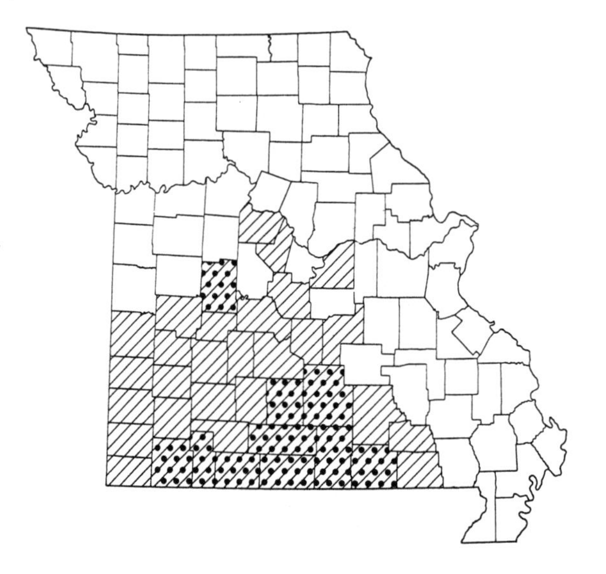

Records:
Fall Migration: All five records are between mid-Oct and late Nov.: 1, photographed and banded, 10 Oct 1978, Kansas City (M. Myers-**BB** 46[1]:25); 1, later died, 11–26 Oct 1979, Kingdom City, Callaway Co. (K. Evans, JR et al.-**BB** 47[1]:16); 1, 9–11 Nov 1972, Columbia (D. Rickett et al.; Goodge 1972); and 1, shot, 22 Nov 1973, Greenwood, Jackson Co. (Rising et al. 1978).
Comments: Like the Common Ground-Dove's fall dispersal pattern, the vast majority of the ani records for the midwest are between late Sept and mid-Dec.

Order Strigiformes: Owls

Family Tytonidae: Barn Owls

Barn Owl (*Tyto alba*)

Status: Rare permanent resident.
Documentation: Specimen: male, road kill, 7 June 1965, near Aquilla, Stoddard Co. (NWMSU, DAE 900).

Habitat: Open country with scattered trees; often nests and roosts in abandoned city and rural buildings.

Comments: Widmann (1907) stated that this owl was "a rather rare resident . . . only in the northern and western prairie, and in the Ozark border regions, but not in the Ozarks and the southeast, which are too densely wooded." Indeed, his statement accurately describes the status of the bird today; except that it is also found in the Mississippi Lowlands and the Ozarks. In fact, due to the nearly complete elimination of forest in the Mississippi Lowlands, this species may be as common there as anywhere in the state. Eighteen nesting reports were filed in the summer of 1983, and all were from the agricultural areas of the southwestern and southeastern sections of the state (**BB** 50[4]:33).

No information exists on the migratory behavior of this species in the state. Undoubtedly, some of the birds found in winter are from more northern populations. High count: 3, 18 Dec 1982, Mingo CBC.

Family Strigidae: Typical Owls

Eastern Screech-Owl (*Otus asio*)

Status: Common permanent resident.

Documentation: Specimen: male, 5 March 1921, Jackson Co. (KU 39082).

Habitat: Open woodland, deciduous forest, and residential areas in towns.

Comments: Appears to be uniformly distributed throughout the state, but no data exists on whether there are differences in abundance among the natural regions of the state. High count in winter: 10, 2 Jan 1983, Kansas City North CBC.

Great Horned Owl (*Bubo virginianus*)

Status: Common permanent resident.

Documentation: Specimen: male, 26 Jan 1917, McDonald Co. (KU 10498).

Habitat: Found in a wide variety of habitats, ranging from mature deciduous forest to scattered trees in agricultural areas.

Comments: Presumably this species is most common in the less forested regions of the state; however, statewide early spring censuses are needed to clarify if this is indeed the case. High winter counts: 33, 16 Dec 1989, Grand River CBC; 27, 20 Dec 1986, Kansas City Southeast CBC.

Fig. 21. The Great Horned Owl is a common and widespread permanent resident. Pen and ink by David Plank.

Snowy Owl (*Nyctea scandiaca*)

Status: Irregular; rare winter resident.

Documentation: Specimen: male, 3 Dec 1949, Callaway Co. (MU 1517).

Habitat: Bare, open country, usually seen sitting on fence posts, hay bales, etc.

Records:

Spring Migration: Only two Mar and an extremely late Apr record of lingering birds: 1, 19 Mar 1985, near L. Contrary (FL-**BB** 52[3]:14); 1, 23 Mar 1963, St. Joseph (m.ob.-**BB** 31[2]:12); and male, D.O.R., 24 Apr 1977, near Craig, Holt Co. (fide DAE-**BB** 44[4]:27; NWMSU, DST 104).

Fall Migration: During invasion years, birds usually reach Missouri by mid-Nov. Most records are from the northern half of the state. Earliest date (and only Oct record): immature, 19 Oct 1983, Ted Shanks (M. Bates-**BB** 51[1]:42). Southernmost records: 1, photos, 5–30 Dec 1989, east of Big Oak Tree SP (S. Baker et al.-**BB** 57[2]:104); and 1, photos, 30 Nov 1984, Essex, Stoddard Co. (C. Fivek-**BB** 52[1]:31).

Winter: Most observations are for Dec and Jan, and all but a couple of sightings involve only single birds. The winter of 1976–77 (8+ reports) and 1980–81 (13+ reports) were two of the largest recorded incursions (Table 6).

Table 6. Invasion years of Snowy Owl in Missouri since the winter of 1959–60.

1959–60	1976–77
1963–64	1980–81
1967–68	1983–84
1974–75	1984–85
1975–76	1986–87

[Northern Hawk Owl (*Surnia ulula*)]

Status: Hypothetical: see below.

Comments: Very dubious records listed by Widmann (1907) and Bennitt (1932), and all probably involve misidentified Short-eared Owls.

Burrowing Owl (*Athene cunicularia*)

Status: Casual transient in west, accidental in east; accidental summer and winter resident.

Documentation: Specimen: o?, 21 Apr 1965, along Missouri R., near Fairfax Bridge, Platte Co. (Easterla 1967a; mount NWMSU).

Habitat: Pastures, short-grass fields, and edge of cultivated fields.

Records:

Spring Migration: At least eight records; all are from the western quarter of the state and range from mid-Apr through the end of the period. At least four of these observations involved birds at badger holes. Earliest date: 1, 10 Apr–29 July 1982, near Watson, Atchison Co. (T. Miller, D. Brickman; Robbins et al. 1986a).

Fig. 22. This adult Burrowing Owl was photographed at an abandoned badger burrow southeast of Mound City, Holt Co., on 18 May 1974 by David Easterla. It and another adult were seen with young at this site in June. It is the only nesting record for Missouri.

Summer: A single definite nest record, where 5+ young were raised at a badger hole near Squaw Creek in May and June 1974 (FL, DAE et al.; Robbins et al. 1986a; Fig. 22). There are two other possible breeding records: two birds (no evidence of breeding activity was observed) were present at a badger hole from mid-Apr to mid-May 1968 in extreme northern Atchison Co. (H. and F. Diggs; Robbins et al. 1986a), and see below.

Fall Migration: Only five observations, including one where a bird appeared in the fall of 1960 and remained at the site until Feb 1962, south of Sikeston, Mississippi Co. (W. Pollock-**BB** 29[2]:14; photo in ref). Other records: 1, 22 Sept 1980, Cass Co. (JJ-**BB** 48[1]:9); 1, 26 Sept 1982, Kansas City (J. Overton-**BB** 50[1]:24); and 2, 8–15 Oct 1966, at a badger hole, near L. Contrary (J. Farlie).

The record involving 4 birds (adult female collected but apparently no longer extant) at a hole in Sept 1934 near Lee's Summit, Jackson Co., opens the possibility that these birds bred there (Teachenor 1940).

Winter: Two records: 1, 19 Dec 1987, Montrose CBC (M. Taylor); and see Fall Migration above.

Barred Owl (*Strix varia*)

Status: Common permanent resident.

Documentation: Specimen: male, 27 Dec 1947, Sullivan Co. (KU 39085).

Habitat: Primarily found in mature forest, especially river bottom.

Comments: Less common and conspicuous than the Great Horned Owl, preferring river- and stream-bottom woods and forest. In view of the available habitat, it is perhaps most common in the Ozarks and Ozark Border regions. However, formerly it was undoubtedly most common in the Mississippi Lowlands. High counts in winter: 17, 19 Dec 1987, Montrose CBC; 11, 1 Jan 1990, Springfield CBC.

Long-eared Owl (*Asio otus*)

Status: Uncommon transient and winter resident. Casual summer resident.

Documentation: Specimen: male, 13 May 1926, Fayette, Howard Co. (CMC 132).

Habitat: Most readily encountered roosting in conifers and cedars but also found roosting on the ground in thick brush. Breeds in forest, usually near water.

Records:

Spring Migration and Summer: Migrants and lingering winter residents are regularly seen until mid-Apr. Birds seen during late Apr or in May possibly represent breeders, since eggs are laid in Mar and Apr. Its true breeding status is undoubtedly underrepresented by the ca. fifteen records (most of these are prior to 1915) because it is easily overlooked. Most nests have been in deep woods bordering rivers or lakes.

Published nesting records since 1950 include the following: nest with 6 eggs, spring 1956, Busch WA (B. Dowling-**BB** 23[6]:3); pair with 5 young in nest, 9 May 1957, Missouri R. bottoms, near McBaine, Boone Co. (DAE, D. Snyder-**BB** 24[6]:4; photographed); "pair nested," Mar–May 1966, Rockwoods, St. Louis Co. (B. Wetteroth-**NN** 38:42); pair by incomplete nest that was eventually abandoned, 6 Mar 1966, 10 miles west of Maryville, Nodaway Co. (DAE); and 1, nearly fledged young, 10 June 1983, St. Charles Co. (fide JW-**BB** 50[4]:33). High count: 15, 6 Mar 1949, Marais Temps Clair (J. Ziervogel et al.-**NN** 21:7).

Fall Migration: Migrants are normally not detected until mid-Nov. No large concentrations have been reported for this season. Earliest dates: female, 2 Oct 1933, Grand Glaize, Camden Co. (MU 454); 1, 10–12 Oct 1975, Independence (KH et al.-**BB** 44[1]:21).

Winter: This is the season when relatively large concentrations may be found roosting in conifers or in dense undergrowth in moist areas; numbers fluctuate considerably from year to year. High counts: 60, no date specified, near Courtney, Jackson Co. (B. Bush; Harris 1919b); 30+, 8 Feb 1990, Dorsett Hill Prairie, Cass Co. (D. Myers, JG, JJ, E. Johnson, J. Leo-**BB** 57[2]:104); 30, 30 Jan 1873, Mississippi R., near St. Louis (J. Hurter; Widmann 1907).

Short-eared Owl (*Asio flammeus*)

Status: Uncommon transient and winter resident; presently a very rare breeder in the Glaciated and Osage plains.

Documentation: Specimen: female, 11 Dec 1973, near Maryville (NWMSU, DLD 83).

Habitat: Open grassy areas, especially tallgrass prairies and fallow fields.

Records:

Spring Migration: The largest concentrations for this period are encountered in Mar, but smaller numbers (< 5 birds/site) are regularly seen into early Apr. One or two birds are casually seen during the last half of Apr. Besides unquestionable breeding records (see below), there is only one May observation (and this probably involved nesting birds): 2, 13–14 May 1986, Taberville Prairie (S. Schoech). Definite late migrant date: 1, 28 Apr 1974, Maryville SL (DAE-**BB** 41[3]:4). High count: 125, 14 Mar 1970, Squaw Creek (FL-**BB** 38[3]:7). This extraordinary concentration remained through Mar (60 were estimated still to be present there on 29 Mar, FL).

Summer: This species formerly bred locally throughout the Glaciated and Osage plains. Widmann (1907) lists definite nesting records for the following counties at the turn of the century: St. Clair, Johnson, St. Charles, and Clark. The following recent observations may pertain to breeding birds: 1, 18–26 June 1966, Squaw Creek (FL, JHA-**AFN** 20:575); 2, 26 June 1982, Taberville (FL). Two birds were observed courting over an uncultivated field in western Bates Co. until the beginning of June 1989, but apparently did not breed (JJ-**BB** 56[4]:148).

Four nesting records were obtained in 1990 as a result of "Breeding Bird Atlas" work. Two pairs were located in Putman Co., 1 pair at Taberville Prairie, and another pair in a natural area on the Vernon/St. Clair Co. border (JW, pers. comm.).

Fall Migration: The first migrants appear at the end of the third week in Oct. Most observations are from mid-Nov through the end of the period. Earliest dates, both of single birds south of Gravois Mills, L. of the Ozarks by DAE: 15 Oct 1954 and 16 Oct 1955. High counts: 22, 25 Oct 1970, 10 miles north of Maryville (MBR); 19, 7 Dec 1961, Callaway Co. (DAE).

Winter: The majority of observations are from this period. Numbers fluctuate considerably from year to year and are presumably related to rodent densities. High counts: 36, 26 Dec 1975, Squaw Creek CBC; 27, 21 Dec 1975, Orchard Farm CBC.

Northern Saw-whet Owl (*Aegolius acadicus*)

Status: Rare transient and winter resident; accidental summer resident.

Documentation: Specimen: o?, 3 Dec 1970, between Warrensburg and Knob Noster, Johnson Co. (CMSU 283).

Habitat: Frequents small, dense cedars and conifers; rarely detected in strictly deciduous woods.

Records:

Spring Migration: Lingering winter residents and presumed migrants are seen until mid- to late Mar. Two Apr records: 1, 11 Apr 1976, Maryville (fide DAE-**BB** 44[2]:17); 1, 16 Apr 1893, St. Louis (Currier; Widmann 1907).

The only breeding record is the one given by Widmann (1907): nest with three young, spring 1904, Bluffton, Montgomery Co. (J. Muller). High counts: at least 3 observations of 2 birds in a day (all from Squaw Creek).

Fall Migration: No definite Sept observations; casually seen from late Oct (six Oct records) through the end of the period. Earliest date: 1, 3 Oct 1951, St. Louis Co. (G. Moore-**BB** 18[10]:3).

Winter: With only a few exceptions (< 5 records), all observations are north of the Missouri R.; however, this is probably related more to the ease in finding birds there than to any real differences in abundance. Cedar and conifer stands are more local and less extensive in northern Missouri than in the south. High count: 2–3, 2 Dec–21 Feb 1981–82, Boone Co. (TB, BG-**BB** 49[2]:16). Southernmost record: 1, specimen (not preserved), 11 Feb 1984, Taum Sauk L., Iron Co. (JW-**BB** 51[3]:20).

Order Caprimulgiformes: Goatsuckers and Allies

Family Caprimulgidae: Goatsuckers

Common Nighthawk (*Chordeiles minor*)

Status: Common summer resident.

Documentation: Specimen: female, 1 Sept 1962, Puxico, Stoddard Co. (NWMSU, DAE 836).

Habitat: During migration can be seen virtually anywhere; now primarily nests on flat, gravel covered roofs of buildings; formerly nested on prairies, cliffs, and rocky ledges.

Records:

Spring Migration: Normally not encountered until the end of Apr, with peak in mid-May. Dramatic concentrations, like those seen in the fall, are almost never encountered in the spring. Earliest date: 1, 14 Apr 1984, at Busch Stadium, St. Louis (M. Andrew-**NN** 56[6]:48). High count: 25, 13 May 1987, Ted Shanks (KM-**BB** 54[3]:14). Widmann (1907) stated that "very large flocks" were seen passing over St. Louis on 25–27 May 1882, but he gave no numbers.

Summer: Common summer resident statewide; the greatest concentrations are present in cities and towns. Dusk and dawn censuses are needed for monitoring its abundance.

Fall Migration: One of the most spectacular migration events during the latter half of Aug and the first half of Sept is the movement of large flocks of nighthawks. Migration is in full swing by mid-Aug, and numbers peak in late Aug or early Sept. Large flocks (sometimes numbering in the hundreds/flock) continue to be seen through the third week in Sept, but numbers drop off drastically thereafter. It is regularly seen, in much smaller numbers (usually only solitary birds are seen after the first week in Oct) through mid-Oct but rarely into the third week of Oct. High counts: 1,000, 28 Aug 1981, St. Louis (MP); 1,000, 28 Aug 1981, between Jefferson City and St. Louis (JW-**BB** 49[1]:14); 820, 9 Sept 1985, Jefferson Co. (MP-**NN** 57:47). Latest dates: 3, 14 Nov 1987, Farmington (SD-**BB** 55[1]:12); 1, 11 Nov 1989, Columbia (G. Perrigo-**BB** 57[1]:49).

Comments: The scarcity of specimen material from the migratory period precludes an assessment of what subspecies pass through the state.

[**Common Poorwill** (*Phalaenoptilus nuttallii*)]

Status: Hypothetical transient and summer resident.

Documentation: Sight and voice records: none with details.

Habitat: Prairie and pastures.

Comments: This species was purported to have bred along the extreme western edge of the state. Widmann (1907) states that H. Nehrling found the bird nesting in Lawrence Co. in 1885. Around the turn of the century breeding was reported in the open areas of Swope Park and Dodson, Jackson Co. (B. Bush; Harris 1919b). Widmann also mentions that Bush heard birds in McDonald and Barry counties. These latter counties do not contain Poorwill habitat, thus these and the Swope Park records of Bush are suspect. They probably pertained to Chuck-will's-widows. The Nehrling record(s) are difficult to assess.

All of the above purported records are prior to 1900. Since then there have been a few reports of heard birds; however, at least three of these reports (perhaps all?) involved Chuck-will's-widows. There is only one observation that we consider probably valid: a single bird was flushed on a virgin prairie in the summer of 1945 (no specific date given) in Vernon Co. (WC-**BB** 12[11]:61). We feel, however, that more concrete evidence is needed before this species is added to the state list.

Chuck-will's-Widow (*Caprimulgus carolinensis*)

Status: Uncommon summer resident in the Ozarks and Ozark Border; rare in the Osage Plains and even rarer in the Glaciated Plains.

Documentation: Specimen: male, 23 May 1964, Gravois Mills, L. of the Ozarks (NWMSU, DAE 569).

Habitat: Common in open, pine-oak or cedar-oak associations, but also found in strictly deciduous woods.

Records:

Spring Migration: In the south the first birds arrive by mid-Apr, but in the north birds are not encountered at breeding sites until the end of Apr or early May. Earliest dates: 1, 4 Apr 1983, St. Louis (fide RK-**BB** 50[3]:13); 1, 11 Apr 1964, Salem, Dent Co. (D. Plank-**BB** 31[2]:12).

Summer: Most common in the Ozark and Ozark Border regions. It is less common and more local in the Osage Plains, and in the Glaciated Plains it is principally found along the bluffs bordering the Missouri and Mississippi rivers, where it is rare and local. The Chuck-will's-widow's status in the north central section of the Glaciated Plains needs to be further clarified. It has been recorded on only one (route 26) of the 14 BBS routes in the Glaciated Plains. However, during 1990 "Breeding Bird Atlas" work it was found near Milan, Sullivan Co. (JHI). It is not known to breed in the Mississippi Lowlands.

Until very recently, abundance data were nonexistent. However, in 1989 the first census routes were established in southern Missouri (Hayes 1989b). Preliminary results for two routes, both with 20 stops/route along 10 miles of road, are as follows: 32, 28 May 1989, Taney Co. (JH, PM), and 15, 9 June 1989, Dallas Co. (JH).

Fall Migration: Almost all records are of breeding birds that are still calling in Aug. Latest date: 1, 10 Sept 1961, Busch WA (K. Stewart-**NN** 33:39). Widmann (1907) states, without giving any details, that birds are recorded until the end of Sept.

Whip-poor-will (*Caprimulgus vociferus*)

Status: Common summer resident statewide.

Documentation: Specimen: female, 3 June 1964, road kill near Round Spring, Shannon Co. (NWMSU, DAE 581).

Habitat: Forest and woodland.

Records:

Spring Migration: The first individuals are heard in early Apr in the south and by mid-Apr in the north. Earliest date: 1, 23 Mar 1986, Buffalo, Dallas Co. (J. Hancock-**AB** 40:479).

Summer: More widespread and common than the Chuck-will's- widow, although there are some areas in the south where Chuck's outnumber Whip's, e.g., in cedar-oak associations.

As mentioned in the previous species account, prior to 1989, no estimates were available on the abundance of caprimulgids in the state. Results from two surveys (Hayes 1989b), both with 20 stops along 10 miles of road, are as follows: 25, 28 May 1989, Taney Co. (JH, PM), and 48, 9 June 1989, Dallas Co. (JH).

Fall Migration: Birds are regularly heard through mid-Sept (residents or migrants?). Widmann mentions that birds are occasionally seen in early Oct, giving 10 Oct as the latest date with no locality. Latest date with full details: 1, calling, 29 Sept 1978, Licking, Texas Co. (D. Hatch-**BB** 46[1]:25).

Order Apodiformes: Swifts and Hummingbirds

Family Apodidae: Swifts

Chimney Swift (*Chaetura pelagica*)

Status: Common summer resident; hypothetical winter visitant.
Documentation: Specimen: male, 20 May 1967, Maryville (NWMSU, NNL 6).
Habitat: Now primarily nests and roosts in chimneys, but formerly bred in hollow trees and cliffs.
Records:
Spring Migration: Arrives at the beginning of Apr in the south and by mid-Apr in the north. Peak is in early May when huge flocks may be seen roosting in chimneys. Earliest date: 1, 20 Mar 1963, Charleston, Mississippi Co. (JH-**BB** 30[2]:26). High count: 3,500–4,000, 7 May 1967, University of Missouri–Columbia campus (M. Entrikin-**BB** 34[2]:8).
Summer: Undoubtedly this species has increased dramatically since the state was settled. Prior to settlement, the species primarily used hollow trees and cliff faces for nesting. Cunningham's observation of a pair nesting in a hollow Basswood in June 1936 in Swope Park, Jackson Co., was one of last reports of this swift nesting in natural sites in Missouri (WC; Rising et al. 1978). Presumably this species still occasionally nests in natural cavities.
Fall Migration: Birds begin congregating at roosts by mid-Aug, and their numbers build through Sept. Relatively large numbers are regularly seen into early Oct, with an occasional large flock seen at roost sites in mid-Oct, e.g., 1,000, 19 Oct 1971, St. Louis (JC-**BB** 39[1]:6). However, normally it is rarely seen in the state after the third week in Oct. High count: 3,000–4,000+, 28 Sept 1975, downtown Kansas City (NJ-**BB** 44[1]:21). Latest dates: 1, 18 Nov 1986, Columbia (TB-**BB** 54[1]:34); 8, 6 Nov and 1, 8 Nov 1925, St. Louis (Widmann 1928; note incorrect year in paragraph 1, p. 151).
Winter: Rising et al. (1978) mention a December 1938 record near Parkville, Platte Co., by two reliable observers (WC, A. Shirling). Unfortunately, no details were given for this observation, thus precluding the elimination of the very similar Vaux's Swift (*Chaetura vauxi*). The latter species is just as likely to occur in Missouri at this time of the year, since it winters much further north (Mexico through Central America) than does the Chimney Swift (South America). We therefore treat this record as hypothetical.

White-throated Swift (*Aeronautes saxatalis*)

Status: Accidental transient.
Documentation: Specimen: o?, 7 Nov 1988, Cape Girardeau (T. Lam-

bert, J. Glubeck; Wilson 1989; specimen preserved as a live mount and deposited at SEMO).

Comments: The above bird was found emaciated on the wall of an elevator at a cement manufacturing plant. There is an early Nov sight record for eastern Kansas (Manhattan; Fretwell 1978).

Family Trochilidae: Hummingbirds

Ruby-throated Hummingbird (*Archilochus colubris*)

Status: Common transient; uncommon summer resident.

Documentation: Specimen: female, 14 Aug 1913, Jackson Co. (KU 39089).

Habitat: Found virtually anywhere there are flowering plants during migration. Breeds in forest and woodland edge and in residential areas and parks.

Records:

Spring Migration: The first birds arrive by mid-Apr in the south and late Apr in the north. Peak is in early to mid-May, but it is not seen in the relatively large concentrations as in the fall. Earliest date: 1, 5 Apr 1986, Buffalo, Dallas Co. (J. Hancock-**AB** 40:479).

Summer: An uncommon summer resident. BBS data indicate that this hummer is most abundant in the Osage Plains (average 1.2 birds/route), whereas all other natural divisions have an average of only 0.2–0.4 birds/route. However, BBS is a poor method for censusing this species. If reliable data were available, it might show that the Ruby-throat is as common, if not more common, in the Ozarks and Ozark Border than in the Osage Plains.

Fall Migration: Migration is in full swing by mid-Aug, with peak in early Sept. It is rarely seen after the first week of Oct. High counts: 200+, 5–6 Sept 1981, Lawrence and Barry counties (KH et al.-**BB** 49[1]:14); 44+, 10–11 Sept 1977, Roaring R. SP, (JG, NJ-**BB** 45[1]:27); 40, 3 Sept 1967, Charleston, Mississippi Co. (JH-**BB** 35[1]:11). Late dates: see below; 1, 2 Nov 1902, New Haven, Franklin Co. (Eimbeck; Widmann 1907); 1, 19 Oct 1975, St. Joseph (FL-**BB** 44[1]:21).

Comments: Observers should use extreme care in identifying late (late Oct through Dec) hummingbirds because the majority of these late sightings in the eastern U.S. have proven to pertain to western North American species (especially *Selasphorus*). Although the following Missouri observations were originally attributed to the Ruby-throated, they may actually represent other species: 1, 24 Nov 1982, St. Louis (C. Hath-**NN** 55:5); 1, 30 Nov 1983, Hermann, Gasconade Co. (**BB** 51[1]:43); and 1, last seen 9 Dec 1978, Stone Co. (A. Gould-**BB** 46[2]:13).

Late hummingbird observations should be documented with detailed field notes and photographs.

Rufous Hummingbird (*Selasphorus rufus*)

Status: Casual late summer and fall transient.

Documentation: Photograph: adult male, early Aug to 29 Oct 1968, Charleston, Mississippi Co. (JH et al.-**BB** 36[4]:1; Pl. 8a).

Habitat: Virtually anywhere there are flowering plants and feeders.

Records:

Summer and Fall Migration: In addition to the above record, which was the first for the state, there are a minimum of seven other observations involving single, adult male Rufous. With the exception of one record, all of these males were initially detected between late July and early Oct. Earliest date: 25–31 July 1985, 4 miles north of Corning, Atchison Co. (E. Wright-**BB** 52[4]:19; photographed). Latest date: 20 Sept–2 Dec 1984, St. Joseph (FL et al.-**BB** 52[1]:31; color photo in *St. Joseph News-Press,* 4 Oct 1984).

There are at least four additional observations that involve immature male or female *Selasphorus:* 3+, 4–11 Sept 1984, Rogersville, Webster Co. (A. Simmerman et al.-**BB** 52[1]:31); 1, 17 Sept–16 Nov 1986, Eldon, Miller Co. (JW et al.); immature male, 19 Sept 1989, New Bloomfield, Callaway Co. (J. Vickery, JW-**BB** 57[1]:49); 1, 15–30 Nov 1986, Salisbury, Chariton Co. (JW-**BB** 54[1]:34).

An increase of hummingbird feeder observers and a better network of communication among bird-watchers in the state has modified considerably our perception of the status of *Selasphorus* hummingbirds during the last decade. Since 1981, with the exception of 1982 and 1983, at least one *Selasphorus* hummer has been recorded annually in the fall. Most observations are from the western half of Missouri; the above Charleston record represents the only sighting for the eastern third of the state.

Comments: Immature or female *Selasphorus* hummingbirds are extremely difficult to identify in the field. Birds must be examined in the hand to determine positive identification. This coupled with the fact that the Allen's Hummingbird (*S. sasin*), a species that is very similar to the Rufous, has been taken in Louisiana and Texas make all field identifications of immature and/or female Rufous highly problematic.

Order Coraciiformes: Kingfishers, Rollers, Hornbills, and Allies

Family Alcedinidae: Kingfishers

Belted Kingfisher (*Ceryle alcyon*)

Status: Uncommon transient and summer resident; uncommon winter resident in south, rare in north.

Documentation: Specimen: female, 5 June 1965, West Eagle Rock, Barry Co. (NWMSU, DAE 894).

Habitat: Primarily breeds in banks along rivers and streams. At other seasons may be found along any clear body of water.

Records:

Spring Migration: Migrants begin supplementing the winter population as streams, rivers, and lakes thaw; usually at the end of Feb in the south and by early Mar in the north. Peak is in early to mid-Apr.

Summer: Although breeding bird data underrepresent this species, the overall trends among the natural divisions are probably correct. It is most common in the Ozarks and Ozark Border (average 0.4 birds/route) and least abundant in the Mississippi Lowlands and Glaciated Plains (average < 0.1 birds/route).

Fall Migration: By early Sept migrants are seen, and their numbers peak at the end of Sept or in early Oct. Numbers remain relatively high through Oct, with a gradual decline through Nov. High count: 25+, 3 Oct 1977, along Niangua R., Dallas Co. (S. Yeskie-**BB** 45[1]:27).

Winter: With the freezing of water (usually in mid-Dec in the north), the last migrants are seen moving south. Numbers vary considerably from year to year, depending on the mildness of the winter. Regardless of the severity of the winter, it is most prevalent in the Ozarks. High counts: 30, 28 Dec 1974, Sullivan CBC; 29, 18 Dec 1976, Springfield CBC.

Order Piciformes: Puffbirds, Toucans, Woodpeckers, and Allies

Family Picidae: Woodpeckers and Allies

Lewis' Woodpecker (*Melanerpes lewis*)

Status: Accidental winter resident and spring transient.

Documentation: Photograph: 1, 23 Dec–24 Mar 1962–63, near Rolla, Phelps Co. (Mrs. Ollar, F. Frame et al.-**BB** 30[1]:28; Pl. 8b).

Records:

There is only one additional record: a bird was carefully studied and photographed on 6 May 1989, just northeast of Gainesville, Ozark Co. (Jacobs 1989; photos in MRBRC file). Apparently this individual was migrating since subsequent searches failed to locate it. An interesting note: the above May record coincides with most extralimital records in adjacent states. Both Illinois sight records are in May (Bohlen and Zimmerman 1989), the single Arkansas record is from late June and early July (James and Neal 1986), and there are at least three records for May through July for Kansas (Thompson and Ely 1989).

Red-headed Woodpecker (*Melanerpes erythrocephalus*)

Status: Common transient and summer resident statewide; an uncommon winter resident in south, rarer and more local in north.

Documentation: Specimen: male, 26 Dec 1961, Lincoln Co. (KU 40311).

Habitat: During the breeding season prefers open areas with scattered trees, forest and woodland edge. During winter, primarily found in mature stands of forest (especially with numerous oaks).

Records:

Spring Migration: Migrants begin reappearing by early Apr, but peak is not until the end of Apr or early May. High count: 30, 2 May 1965, St. Joseph/Squaw Creek (FL).

Summer: During the breeding season this species is most common in relatively open areas with scattered trees or nearby woodland. It is decidedly more common in the Glaciated Plains than in any other region (Map 21). Graph 4 depicts the pronounced changes in summer abundance of this species over the past twenty-three years. Presumably, these differences are related to the availability of acorns during the fall and winter seasons, i.e., woodpecker breeding density is greatest following years of high mast productivity.

Fall Migration: Birds begin leaving breeding areas by the latter part of Aug. Migration peak is normally during early to mid-Sept. By mid-Oct most birds have left the state, and those that remain are on winter territories. High counts: 290+, 2 Sept 1942, Fairfax, Atchison Co. (A. White-**BB** 9[10]:66); 284, in one hr, 15 Sept 1884, along Mississippi R., St. Louis (Widmann 1907).

Winter: More widespread and common in the Ozarks than any other region, although it can be locally common during some winters in other areas. Numbers fluctuate considerably from year to year statewide in relation to acorn mast availability (Smith 1986). High counts: 592, 2 Jan 1989, Mingo CBC; 207, 28 Dec 1974, Sullivan CBC.

Comments: The Red-headed Woodpecker is an example of a species that has actually benefited from man's alteration of the environment. It is probably much more common now than it was prior to settlement, as a result of the opening of the Ozarks and the planting of trees in the former prairie regions of the state.

Red-bellied Woodpecker (*Melanerpes carolinus*)

Status: Common permanent resident.

Documentation: Specimen: male, 12 Dec 1920, Jackson Co. (KU 39103).

Map 21.
Relative breeding abundance of the Red-headed Woodpecker among the natural divisions. Based on BBS data expressed as birds/route.

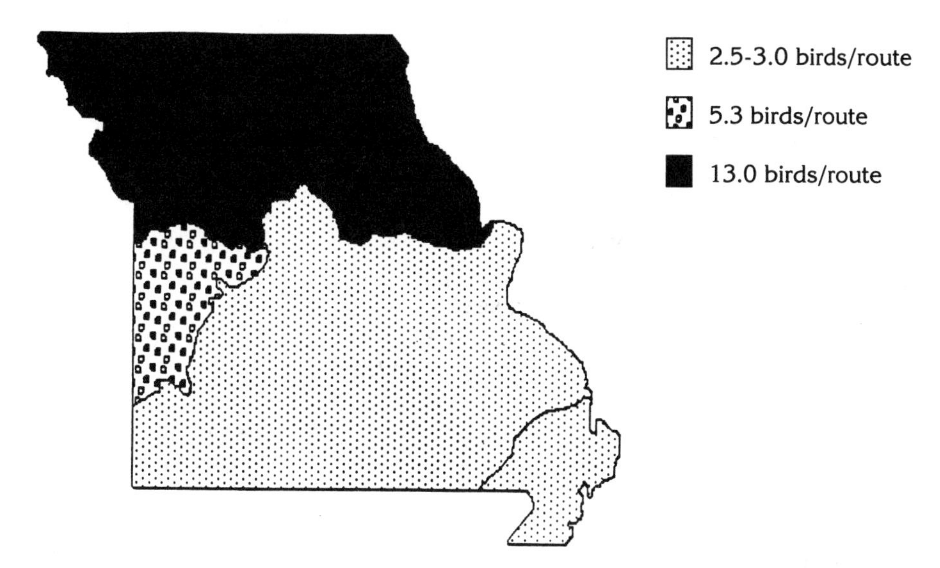

Graph 4.
Missouri BBS data from 1967 through 1989 for the Red-headed Woodpecker. See p. 23 for data analysis.

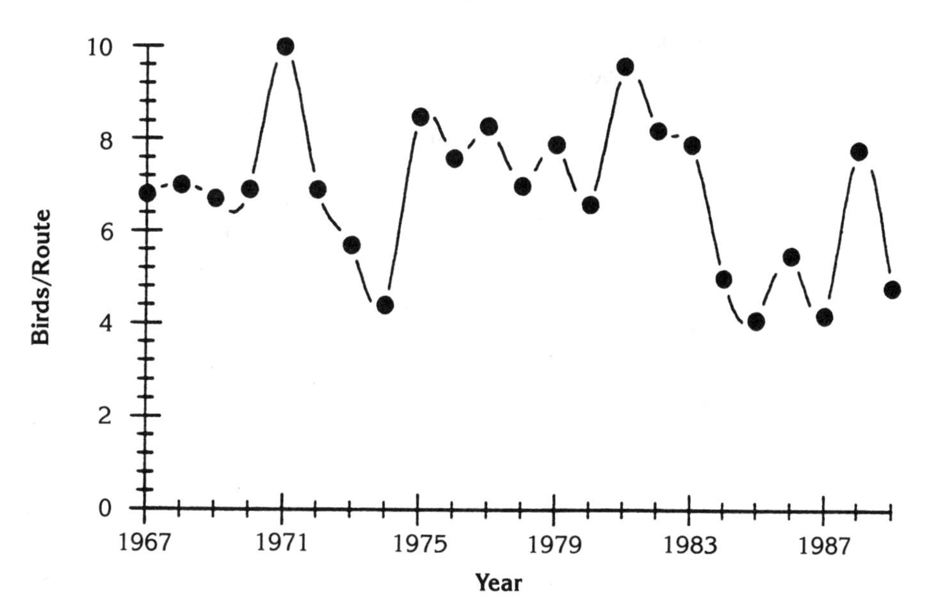

Map 22.
Relative breeding abundance of the Red-bellied Woodpecker among the natural divisions. Based on BBS data expressed as birds/route.

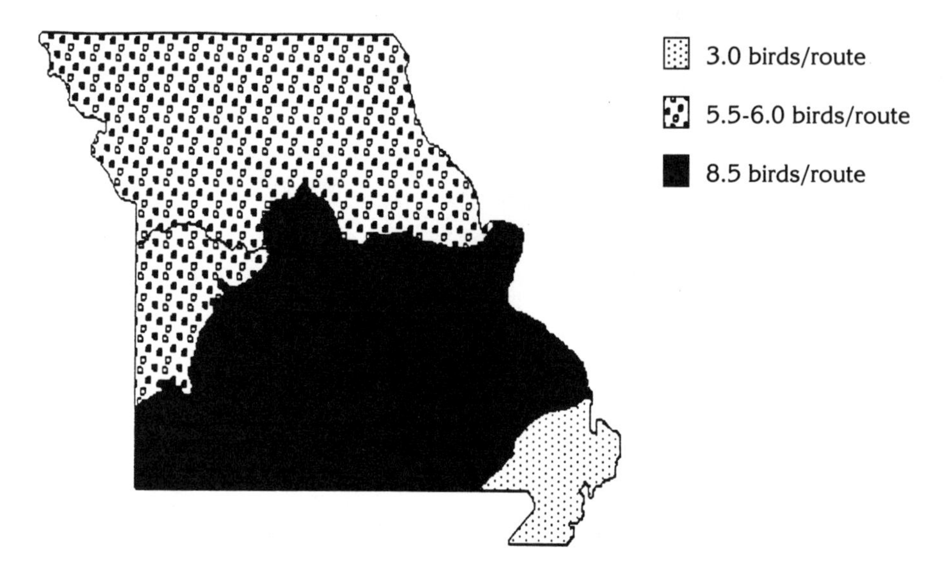

Habitat: Principally found in open woodland, but also found in sparsely wooded residential areas.

Records:

Summer: The importance of woodland and forest to this species is apparent when BBS data are examined. It is most abundant in the forested Ozarks and Ozark Border, and least so in the largely denuded Mississippi Lowlands (Map 22). In addition, although the overall mean for this species in the Glaciated Plains is similar to that of the Osage Plains, it is less common in the western (less forested) than the eastern half of the Glaciated Plains (west, 4.4 birds/route vs. east, 6.7 birds/route). A similar pattern would undoubtedly emerge for the Osage Plains if data were available.

Winter: CBC data show that it is most common in the Ozark Border (average > 10.0 birds/10 pa hrs) and least common in the Osage Plains (average < 6.0 birds/10 pa hrs). High counts, both on the Weldon Spring CBC: 238, 29 Dec 1957; 159, 2 Jan 1971.

Yellow-bellied Sapsucker (*Sphyrapicus varius*)

Status: Uncommon transient; uncommon winter resident in Mississippi Lowlands and extreme southern Ozarks, rare elsewhere; former accidental summer resident along upper Mississippi River.

Documentation: Specimen: female, 20 Mar 1967, 10 miles west of Maryville (NWMSU, DAE 1772).

Habitat: Primarily seen in deciduous forest, but frequently seen in conifers, orchards, etc.

Records:

Spring Migration: There is an influx of birds into the state, beginning in Mar and peaking in late Mar or early Apr. Relatively large numbers (6–7/day) may be encountered through the third week in Apr, but numbers decline rapidly thereafter. Stragglers are casual after the first week of May. High count: 10, 2 Apr 1980, Big Oak Tree SP (JH). Latest date: 1, 29 May 1979, Springfield (fide CB-**BB** 46[3]:10).

Summer: There are no recent records, but according to Widmann (1907) this species was once a local, very rare breeder in the Mississippi R. floodplains from St. Louis northward to Clark Co. Recent searches of this area of the state have failed to uncover any breeding birds. Furthermore, this species has declined considerably as a breeder since the turn of the century in Iowa (Dinsmore et al. 1984), and there have been few recent Illinois records (Bohlen and Zimmerman 1989).

Fall Migration: Very few are seen until the last week of Sept, but there is a dramatic increase in numbers during the first week of Oct that peaks about a week later. By the end of Oct the majority have passed through (except winter residents), but a few stragglers (mainly immatures) are still moving through until the end of the period. Earliest date: 1, 17 Sept 1972, St. Joseph (FL-**BB** 40[1]:8).

Winter: CBC data indicate the largest concentrations are encountered in the few remaining stands of mature forest in the Mississippi Lowlands (average 5.3 birds/10 pa hrs), whereas it is rare in all of the other natural divisions (average 0.3–0.5 birds/10 pa hrs). However, newly established counts in the White River Cedar Glade section of the southwestern Ozarks suggest that the sapsucker is not uncommon there, e.g., on the Taney Co. CBC an average of 3.6 birds/10 pa hrs has been recorded (n=7 years; 1982–88). High counts, both on the Taney Co. CBC: 38, 2 Jan 1988; 24, 27 Dec 1986.

Comments: Observers should be cognizant of the possibility of the occurrence of the Red-naped Sapsucker (*S. nuchalis*), as this species has been recorded east of the Great Plains. Until recently, this Great Basin woodpecker was treated as a subspecies of the Yellow-bellied Sapsucker (*A.O.U. Check-list 1985*). Occasionally, adult Yellow-bellieds do have red on the nape; pay particular attention to head, throat and back pattern of any sapsucker with a red nape. See Kaufman (1990) for a thorough treatment on sapsucker identification.

Downy Woodpecker (*Picoides pubescens*)

Status: Common permanent resident.

Documentation: Specimen: female, 24 Aug 1920, Taney Co. (KU 39124).

Habitat: Found in virtually every habitat with trees.

Records:

Summer: Although BBS data are not appropriate for determining the abundances of this species and the Hairy Woodpecker (because these species vocalize very infrequently and are relatively inconspicuous at this time of year), they nevertheless exhibit a similar overall pattern for both species. The data reveal that the Downy, like the Hairy, is least common in the Mississippi Lowlands (average 1.0 birds/route), with the Ozarks, Glaciated and Osage plains exhibiting a similar relative abundance (average 2.0–2.8 birds/route). The Downy is more abundant than the Hairy at all seasons statewide.

Winter: CBC data reveal that it is most common in the Mississippi Lowlands, Glaciated Plains and Ozark Border (average 12.0–17.0 birds/10 hrs), whereas the lowest numbers (average 7.0–8.0 birds/10 hrs) are recorded in the Osage Plains. High counts: 197, 15 Dec 1979, Kansas City Southeast CBC; 176, 21 Dec 1980, Columbia CBC.

Hairy Woodpecker (*Picoides villosus*)

Status: Common permanent resident.

Documentation: Specimen: male, 25 Nov 1917, Jackson Co. (KU 39112).

Habitat: Similar to Downy Woodpecker, except prefers more extensive mature woodland.

Records:

Summer: As mentioned under the Downy Woodpecker species account, BBS data depict a similar pattern of relative abundance among natural regions for both species in the state. Like the Downy, this species is least common in the Mississippi Lowlands (average 0.1 birds/route) than in the other regions, where an average of 0.5 birds/route is found.

Winter: The average ratio of Hairy to Downy woodpeckers is similar among the Glaciated (1:10), Ozark Border (2:10), and Ozark (2:10) regions, i.e., for every 1–2 Hairies seen per 10 pa hrs, one would expect to see 10 Downies. High counts, both on the Sullivan CBC: 37, 31 Dec 1977; 30, 29 Dec 1979.

Red-cockaded Woodpecker (*Picoides borealis*)

Status: EXTIRPATED; formerly a local, uncommon permanent resident of the Shortleaf Pine region of the Ozarks.

Documentation: Specimen: male, 12 May 1907, Shannon Co. (AMNH 229479).

Habitat: Was restricted to relatively large, open mature stands of Shortleaf Pine.

Comments: The last colony disappeared shortly after the last stand of virgin Shortleaf Pine was cut in the spring of 1946 along highway 19, just south of Round Spring, Shannon Co. Cunningham's observation of 5 at this locality on 16 June 1946 appears to be the final published sighting (**AM** 48:125). Searches of this area in the late 1950s failed to find any remnant populations (DAE).

Even at the time of its initial discovery in the state, it was predicted that this species would soon disappear, as a result of the relentless lumber industry (Woodruff 1907, 1908; specimens, AMNH 229479–229483). Although there are definite records for only Shannon and Carter counties, this woodpecker very likely was present in Butler, Reynolds, Ripley, Oregon, Texas, and Wayne counties before mature stands of pine were eliminated by man (Eddleman and Clawson 1987). A recent, extensive survey of the former range of this species delineated potential sites where it might be reintroduced (Eddleman and Clawson 1987).

Similar extirpations and declines have been noted throughout the range of this species, and it is now listed on the U.S. Fish and Wildlife Service's Endangered Species List.

[**Black-backed Woodpecker** (*Picoides arcticus*)]

Status: Hypothetical.

Documentation: Sight records only: see below.

Comments: There is a single, reliable sighting with minimal details. A male was seen at Kansas City on 9 Apr 1946 (Williamson 1946). The bird was apparently quite tame, and the observer was able to note the all black back and yellow "head-patch" at a distance of three feet, as it fed in an apple tree.

Certainly a less reliable report is of a bird on 22 Feb 1921, St. Charles Co. (J. Jokerst; Bennitt 1932). Jokerst reported a number of extraordinary sightings for the state, so many of his observations are questionable (see comments of Comfort 1972; Easterla 1976). This species has also been on the Squaw Creek checklist since at least the mid-1950s (FL, pers. comm.); however, no details are available on the record.

Northern Flicker (*Colaptes auratus*)

Status: Common permanent resident.

Documentation: Specimen: male, 27 Dec 1964, Gravois Mills (NWMSU, DAE 818).

Habitat: Prefers woodland and forest edge and open areas with scattered trees.

Records:

Spring Migration: Obvious migrants begin appearing by mid-Mar and

Map 23.
Relative breeding abundance of the Northern Flicker among the natural divisions. Based on BBS data expressed as birds/route.

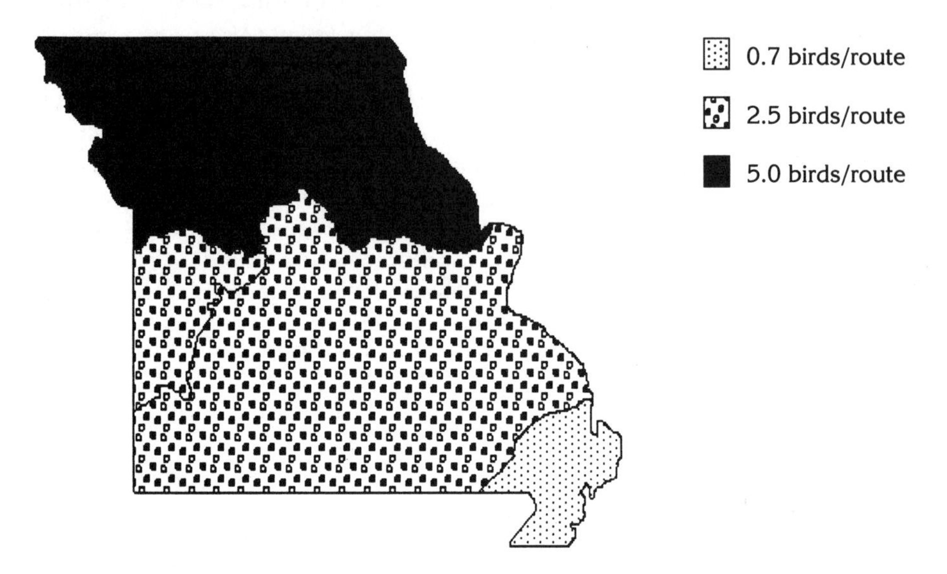

peak in early to mid-Apr. High counts: 100 seen flying NE in ca. 1.5 hrs of observation, 23 Mar 1895, St. Louis (Widmann 1907); 75, 5 Apr 1964, Trimble/Squaw Creek (FL); 65+, 12 Apr 1975, Bigelow Marsh (MBR, DAE).

Summer: This open country bird is twice as common in the Glaciated Plains than it is in the Ozarks, Ozark Border and Osage Plains (Map 23).

Fall Migration: Migrants begin supplementing the summer resident population in early Sept. Peak is in late Sept and early Oct. High count: 496, 5 Oct 1983, Forest Park (MP-**BB** 51[1]:43).

Winter: In the heart of winter it is probably least common in the Glaciated Plains, but CBC data show that in early winter the Ozark region has the fewest number seen on average, > 7.0/10 hrs, while the Mississippi Lowlands has the highest average, 19.0 birds/10 hrs. High counts, both on the Weldon Spring CBC: 171, 20 Dec 1981; 152, 2 Jan 1971.

Comments: The western, Red-shafted race (*C. a. cafer*) is a rare migrant and winter resident and is more common in the western half of the state. The vast majority, if not all, of the birds seen are hybrids (n=5 specimens). Most have left by mid-Apr, with the latest spring record being 1, 25 May 1971, Busch WA (K. Boldt, JEC et al.-**NN** 43:82). Earliest fall records: 1, 20 Sept 1968, St. Joseph (FL-**BB** 36[4]:12); 1, 21 Sept 1976, St. Louis (K. Boldt-**NN** 48:71). High count: 4, 21 Dec 1969, Squaw Creek CBC.

Pl. 1. *Top,* Harris' Sparrow; *middle,* Bell's Vireo; and *bottom,* Lark Sparrow. All three were first made known to science through their discovery in Missouri. Watercolor by David Plank.

Pl. 2. *Upper left,* Hudsonian Godwit; *upper right,* Stilt Sandpiper; *lower left,* Buff-breasted Sandpiper; and *lower right,* Lesser Golden-Plover. These are shorebirds that primarily migrate during the spring through the interior of North America. Watercolor by David Plank.

Pl. 3. The Greater Roadrunner is a relatively recent immigrant to Missouri. It is most frequently observed in the "cedar glades" of the southwestern Ozarks. Watercolor by David Plank.

Pl. 4. The Scissor-tailed Flycatcher is primarily found in the Osage Plains. Hundreds may be seen migrating through this region of the state in fall. Watercolor by David Plank.

Pl. 5. The first nest and eggs of the Bachman's Warbler (male *top,* female *bottom*) were discovered in Missouri by Otto Widmann in May 1897 along the St. Francois River in Dunklin Co. This species may now be extinct. Watercolor by David Plank.

Pl. 6. A male Painted Bunting singing from an Eastern Red Cedar. This colorful summer resident is principally found in the "cedar glades" of the southwestern Ozarks. Watercolor by David Plank.

Pl. 7. The once locally common Bachman's Sparrow is now on the Missouri Endangered Species List; the reasons for its decline are unknown. This male was observed singing on the ground in an abandoned field on 18 April 1965 by David Plank and David Easterla near Salem, Dent Co. Watercolor by David Plank.

Pl. 8. *a,* male Rufous Hummingbird, 7 September 1969, at Charleston, Mississippi Co., photograph by Jim Haw; *b,* Lewis' Woodpecker, 9 February 1963, at Rolla, Phelps Co., photograph by David Easterla; *c,* Say's Phoebe, 2 December 1983, Maryville Sewage Lagoons, Nodaway Co., photograph by David Easterla; *d,* male Mountain Bluebird, 10–14 April 1990, near Cedar Springs, Cedar Co., photograph by JoAnn Garrett; *e,* immature male Black-headed Grosbeak, 28 November–3 December 1984, Roaring River State Park, Barry Co., photograph by Merle Rogers; *f,* adult Bronzed Cowbird, 5 January 1979, Squaw Creek NWR headquarters, Holt Co., photograph by Berlin A. Heck, courtesy of USFWS.

Map 24.
Relative breeding abundance of the Pileated Woodpecker among the natural divisions. Based on BBS data expressed as birds/route.

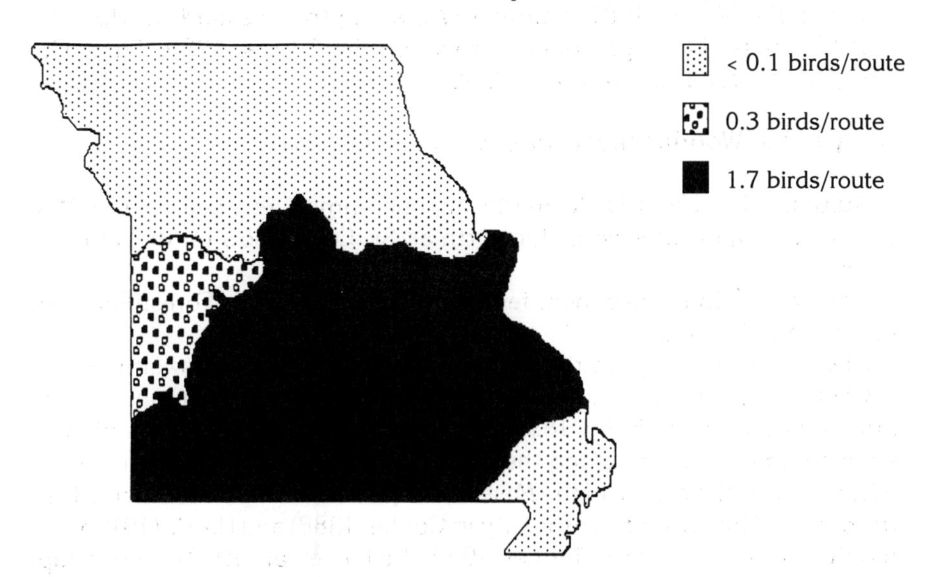

Pileated Woodpecker (*Dryocopus pileatus*)

Status: Uncommon permanent resident in south and east, casual in the northwest.

Documentation: Specimen: male, 8 June 1937, Bethany, Harrison Co. (MU 434).

Habitat: Relatively large stands of mature forest, especially bottomland forest.

Records: It is most common in the Ozarks and Ozark Border and in the remaining tracts of forest in the Mississippi Lowlands (Map 24). In the Glaciated Plains region it is rarer and more local, being primarily restricted to forest bordering the larger river systems. It is rarest in the northwestern corner of this latter region, where small populations are known from along the Platte R., just east of St. Joseph (FL et al.), and along the Grand R., as far upstream as Evona, Gentry Co. (MBR-**BB** 56[3]:91). Also, it is very rare and local in the Osage Plains, especially in the westernmost tier of counties, where it is primarily found along the Osage and Grand rivers and their tributaries.

Formerly, it was more common and widespread across the state. Widmann (1907) mentions that it was most common in the Mississippi Lowlands, an area largely denuded now. Although there are no specific records, it probably once occurred along the Missouri R. in the northwestern corner. Harris (1919b) states that it was a regular breeder along the Missouri R. bottoms in the Kansas City area until the early 1890s.

Occasionally, wandering birds are seen in areas not normally inhabited, e.g., female, 30 Jan 1987, Thurnau WA, along the Missouri R., Holt Co. (MBR-**BB** 54[2]:14). High counts for winter: 44, 30 Dec 1989, Taney Co. CBC; 42, 27 Dec 1986, Taney Co. CBC.

Ivory-billed Woodpecker (*Campephilus principalis*)

Status: EXTIRPATED; formerly an uncommon resident in the Mississippi Lowlands at least as far north as St. Louis. Now on the verge of extinction.

Documentation: Specimen: female, 8 May 1886, near Forest Park, St. Louis (DMNH 27343).

Habitat: Apparently restricted to mature bottomland and swamp forest.

Comments: The last record is of a male that was taken 8 miles SW of Morley, Scott Co., on 8 Nov 1895 (Widmann 1907). In the 1820s it still was an uncommon resident along the Mississippi R. bottomland from at least as far north as the mouth of the Missouri R. (see review by Hasbrouck 1891). Records for Fayette and Kansas City in Cooke (1888) and Harris (1919b) are highly questionable (see Tanner 1942; Rising et al. 1978). The disappearance of this species in the state coincided with the logging of the southeastern lowlands.

Order Passeriformes: Passerines

Family Tyrannidae: Tyrant Flycatchers

Olive-sided Flycatcher (*Contopus borealis*)

Status: Uncommon transient.

Documentation: Specimen: male, 1 June 1976, north of Duck Creek (SEMO 422).

Habitat: Usually seen perched on top of dead snags above forest or woodland canopy or at forest edge.

Records:

Spring Migration: Less common at this season than in the fall. It is a late migrant, with the first arrivals appearing by the middle of the second week of May. Peak is during the last week of May. Earliest dates: 1, 27 Apr 1985, St. Francois SP, St. Francois Co. (BRE, SD, BL); 1, 2 May 1978, Maryville (DAE, TB-**BB** 45[3]:18). High counts: 4, 29 May 1984, Bluffwoods SF (FL); 4, 24 May 1986, Nodaway, Atchison, and Holt counties (MBR).

Summer: Most observations are clumped at the beginning and at the end of this period, representing late spring and early fall migrants, respectively. Birds heading north are readily seen through the first week of June. The earliest fall transients are usually not seen until the end of July. The following represent the only records between mid-June and late July: 1, 22 June 1989, Higginsville, Lafayette Co. (F. Young-**BB** 56[4]:148); 1, 11 July 1987, Hermitage, Hickory Co. (H. John-**BB** 54[4]:28).

Fall Migration: Most observations during the first three weeks of Aug involve single birds, but by the end of the month and in early Sept it is not unusual to see 3–4 birds/day. By mid-Sept only single birds are seen, and virtually all have passed through the state by the end of Sept. High count: 11, 6 Sept 1979, Warren Co. (TB-**BB** 47[1]:25). Latest date: 1, 11–12 Oct 1974, Brickyard Hill WA (DAE, MBR-**BB** 42[2]:6).

Eastern Wood-Pewee (*Contopus virens*)

Status: Common summer resident.

Documentation: Specimen: male, 23 Aug 1920, Taney Co. (KU 39145).

Habitat: Primarily found breeding in woodland and forest openings and edge.

Records:

Spring Migration: Extremely early migrants are casually seen in the bootheel in early Apr, but normally the first arrivals are not seen until the last week of Apr. The majority arrive in mid-May. Earliest dates, both in Mississippi Co.: 2, heard, 25 Mar 1967 (JH); 1, 31 Mar 1974 (JH).

Summer: This woodland inhabiting bird is most common in the Ozarks and Ozark Border (Map 25). The low value for the Mississippi Lowlands is a reflection of the lack of remaining woodland there.

Fall Migration: The largest numbers are seen during the first two weeks of Sept. By the end of the first week of Oct only a few scattered individuals are seen. Latest dates: 1, 11 Nov 1987, Busch WA (BRE, BL-**BB** 55[1]:12); 1, 25 Oct 1975, Montgomery City, Montgomery Co. (RW-**BB** 44[1]:21).

Comments: Observers should be alert to any "atypical" vocalizing pewees, as the Western Wood-Pewee (*Contopus sordidulus*) has been recorded in adjacent states. Vocalizations are the only means of separating these species under field conditions. Therefore, attempts should be made to record the voice of any suspect pewee. See Kaufman (1990) for details on pewee identification.

Yellow-bellied Flycatcher (*Empidonax flaviventris*)

Status: Rare transient.

Documentation: Specimen: male, 16 May 1907, Grandin, Carter Co. (AMNH 229528).

Habitat: Often seen in damp, shady areas in woodland and forest; frequently encountered in conifers.

Map 25.
Relative breeding abundance of the Eastern Wood-Pewee among the natural divisions. Based on BBS data expressed as birds/route.

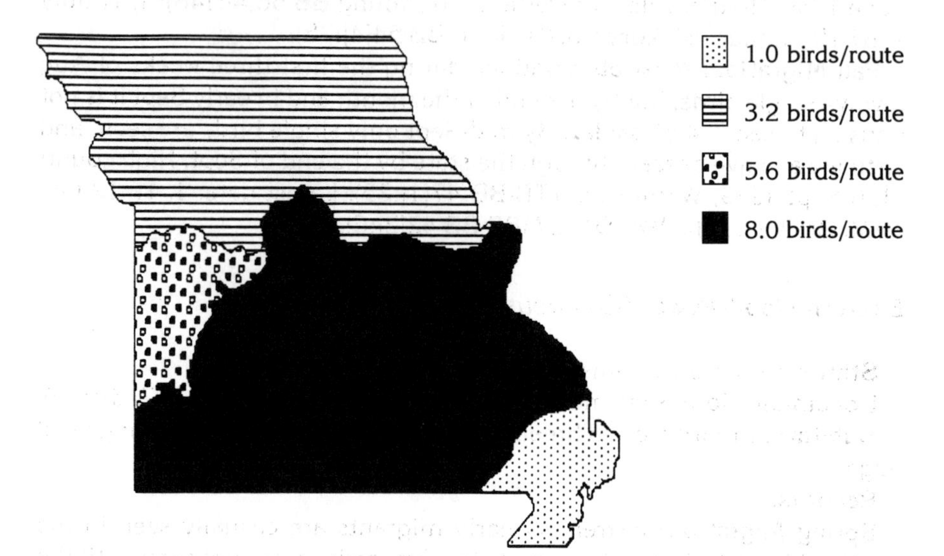

Graph 5.
Missouri BBS data from 1967 through 1989 for the Eastern Wood-Pewee. See p. 23 for data analysis.

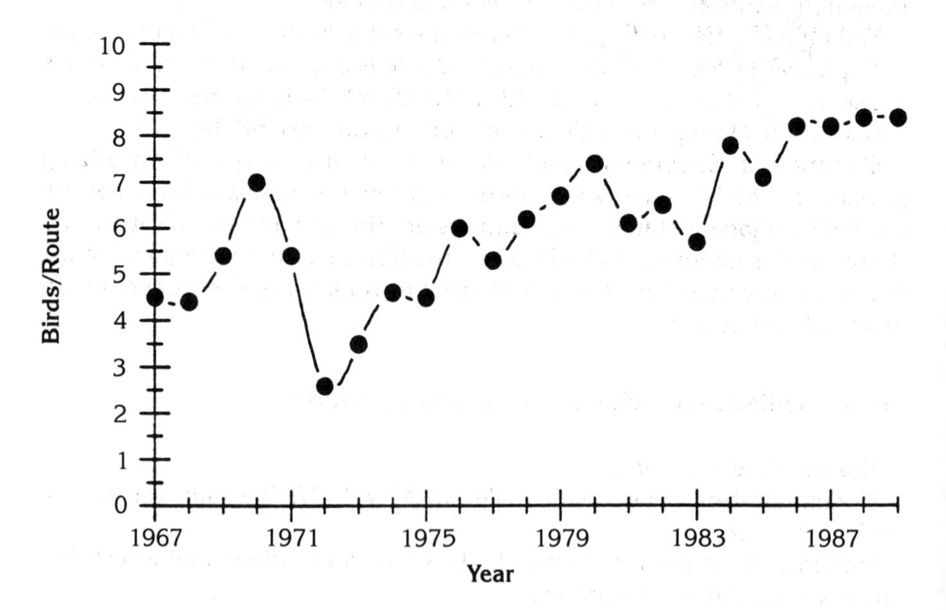

Records:

Spring Migration: Most records are for this period. It is rarely seen before mid-May, with ca. 6 observations involving more than a single bird/day. Peak is at the beginning of the fourth week of May. Earliest dates: 1, 7 May 1982, Mingo (K. Adams-**BB** 49[3]:20); 1, 7 May 1988, Mingo (BRE, SD-**BB** 55[3]:89). High counts: 5, 24 May 1990, Tower Grove Park (JZ-**BB** 57[3]:139); 4, 27 May 1973, Trimble (SP-**BB** 40[3]:6).

Summer: Late migrants are casually seen during the first few days of June. Latest date: 1, 4 June 1990, Tower Grove Park (JZ).

Fall Migration: Only a couple of records exist prior to the third week in Aug, with most in early to mid-Sept. Earliest date: 1, 1 Aug 1970, Maryville (MBR). High counts: three different counts involving two birds, all from St. Louis, between 5–22 Sept. Latest dates (and only Oct records): 1, 11 Oct 1973, Busch WA (PS, JEC-**NN** 45:123); 1, 3 Oct 1971, L. of the Ozarks SP (MBR, RA, FL-**BB** 39[1]:6).

Comments: All late fall (Oct–Dec) *Empidonax* should be thoroughly documented. Details on morphology (e.g., mandible [lower bill] color, eye-ring characteristics), behavior (note if tail is jerked when calling), and vocal characters should be noted. Efforts should be made to tape record a calling bird, as vocalizations, song, and call notes are particularly important in ascertaining an identification. Often birds can be stimulated to call by playing prerecorded *Empidonax* vocalizations. Kaufman (1990) offers an in-depth discussion on identification of this difficult group.

Acadian Flycatcher (*Empidonax virescens*)

Status: Common summer resident in the Ozarks and Ozark Border, uncommon in the Glaciated and Osage plains and the Mississippi Lowlands.

Documentation: Specimen: male, 29 Apr 1907, Spring Valley, Shannon Co. (Woodruff 1908; AMNH 229534).

Habitat: Prefers heavily shaded forest and woods along streams, rivers, and swamps.

Records:

Spring Migration: Arrives in early May in the south and about a week later in the north. Most common along the Ozark streams and rivers. Earliest dates: 1, 25 Apr 1986, Creve Coeur L. (RG); 1, 25 Apr 1989, Taney Co. (J. Hayes-**BB** 56[3]:91). High count: 60+, 20 May 1978, along 13 mile stretch of Niangua R., Dallas Co. (MBR, FL-**BB** 45[3]:18).

Summer: The Acadian reaches its greatest abundance along the streams and rivers of the Ozarks and Ozark Border. Although found throughout the Glaciated and Osage plains, it is more local in these areas as a result of less extensive habitat. As with other riparian species in Missouri, BBS data underrepresent this species. Eighteen were counted on a float census along 3.7 miles of Center Creek, just outside Joplin in Jasper

Co. on 8 July 1990 (L. Herbert, pers. comm.).

This species all but disappeared from Swope Park in the interval between 1916 (n=27 males; A. Shirling 1920) and 1973 (n=0; Branan and Burdick 1981), even though the number of hectares of floodplain forest apparently did not change.

Fall Migration: Very little information exists on the fall movement. Almost all observations are of birds in Aug and early Sept at breeding sites. Latest date: 2, heard, 12 Sept 1964, Big Oak Tree SP (JH).

Comments: See remarks under Comments of Yellow-bellied Flycatcher account.

Alder Flycatcher (*Empidonax alnorum*)

Status: Common transient.

Documentation: Specimen: male, 26 May 1985, 10 miles north of Maryville (ANSP 178011).

Habitat: Principally encountered in willows along streams and rivers, but also found at edge of deciduous woods away from water.

Records:

Spring Migration: One of the latest passerine migrants, with the first individuals not appearing until the middle of the second week in May. The bulk of them pass through during the final week of May. Earliest date: 1, 7 May 1988, Mingo (BRE, SD-**BB** 55[3]:89); 1, 10 May 1987, Rock Bridge SP, Boone Co. (BG). High count: 14, 27 May 1982, along Missouri R., Atchison and Holt counties (MBR).

Summer: Commonly encountered during the first week of June. Purported nesting records are erroneous and either pertain to late spring or early fall migrants or to the Willow Flycatcher. Latest dates: 1, 3 June 1978, BBS, Ray Co. (MMH); 3, 3 June 1986, 11 miles north of Maryville, Nodaway Co. (MBR; ANSP 178063).

Fall Migration: Virtually no information exists for this period. Presumably it passes through during the later half of Aug and early Sept.

Comments: See Willow and Yellow-bellied Flycatcher accounts.

Willow Flycatcher (*Empidonax traillii*)

Status: Uncommon summer resident.

Documentation: Specimen: male, 26 May 1986, Bigelow Marsh (ANSP 178064).

Habitat: Virtually identical to Alder Flycatcher.

Records:

Spring Migration: Less common than the preceding species. Like the Alder Flycatcher it is a late migrant, with very few records before mid-May. Earliest date: 1, 7 May 1988, Mingo (BRE, SD-**BB** 55[3]:89). High count: 3, 19 May 1981, Springfield (CB-**BB** 48[3]:19).

Summer: A locally uncommon breeder statewide, almost always associated with willows. Less common and more local in the Ozarks and Ozark Border.

Fall Migration: Virtually no information exists, except for that provided by Widmann (1907), and he notes that it is usually still present through mid-Sept. We treat Widmann's Oct dates for this species and the Least Flycatcher as suspect.

Comments: This and the Alder Flycatcher were considered to comprise a single species (referred to as the Traill's Flycatcher) until recently. Morphologically the two are very similar (often they cannot be distinguished even in the hand), and the only reliable way to separate them is by voice (both call notes and song are distinct). Refer to the recordings associated with the various North American field guides for differences in song. The Willow Flycatcher call notes are an unemphatic "whit," whereas the Alder's call notes are higher pitched, best portrayed as "pip." See remarks under Comments of Yellow-bellied Flycatcher account.

Least Flycatcher (*Empidonax minimus*)

Status: Common transient; accidental summer resident.

Documentation: Specimen: male, 24 May 1985, 10 miles north of Maryville (ANSP 178008).

Habitat: Woodland edge, hedgerows, scattered trees in country and residential areas.

Records:

Spring Migration: The most common *Empidonax* during migration. Arrives in late Apr in the south and about a week later in the north. Peak, in mid-May, is earlier than that of all the other *Empidonax*'s, except for the Acadian. It is regularly seen through May. Earliest date: 1, 21 Apr 1964, St. Joseph (FL). High count: 40+, 18 May 1982, Nodaway, Holt, and Atchison counties (MBR); 40, 5 May 1963, St. Joseph/Squaw Creek (FL).

Summer: Late migrants are regularly seen during the first week of June. A bird taken on 11 June 1965, along the Current R., near Eminence, Shannon Co., had small gonads (DAE; NWMSU, DAE 906). However, there are observations during June and July that possibly pertain to breeding birds, and there is one definite nesting record for the state: nest with 3 eggs was collected in Jackson Co. on 16 June 1891 (Harris 1919b). The breeding status of the following birds is unknown: 1, singing, 22 June 1938, Joplin (G. Banner-**BB** 5[7]:69); 1, singing, 26 June 1976, BBS route 24, Jackson/Cass counties (SP); 1, 10 July 1976, Trimble (NJ et al.-**BB** 44[2]:21).

Fall Migration: The earliest migrants begin reappearing by early Aug, and their numbers peak at the very end of Aug or early Sept. It is rarely seen after the third week of Sept. Earliest dates: 1, calling, 1 Aug 1982, Lonedell, Franklin Co. (P. Hoell-**NN** 54:68); 1, 7 Aug 1963, Carter Co. (DAE).

Map 26.
Relative breeding abundance of the Eastern Phoebe among the natural divisions. Based on BBS data expressed as birds/route.

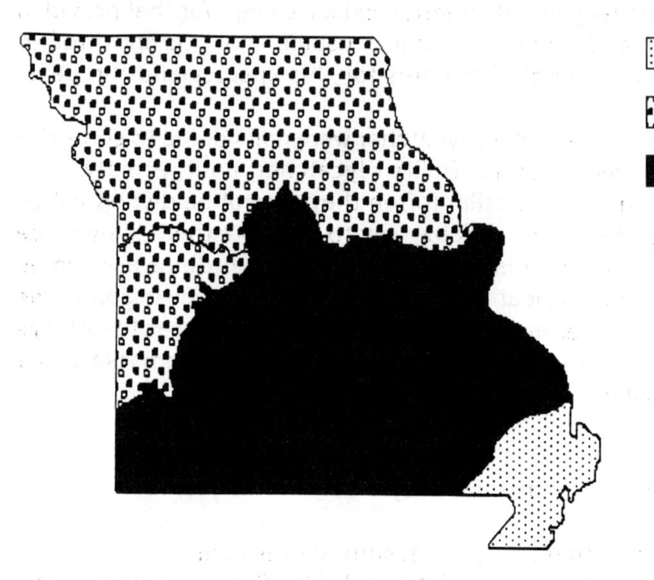

Comments: See Comments of Yellow-bellied Flycatcher account.

Eastern Phoebe (*Sayornis phoebe*)

Status: Common transient; uncommon summer resident; rare winter resident, primarily in the southeast.

Documentation: Specimen: male, 13 Apr 1968, near Sam A. Baker SP, Wayne Co. (NWMSU, SM 40).

Habitat: Usually associated with water during migration and breeding season; nests are often under bridges, cliff overhangs, abandoned buildings, and cave entrances.

Records:

Spring Migration: In the south the first migrants appear in early Mar and about a week later in the north. Peak is in early to mid-Apr.

Summer: Because this flycatcher uses manmade structures such as bridges, barns, and sheds for nest placement, it is better represented on BBS than other riparian species, although data are still poor. It is most common in the Ozarks and Ozark Border and least abundant in the Mississippi Lowlands (Map 26).

Fall Migration: Peak numbers of migrants are seen in Sept, with 2–3 birds/day encountered until mid-Oct. Thereafter usually only solitary birds are observed, with very few seen in Dec.

Winter: The majority of winter records are from the southeastern section of the state, and most of these are seen on CBCs; for example, at least one bird was recorded on 7 of the last 19 Mingo CBCs. There are an additional 15+ records for this season, from as far north as St. Louis and west to the Branson area. High counts, three birds on each of the following Mingo CBCs: 27 Dec 1972, 19 Dec 1981, and 2 Jan 1988. Northernmost record: 1, 7 Feb 1984, St. Louis (DC-**BB** 51[2]:21).

Say's Phoebe (*Sayornis saya*)

Status: Casual transient; accidental winter resident.
Documentation: Specimen: immature o?, 2–3 Dec 1983, found dead at Maryville SL (DAE; NWMSU, DAE 3100; Pl. 8c).
Habitat: Perches atop or on outer part of bushes, and on fences in open areas.
Records:
Spring Migration: Single observation: 1, 6 May 1984, Squaw Creek (B. Clark-**BB** 51[3]:13).
Fall Migration: Two additional observations besides the one given above: 1, 27 Sept 1981, Washington Co. (SD-**BB** 49[1]:14); and an immature, 11 Nov 1984, Independence (R. Franklin-**BB** 52[1]:31).
Winter: An individual that was first detected on 28 Dec 1952, Busch WA, stayed for a month (E. Atkins et al.-**BB** 20[3]:2).
Comments: The Maryville SL bird was observed feeding on insects frozen in the ice covering the lagoon (photo DAE; Pl. 8c). The bird was found dead the next morning following a severe snow and ice storm. Widmann (1907) mentions a sighting for 1886 (with no specific date) at Butler, Bates Co., by H. Clark.

Vermilion Flycatcher (*Pyrocephalus rubinus*)

Status: Casual transient.
Documentation: Specimen: immature male, found dead, 13 Nov 1967, Busch WA (Easterla and Anderson 1969; NWMSU, DAE 1474).
Habitat: Prefers open country where it perches atop small trees and bushes, often near water.
Records:
Spring Migration: Easterla and Anderson (1969) report two spring observations: male, 13 Mar 1955, Gasconade R., near Hartville, Wright Co. (H. Hedges, O. Hawksley); and 1, 5 May 1958, Forest Hill Cemetery, Jackson Co. (Tatum).
Fall Migration: There are a total of five records ranging from Sept through mid-Nov as follows: 1, Sept 1958, Portageville, New Madrid Co. (C. Moody-**BB** 25[8]:2); male, 28 Sept 1972, Busch WA (m.ob.-**BB** 40[1]:8); 4 (3

females and 1 male; females disappeared after a few days, and the male was found dead on 13 Nov; see above), 30 Sept–13 Nov 1967 (MS et al.; Easterla and Anderson 1969); adult male, 27 Oct 1945, Pleasant Hill, Cass Co. (A. Shirling-**BB** 12[11]:60); and an immature male, 11–23 Nov 1952, Busch WA (A. Bennitt et al.-**BB** 20[3]:1).

Great Crested Flycatcher (*Myiarchus crinitus*)

Status: Common summer resident.

Documentation: Specimen: male, 30 May 1964, Gravois Mills, L. of the Ozarks (NWMSU, DAE 801).

Habitat: Most common in forest and woodland clearings and edge, but also parks and other relatively open areas with trees.

Records:

Spring Migration: A few individuals arrive in the south by mid-Apr and about a week later in the north. The bulk of the birds arrive in early to mid-May. Earliest dates: 1, 7 Apr 1977, Big Oak Tree SP (JH); 1, 13 Apr 1981, Farmington (BL-**AB** 35:830).

Summer: Not surprisingly, this woodland and forest inhabitant is most common in the Ozarks and Ozark Border and least common in the Mississippi Lowlands (Map 27).

Fall Migration: Migrants are common in late Aug through mid-Sept, thereafter it is rarely seen. Latest dates: 1, 6 Nov 1965, Little Dixie L. (J. Roller-**BB** 33[1]:7); 1, 15 Oct 1971, Columbia (L. and J. Falch-**BB** 39[1]:6).

Comments: Any *Myiarchus* seen after late Sept in the fall should be carefully scrutinized, as other *Myiarchus* species have been recorded in adjacent states. Pay particular attention to the amount and distribution of rufous in the tail (most easily seen in the underside), mandible (lower bill) color, and vocalizations.

Western Kingbird (*Tyrannus verticalis*)

Status: A rare and local summer resident in the northwest; very rare transient (less common than in the northwestern corner) and casual summer resident elsewhere.

Documentation: Specimen: female, 16 May 1964, Mud Lake, Buchanan Co. (NWMSU, DAE 556).

Habitat: In migration usually seen perched on fences and utility wires in open country. Nests on telephone poles and in small, isolated groves of trees.

Records:

Spring Migration: Given the paucity of records (about twenty, most for the St. Louis region) for areas outside the extreme northwestern corner, it appears that this species has a narrow corridor for which it enters the state.

Graph 6.
Missouri BBS data from 1967 through 1989 for the Great Crested Flycatcher. See p. 23 for data analysis.

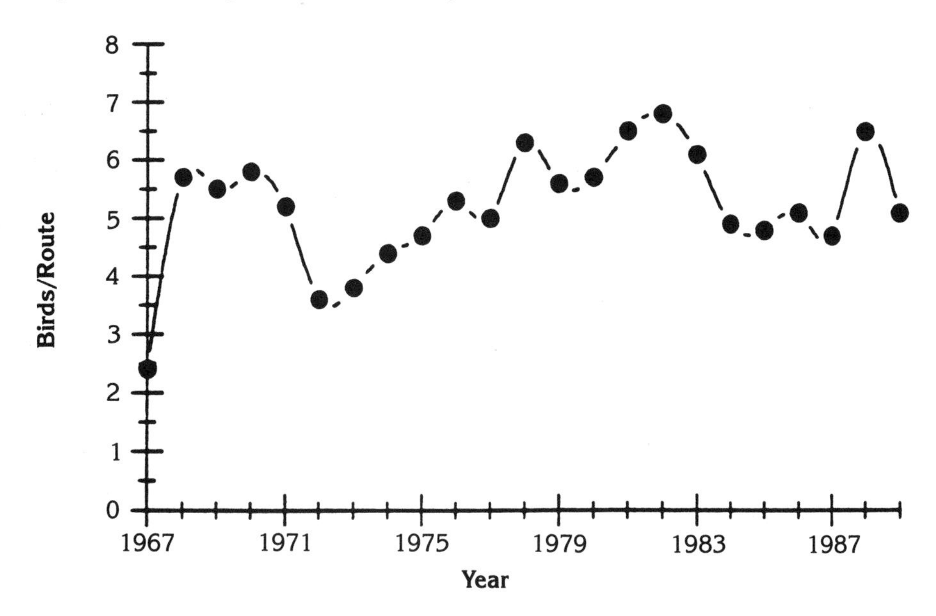

Map 27.
Relative breeding abundance of the Great Crested Flycatcher among the natural divisions. Based on BBS data expressed as birds/route.

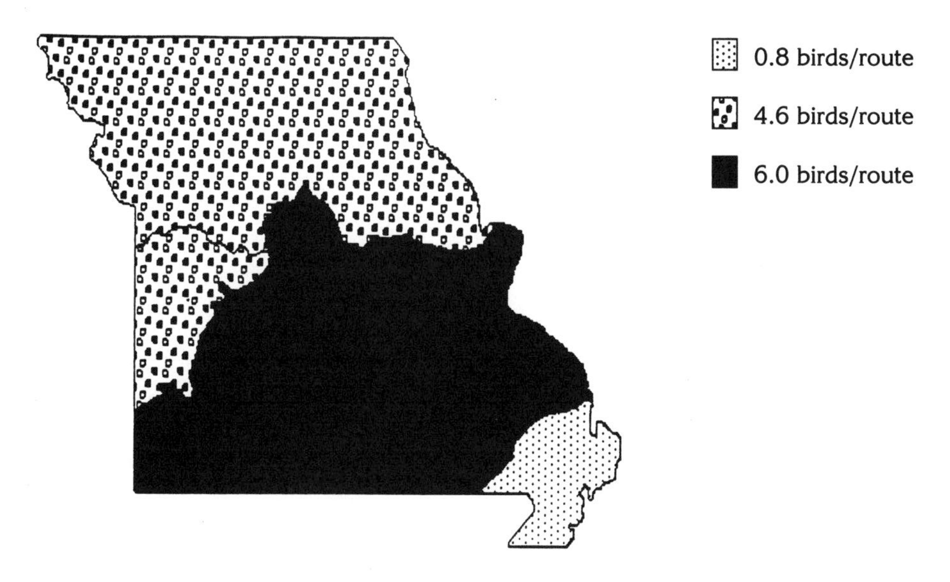

This corridor extends from Kansas City to the Iowa border, as there are only about seven spring observations from the Arkansas border north to Kansas City. The first arrivals are usually not detected until the second week in May, and during some years they are not seen until the final week of May. Earliest dates: 1, 15 Apr 1894, Stotesburg, Vernon Co. (T. Surber; Widmann 1907); three recent records for 29 Apr. High counts for western and eastern Missouri, respectively: 9, 27 May 1962, St. Joseph (FL); 5, 10 May 1989, St. Louis (RK-**NN** 61:53).

Summer: It is a rare and local breeder in the western tier of counties from the Missouri R. at Kansas City to the Iowa border. It is primarily restricted to the Missouri R. floodplain. One to 2 pairs have bred in extreme northern Cass Co. since 1979 (fide JG).

There are the following verified extralimital breeding records: 2 pair, 26 June 1977, near Pulaskiifield, Newton Co. (JE-**BB** 44[4]:30); pair, summer 1978, near Waverly, Lafayette Co. (Hobbs 1981); one or more pairs, 1986–90, St. Louis (RK et al.-**BB** 53[3]:12); 2 adults and 4 young, 29 June 1981, Columbia (JR, BG, SS-**BB** 48[3]:24); pair, summer 1988, Columbia (B. Clark-**BB** 55[4]:169). There are a few other sightings involving single birds away from the northwestern corner. High counts: 12 (includes fledged young), 28 July 1973, Kansas City Municipal Airport, Clay Co. (SP-**BB** 40[4]:30); a few records with 7 adult birds in northwest Missouri (FL, DAE).

This species has been recorded breeding in the northwestern corner since at least the late 1890s; a nest with 3 eggs was collected in Tarkio, Atchison Co., on 17 June 1898 (CM 6390).

Fall Migration: Most have left the nesting sites by early Sept. Away from the breeding areas it is casually seen from Aug through early Oct. High count: 8, 6 Sept 1965, Gasconade R., Gasconade Co. (RA-**BB** 33[1]:7). Latest date: 1, 12 Oct 1938, Fairfax, Atchison Co. (ob.?-**BB** 6[1]:16). The report of a bird on 27 Oct 1985 (**BB** 53[1]:28) actually pertained to a Scissor-tailed Flycatcher.

Eastern Kingbird (*Tyrannus tyrannus*)

Status: Common summer resident.

Documentation: Specimen: female, 13 May 1968, near St. Joseph (NWMSU, SM 42).

Habitat: Prefers relatively open country where it is commonly seen perched atop trees, on fences, and telephone wires.

Records:

Spring Migration: Arrives earlier than the preceding species, with birds first appearing in the south at the end of the second week of Apr and at the beginning of the fourth week in the north. Peak is in early to mid-May. Earliest date: 1, 28 Mar 1983, St. Francois Co. (BL-**BB** 50[3]:13).

Summer: This open-country bird is most common in the Glaciated and

Fig. 23. During the breeding season the Eastern Kingbird is most abundant in the Glaciated and Osage plains. This bird is perched on a Sycamore. Pen and ink by David Plank.

Map 28.
Relative breeding abundance of the Eastern Kingbird among the natural divisions. Based on BBS data expressed as birds/route.

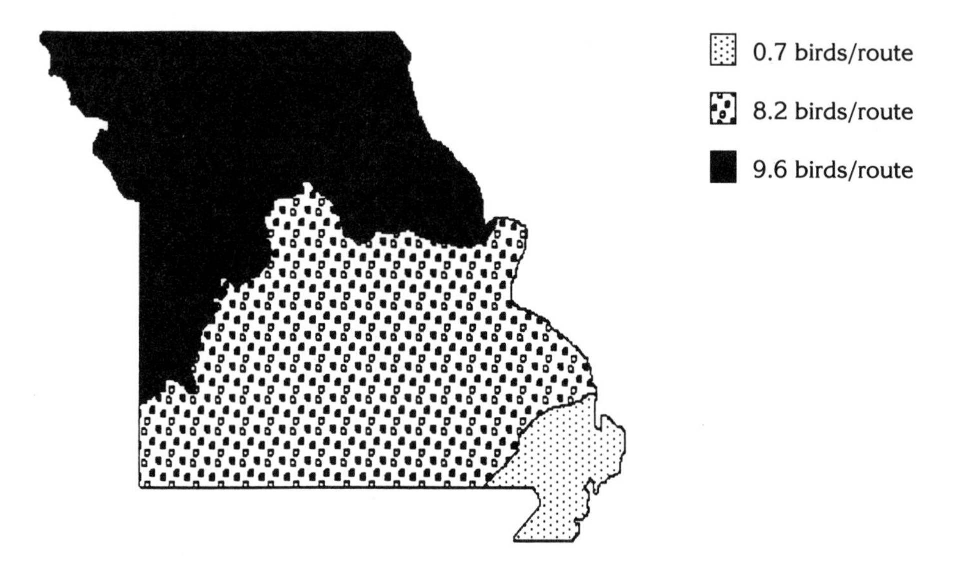

Osage plains, and surprisingly (at least based on BBS data), it has a similar relative abundance in the Ozarks (Map 28). We believe that the value for the Ozarks is somewhat inflated (relative to other natural divisions) as a result of an inherent bias in the BBS census technique; for example, by surveying along roads, one is in effect censusing an edge habitat, which is favorable to this species. Nevertheless, the relatively high value for the Ozarks is not solely an artifact of this censusing technique, as this species is indeed more abundant now in the Ozarks than it was at the turn of the century. Before the Ozarks was "opened up," Widmann (1907) stated that it was "much less common in the Ozarks [than the prairie and Ozark Border]."

Fall Migration: Migration is evident by the second week of Aug, with groups of > 30 birds observed. The impressive peak (when birds appear to be nearly everywhere) is during the last ten days of Aug and the first week of Sept. After mid-Sept numbers drop off considerably, with only small groups or individuals seen thereafter. It is only casually seen after the first few days of Oct. High counts: 300, 11 Sept 1983, Jackson Co. (CH, KH, MMH-**AB** 38:210); 150, 21 Aug 1971, St. Joseph/Squaw Creek (FL). Latest date: 1, 24 Oct 1964, near St. Joseph (FL-**BB** 31[4]:17).

Scissor-tailed Flycatcher (*Tyrannus forficatus*)

Status: Uncommon spring transient and locally common fall transient in Osage Plains and western edge of Ozarks; uncommon summer resident

Graph 7.
Missouri BBS data from 1967 through 1989 for the Eastern Kingbird. See p. 23 for data analysis.

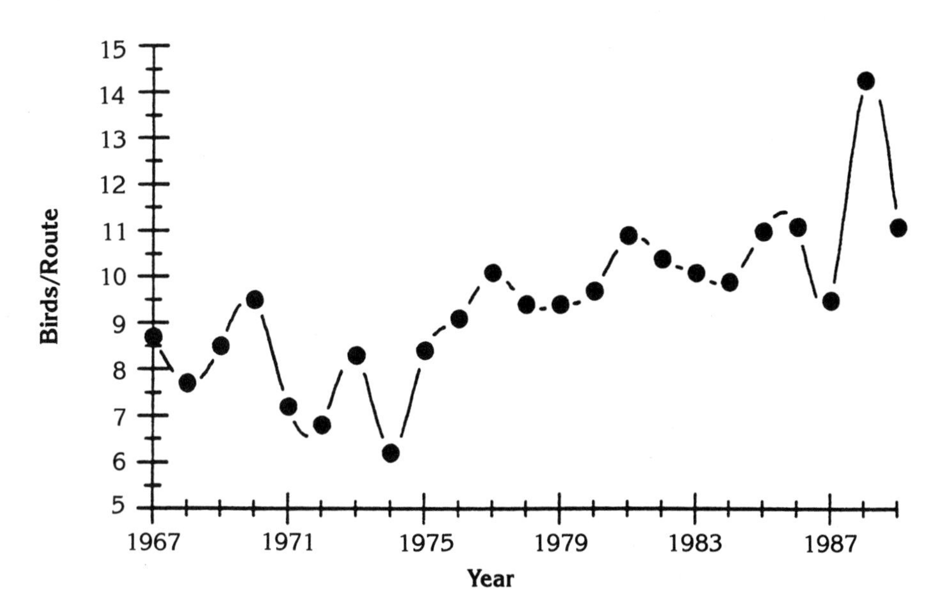

in Osage Plains and the western edge of the Ozarks; rare transient and casual summer resident elsewhere.

Documentation: Specimen: immature female, 11 Oct 1961, south of Columbia (DAE et al.-**BB** 28[4]:26; MU 1781).

Habitat: Prefers prairie and pastures with scattered trees.

Records:

Spring Migration: Concentrations at this season never approach those of the fall. The first males usually appear during the second week of Apr. Peak is at the end of Apr and early May. Earliest dates: 1, 22 Feb 1980, Springfield (W. Carras-**BB** 47[3]:17); 1, 12 Mar 1986, Springfield (B. Dyer-**BB** 53[3]:12). High count: 12, 17 Apr 1979, Reed WA (B. Brown-**BB** 46[3]:10).

Summer: It is apparent that this species was initially recorded in Missouri in the late 1860s or early 1870s. Baird et al. (1874) were the first to mention its presence "occasionally to southwestern Missouri." Coues (1874) credited Baird as the source of this new information (although the person who actually made the original observation remains a mystery), but neither one of them included Missouri as part of the Scissor-tails' range in their earlier works (Baird et al. 1860; Coues 1872).

The initial *A.O.U. Check-list* (1886) echoes the above earlier works in its treatment of the species as "casual to Missouri." The Scissor-tails' status in the state was clarified or updated in the second edition of the checklist

(1895) when "southwestern Missouri" was included in the breeding range. It remained so until the 5th edition (1957) entirely excluded Missouri from the breeding range. This was the result of E. Reilly, Jr., who compiled range information for the 5th edition. Reilly in a letter to Warner (1966) stated the following, "The inclusion of this species in southwestern Missouri as a breeding bird in the early editions must have been an error, since diligent search of the literature failed to uncover the source of earlier statements." As we state above, Baird was the source of this information, and it was surely not included by accident, as Baird was quite cognizant of its significance. Moreover, our review of the literature uncovered a few records which were missed by both Reilly and Warner (see below).

As the above indicates, Scissor-tails were already breeding in the southwesternmost counties of the Osage Plains of Missouri prior to the turn of the century. The paucity of records from this area during the 1800s and early 1900s was undoubtedly more of a reflection of the dearth of observers than of the absence of birds.

There nevertheless has been an extension in its Missouri range over the past fifty years. Warner (1966) summarized this expansion; however, as we state above, he overlooked a few significant observations that predate records included in his review. The first nest record with details was of the four young found on 29 June 1940, Garden City, Cass Co. (E. Markward et al.-**BB** 7[8]:53). Breeding was first noted near Lowry City, St. Clair Co., in the summer of 1946 (B. Hilty, A. Haverland, fide SH), and a pair was found nesting at Lee's Summit, Jackson Co., in 1954 (BK-**AFN** 8:348). A single bird was observed at Squaw Creek as early as 31 Aug 1947 (H. Hedges, JB, HH-**BB** 15[1]:3). Warner's review chronicles the spread between 1955 and 1964.

"Breeding Bird Atlas" work has recently confirmed the first nesting records north of the Missouri R.; these records are from just north of the river in central Missouri (JW, BJ, pers. comm.). It is surprising that the Scissor-tail has not become established as a breeder in the northwestern corner of the state, given that it nests in adjacent Kansas and Nebraska. Easternmost nesting records: at least one pair has nested in the Salem, Dent Co. area 8 of the past 14 years (D. Plank, pers. comm.); pair, summers 1980–82, near Belgrade, Washington Co. (SD, BL, BRE).

Fall Migration: As early as late Aug relatively large groups (> 40 birds) may be observed; however, the spectacular peak is not until the last week of Sept or the first week of Oct when over 100 birds may be seen at a single locality in the Osage Plains. Smaller numbers (5–20 individuals/day) are readily observed during the second week and the beginning of the third week of Oct, with only 1–2 birds/day casually seen at the end of Oct or early Nov. Thereafter it is accidental. High counts: several observations involving about 100 birds, all during late Sept and early Oct. Latest dates: 1, 2 Dec 1985, Springfield (B. Dyer-**BB** 53[2]:9); 1, 20 Nov 1986, Jasper Co. (GD, W. Holloway-**BB** 54[1]:34).

Map 29.
Relative breeding abundance of the Horned Lark among the natural divisions. Based on BBS data expressed as birds/route.

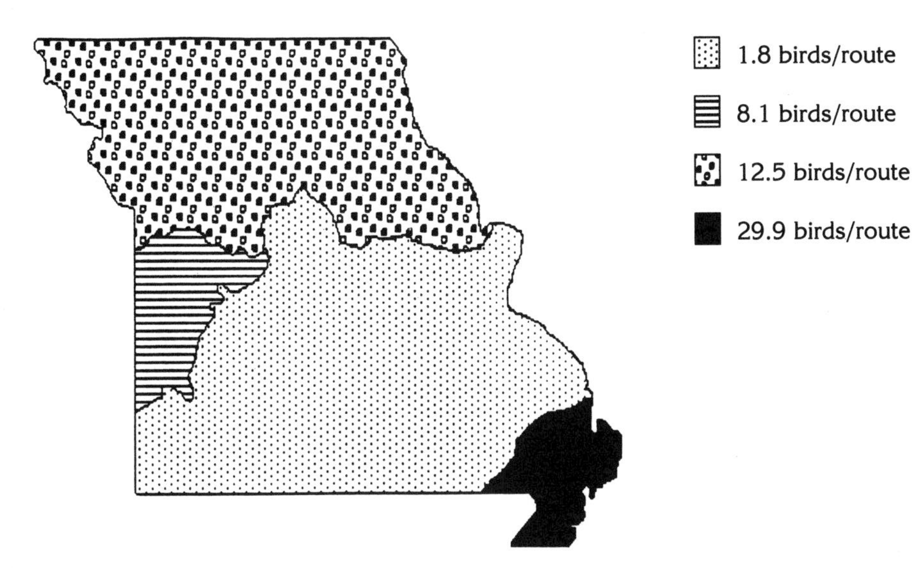

Family Alaudidae: Larks

Horned Lark (*Eremophila alpestris*)

Status: Common permanent resident.
Documentation: Specimen: male, 3 Apr 1968, Maryville (NWMSU, RAR 286).
Habitat: Grazed pasture, roadsides, plowed, and cultivated fields.
Records:
Spring Migration: The largest concentrations for this period are observed in late Feb and early Mar. Most migrants have passed through the state by mid-Mar.
Summer: This species is abundant in the Mississippi Lowlands, and as one would predict, it is least common in the Ozarks (Map 29).
Fall Migration: Migrants in relatively small groups are seen by mid-Oct. Flock-size increases through Nov, with the largest concentrations observed at the end of Nov and early Dec.
Winter: The largest early winter concentrations are found in the cleared Mississippi Lowlands, and as in summer, it is least common in the forested Ozarks (Map 30). High counts: 1,609, 18 Dec 1983, Orchard Farm CBC; 1,300, 27 Dec 1980, Big Oak Tree SP CBC.

Map 30.
Early winter relative abundance of the Horned Lark among the natural divisions. Based on CBC data expressed as birds/pa hr.

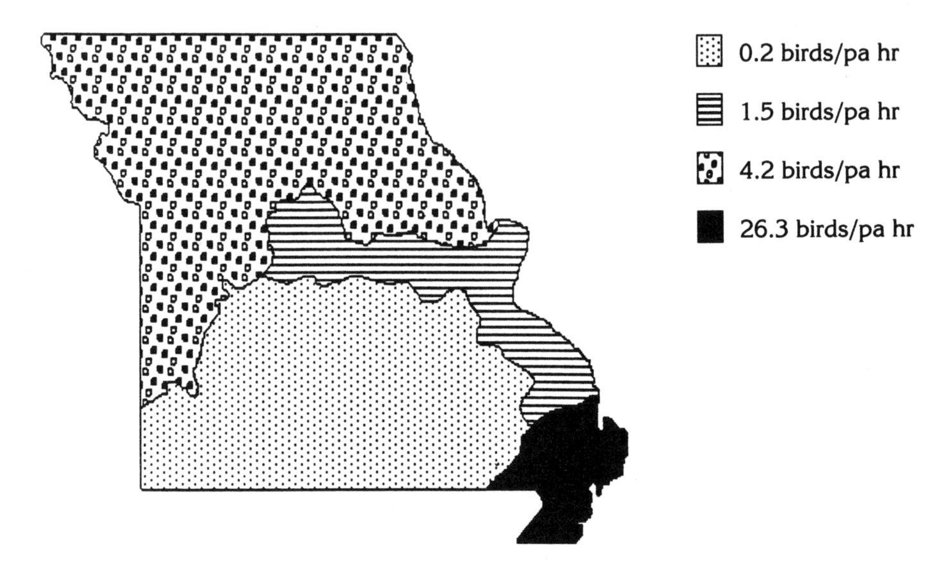

Comments: *Eremophila a. praticola* is the summer resident and the predominant bird in winter. The northeastern subspecies, *E. a. alpestris,* is a rare transient and winter resident, although it probably is more common than the records indicate, since few observers have made racial determinations of wintering larks. Earliest fall records for *alpestris:* 3, 20 Oct 1989 (male, taken on 23 Oct; ANSP 181884), St. Joseph (DAE, MBR-**BB** 57[1]:49); 1, 19 Nov 1961, Callaway Co. (WG, DAE). Latest spring date: 1, 4 Apr 1896, St. Joseph (S. Wilson; Widmann 1907). Specimen confirmation is needed for the occurrence of other subspecies.

Family Hirundinidae: Swallows

Purple Martin (*Progne subis*)

Status: Locally common summer resident.
Documentation: Specimen: female, 21 Aug 1967, Cole Co. (KU 59007).
Habitat: Now largely confined to artificial nest houses in residential areas. During migration usually associated with water.
Records:
Spring Migration: A few males may arrive as early as late Feb in the south in years when the late winter is mild. But in more average years, birds

Map 31.
Relative breeding abundance of the Purple Martin among the natural divisions. Based on BBS data expressed as birds/route.

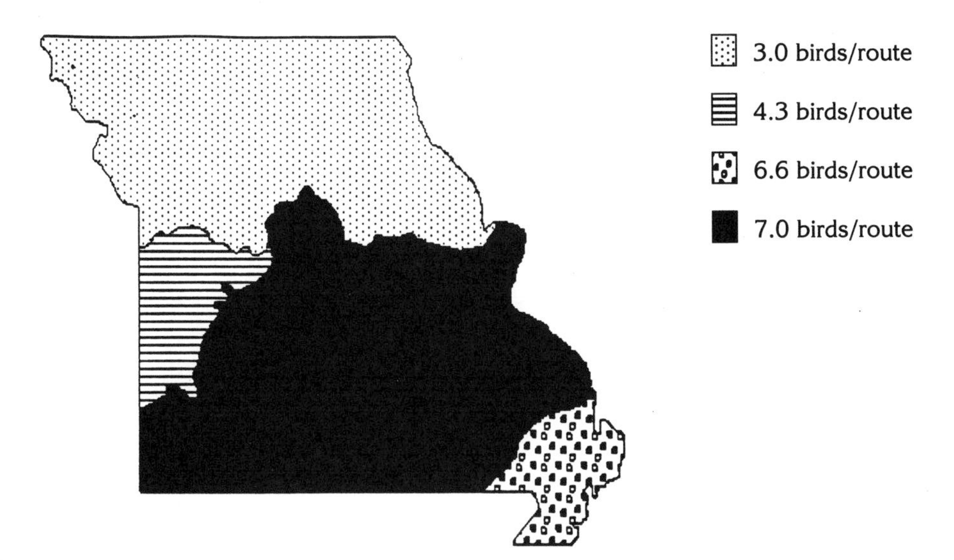

do not appear until early to mid-Mar. The bulk of the birds arrive during the first half of Apr. Earliest dates: 1, killed by cat, 11 Feb 1937, north of Springfield (P. Smith-**BB** 4[4]:43); 1, 19 Feb 1983, St. Louis (G. Dreyer-**NN** 55:36); 3, 22 Feb 1986, near Forsyth, Taney Co. (PM-**BB** 53[2]:9).

Summer: BBS data reveal that the martin is most common in the Ozarks and Mississippi Lowlands and least common in the Glaciated Plains (Map 31). By early July huge concentrations, numbering in the thousands of birds, may be encountered along rivers and lakes. These roosts continue to build in numbers through July. High counts: 30,000, 10 July 1965, roost on Missouri R. at St. Louis (RA-**BB** 32[3]:14); 16,000, 24 July 1963, Busch WA (JC-**BB** 30[3]:22); 5,000–10,000, 3–31 July 1980, Jefferson City (JW-**BB** 47[4]:12); 6,000–7,000, 11 July 1982, Zenith Pond, Springfield (CB-**BB** 49[4]:18).

Fall Migration: The above roosts usually peak in early to mid-Aug, but occasionally spectacular concentrations may remain until the end of Aug. However, by the end of the first week in Sept relatively few birds remain. After mid-Sept only 1–2 birds/day are rarely encountered and it is only casually observed in early Oct. High counts: 500,000, 18 Aug 1981, Springfield (fide CB-**BB** 49[1]:14); the following two observations are at the same site along the Missouri R. at St. Louis: 130,000+, 14 Aug 1965 (RA); 90,000–100,000, 8–10 Aug 1964 (RA-**BB** 31[3]:19). Latest date: 1, 24 Oct 1983, Tyson Valley RC (DC-**NN** 55:91).

Graph 8.
Missouri BBS data from 1967 through 1989 for the Purple Martin.
See p. 23 for data analysis.

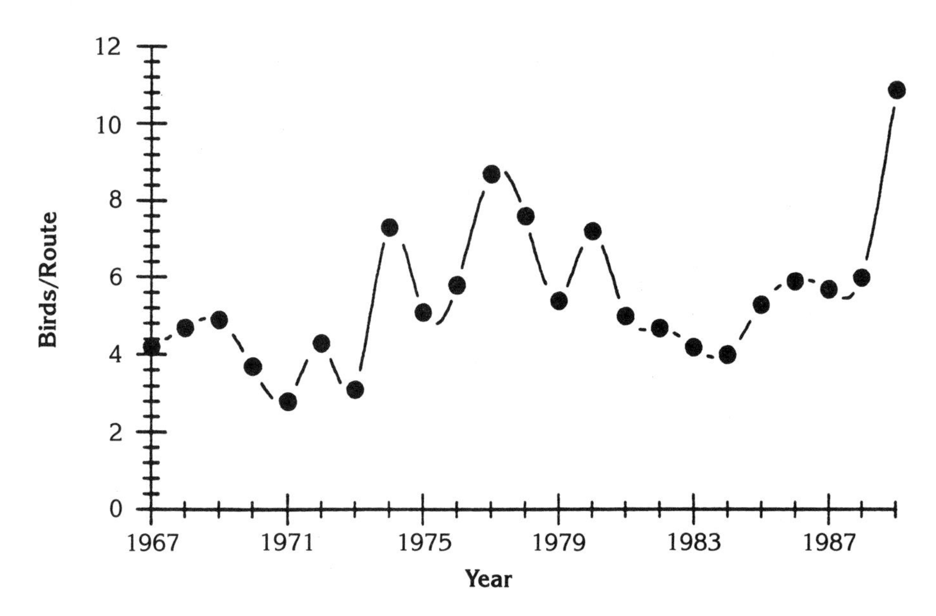

Comments: For fifteen years, the Purple Martin has either been listed on the National Audubon Society's Blue List or their Special Concern List (Tate 1986). The BBS data for Missouri (Graph 8) show that there has not been any major declines in the abundance of this species over the past 23 years. This martin is well known for having great fluctuations in abundance from year to year.

Tree Swallow (*Tachycineta bicolor*)

Status: Common transient; rare to locally uncommon summer resident.

Documentation: Specimen: female, 12 June 1926, Buchanan Co. (KU 39157).

Habitat: Usually associated with water. During breeding season nests in holes of dead tree snags over water.

Records:

Spring Migration: Like the Purple Martin, a few individuals may be seen as early as late Feb during years of mild late winter and "early" spring weather. However, during average years it is not readily encountered in the south until late Mar. The largest numbers are seen during the last ten days of Apr and the first week of May. Earliest dates: 1, 10 Feb 1990, Clarence Cannon NWR, Pike Co. (N. Rattenborg-**BB** 57[2]:105); 1, 16 Feb 1975, Little

Dixie L. (L. Falch, fide BG-**BB** 42[3]:14); 1, 23 Feb 1971, Piedmont, Wayne Co. (S. O'Kelly-**BB** 38[4]:10). High count: 1,500, 14 May 1967, Squaw Creek (FL).

Summer: A local, relatively rare to uncommon breeder statewide, being most common in areas where numerous dead trees are standing in water. The largest breeding concentration is at Duck Creek in the southeast. High count (includes nonbreeding birds): 150+, 29 July 1978, Squaw Creek (MBR, TB-**BB** 45[4]:13).

Fall Migration: Migrants begin appearing in the low hundreds in early Aug, but their numbers do not increase dramatically until mid-Sept. From mid-Sept through mid-Oct (especially late Sept and early Oct) huge concentrations, numbering in the thousands, may be encountered. At the end of Oct numbers drastically drop off, and only a few birds are seen during early Nov. High counts: both at Squaw Creek by FL; 5,000, 12 Sept 1965; 5,000, 21 Oct 1978. Latest dates: 1, 24 Nov 1972, Squaw Creek (SP-**BB** 40[1]:8); 1–3, 13 Nov 1983, Farmington (BL-**BB** 51[1]:43).

[**Violet-green Swallow** (*Tachycineta thalassina*)]

Status: Hypothetical; spring and fall transient.

Documentation: Sight records only: see below.

Comments: There is only a single record with written documentation: three birds were observed on 7 Oct 1974, at Lewis and Clark SP, Platte Co. (M. Myers-**BB** 42[2]:6). Although this report was accompanied by convincing details, only a single observer was involved and the birds could not be located later that same day. There is an additional possible sighting: 3, 9–11 Apr 1966, LaPlata, Macon Co. (R. Luker).

Northern Rough-winged Swallow (*Stelgidopteryx serripennis*)

Status: Common summer resident.

Documentation: Specimen: male, 28 May 1938, Ashland, Boone Co. (MU 429).

Habitat: Nests primarily in banks along watercourses and roadside cuts.

Records:

Spring Migration: First appears during the last ten days of Mar in the south and the first week of Apr in the north. By the end of the third week of Apr it is common statewide. Large congregations, like those seen in the fall, have not been reported during this season. Earliest dates: 10 Mar 1902, Festus, Jefferson Co. (Widmann 1907); 2, 14 Mar 1987, Taney Co. (PM-**BB** 55[1]:14); 14 Mar 1985, Eureka, St. Louis (DC-**BB** 57[5]:38).

Summer: It probably has increased since settlement, primarily as a result of roadside cuts that have provided an increase in nesting sites. On the other hand, the channelization of the larger rivers may have eliminated many sites.

Fall Migration: Birds begin congregating, numbering in the high hundreds or low thousands, over large pools of water and along the larger rivers in Aug, especially during the latter part of the month. The largest concentrations are present during the last two weeks of Sept when groups composed of several thousand birds are seen. Numbers are much reduced after the end of Sept, but it is still readily encountered through mid-Oct. There are no Nov observations. High counts, both at St. Joseph by FL: 7,500, 21 Sept 1985; 6,000, 21 Sept 1980 (**BB** 48[1]:9). Latest dates: a single bird was watched for about 5 min at very close range, 1 Dec 1963, Browning L. (FL-**BB** 31[1]:33); 1, 25 Oct 1980, St. Joseph (FL).

Bank Swallow (*Riparia riparia*)

Status: Common transient; locally common summer resident north of and including the Missouri R. and along Mississippi R. drainage; rare and even more local in the Ozarks and the Osage Plains.

Documentation: Specimen: male, 26 May 1968, near Bigelow (NWMSU, DAE 1921).

Habitat: Nests in similar situations as the preceding species.

Records:

Spring Migration: Arrives later than the N. Rough-winged Swallow, with birds reaching southern Missouri at the beginning of the second week in Apr and about a week later in the north. Peak is early to mid-May. Earliest dates: 1, 31 Mar 1990, L. Taneycomo, Taney Co. (PM, JH-**BB** 57[3]:139); 6, 4 Apr 1981, L. Contrary (FL). High counts, both at Squaw Creek, by FL: 2,500, 15 May 1966; 2,000, 14 May 1967.

Summer: A local colonial breeder along eroded banks of almost all rivers north of and including the Missouri R. In addition, occurs locally along the entire Mississippi R. drainage. Colonies are much more spotty in the Ozarks and Osage Plains. Like the preceding species, it probably has increased as a breeder since presettlement, as a result of an increase in nesting sites, especially roadside cuts. Nesting colonies in south (away from Mississippi R. valley): 500+ birds, 17 May 1980, Bonne Terre Marsh, St. Francois Co. (BRE); 350 nests, 22 June 1982, Willard, Greene Co. (J. Horton-**BB** 49[4]:18). High count (possibly includes migrants): 2,000+, 29 July 1973, Squaw Creek (MBR).

Fall Migration: The autumn buildup in numbers not only occurs earlier than that of the other swallows (except the martin), but it is composed of lesser numbers than the other species. Peak occurs from mid-Aug through early Sept. Very few birds are seen after the third week of Sept. High counts: 1,000, 11 Aug 1946, Sugar L. (H. Hedges-**BB** 13[11]:2); 1,000, 1 Sept 1963, Squaw Creek (FL). Latest date: 5, 7 Oct 1974, St. Joseph (FL).

Cliff Swallow (*Hirundo pyrrhonota*)

Status: Locally common summer resident.

Documentation: Specimen: female, 3 June 1970, near Rock Port, Atchison Co. (NWMSU, DAE 2488).

Habitat: Most often seen near water. Breeds on cliffs, bridges, buildings, and dams.

Records:

Spring Migration: Normally the first arrivals are seen during the second week of Apr. However, it does not appear in numbers until the last few days of Apr and early May. Earliest dates: 16 Mar 1963, Rolla (F. Frame-**BB** 30[2]:26); 1, 27 Mar 1976, Little Dixie L. (BG et al.-**BB** 44[1]:27). High count: 500, 10 May 1981, Callaway Co. (RW-**AB** 35:830).

Summer: Local but widespread throughout the state. Undoubtedly has increased significantly since presettlement as a result of an increase in manmade structures for nesting. For example, of the 50 colonies reported in a 1984 public survey by the MDC, the distribution of nest sites was as follows: 33 on bridges, 9 on buildings, 6 on natural sites, and 2 on dams (JW-**BB** 51[4]:16). The largest colony reported was in Benton Co. and contained nearly 400 nests. By the end of July large concentrations are noted at large water impoundments. High count: 2,500, 19 July 1964, St. Joseph (FL).

Fall Migration: The largest congregations (numbering in the thousands) are seen during Aug. Relatively large groups may be seen through the first half of Sept, but by the end of the month almost all have left. It is casual during the first week of Oct. High counts: 5,000+, 2 Aug 1964, St. Joseph/Squaw Creek (FL); 5,000, 10 Aug 1946, Sugar L. (E. Newton-**BB** 13[11]:2); 5,000, 1 Sept 1968, Squaw Creek (FL-**BB** 36[4]:12). Latest dates: 1, 19 Oct 1954, St. Joseph (FL); 1, 7 Oct 1987, Horseshoe L., Buchanan Co. (MBR-**BB** 55[1]:12).

Barn Swallow (*Hirundo rustica*)

Status: Common summer resident; accidental winter visitant.

Documentation: Specimen: female, 17 June 1932, Jackson Co. (KU 19370).

Habitat: Prefers open areas, breeding primarily in buildings, bridges, etc.

Records:

Spring Migration: The first migrants are seen at the end of Mar. Numbers gradually increase through Apr, with peak in early to mid-May. Earliest date: 1, 1 Mar 1981, Ted Shanks (BG, SS-**BB** 48[3]:19). High count: 5,000, 14 May 1967, Squaw Creek (FL).

Summer: Our most common and most uniformly distributed swallow.

Map 32.
Relative breeding abundance of the Barn Swallow among the natural divisions. Based on BBS data expressed as birds/route.

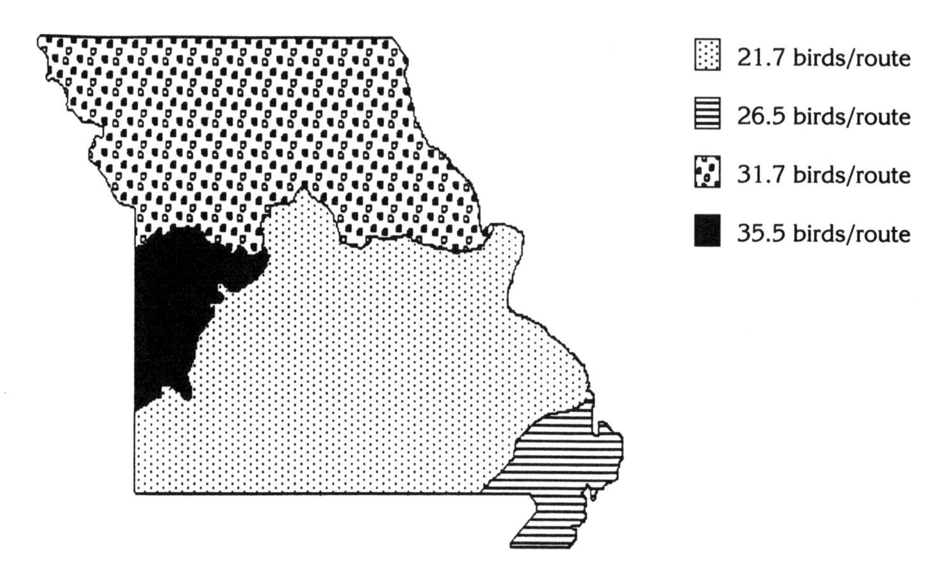

Like the other swallows it has benefited from manmade structures. BBS data show that this species is most abundant in the Osage and Glaciated plains, with the fewest found in the Ozarks (Map 32).

Fall Migration: Although relatively large concentrations (in the hundreds or low thousands) may be encountered at the end of Aug and early Sept, the most impressive gatherings are seen between the latter half of Sept and early Oct. By the end of Oct, normally only small-sized groups (< 10 birds) or solitary birds are seen. It is casual during the first week of Nov. High counts: 10,000, 4 Oct 1978, with 5,000+ still present on 21 Oct, Squaw Creek (MBR, FL-**BB** 46[1]:26); 6,000+, 1 Oct 1990, Squaw Creek (MBR). Latest dates: 1, 14 Nov 1985, Maryville SL (DAE); 1, 13 Nov 1983, Farmington (SD-**BB** 51[1]:43).

Winter: Only one record; 1, 17 Dec 1988, Popular Bluff CBC (V. Moss).

Family Corvidae: Jays, Magpies, and Crows

[**Gray Jay** (*Perisoreus canadensis*)]

Status: Hypothetical.

Documentation: Sight record only: see below.

Comments: There is a single sighting of a bird feeding at a deer carcass near Craig, Holt Co., on 21 Nov 1985 (T. Bell, M. Hoover-**BB** 53[1]:28).

Unfortunately, no photographs were taken nor was the description detailed enough to ascertain subspecific identification, thus precluding determination of the origin of the bird, i.e, whether it belonged to the boreal Canadian population (*P. c. canadensis*) or to the Rocky Mountain race (*P. c.* capitalis). Further clouding the validity of this record is the fact that no other extralimital movements of either *canadensis* or *capitalis* were recorded in the fall and winter of 1985–86. It is possible that the bird was an escapee.

Blue Jay (*Cyanocitta cristata*)

Status: Common permanent resident.

Documentation: Specimen: female, 3 Sept 1933, Jasper Co. (KU 20216).

Habitat: Found in a wide range of habitats with trees.

Records:

Spring Migration: Flocks of "treetop" migrants are most conspicuous from mid-Apr into the third week of May. Peak is during the last week of Apr and the first week of May. High count: "several hundred migrating," 26 Apr 1961, Table Rock L. area (FL); 150, 26 Apr 1964, St. Joseph area (FL).

Summer: The Ozarks is the region where this species is most abundant, and the Mississippi Lowlands is the area where it is least common (Map 33). Within the Glaciated Plains, it is decidedly more common in the eastern half.

This species has been very successful in adapting to man's surroundings (parks, residential areas, etc.). Branan and Burdick (1981) noted a 273% increase of this species at Swope Park between 1916 (n=73 birds; A. Shirling 1920) and 1973 (n=199 birds).

Fall Migration: Southbound movement occurs as early as late Aug: 38 (29 in one string), 30 Aug 1984, St. Joseph (FL). There is a gradual increase in numbers, which peaks at the end of Sept and early Oct, when large flocks (in the hundreds) are noted throughout the state. Numbers gradually decrease until the end of Oct, and migration appears to cease, or is at least very inconspicuous. High counts: 1000+, 30 Sept 1978, L. of Ozarks SP (m.ob.-**BB** 46[1]:26).

Winter: CBC data reveal that it is most common in the Ozarks and Ozark Border (average of 4.0 birds/pa hr) and least common in the Mississippi Lowlands (average of 1.8 birds/pa hr). Smith (1986) showed that winter abundance positively correlates with mast availability. It appears that this species has increased in abundance since presettlement as a result of man's activities, e.g., providing winter food at feeding stations, planting of trees (both native and ornamental), etc. High counts: 831, 27 Dec 1987, Columbia CBC; 678, 3 Jan 1988, Dallas Co. CBC.

Comments: A highly migratory species. Undoubtedly, many of the

Map 33.
Relative breeding abundance of the Blue Jay among the natural divisions. Based on BBS data expressed as birds/route.

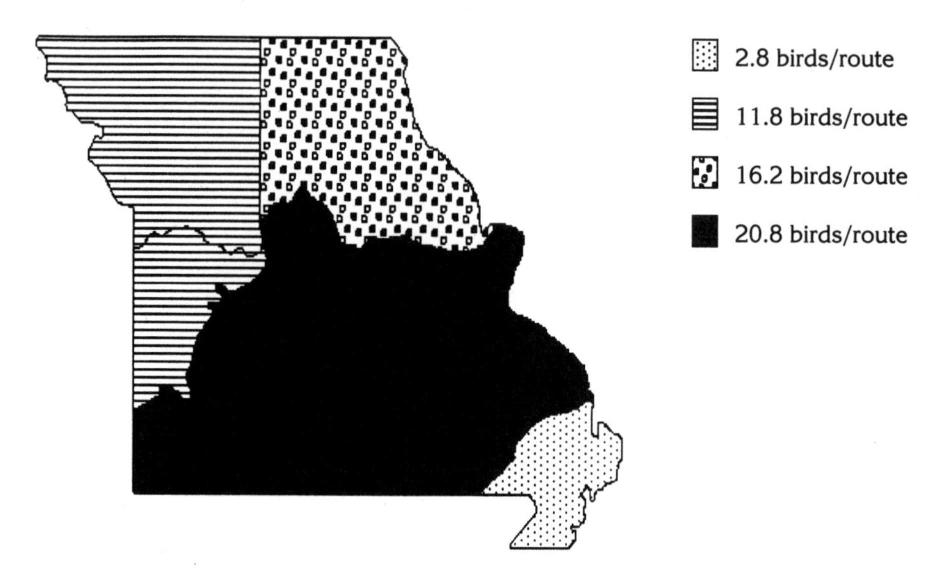

state's nesting birds are replaced or at least supplemented in winter by birds from the north.

Clark's Nutcracker (*Nucifraga columbiana*)

Status: Accidental fall transient and winter resident.

Documentation: Photograph: 3, 5 Nov–late Dec 1972, North Kansas City, Clay Co. (W. Rodgers et al.-**BB** 40[1]:9; Fig. 24).

Habitat: Usually seen at bird feeders; but should be looked for in stands of coniferous trees.

Records:

Fall Migration: The above record was part of a major movement of birds to the Great Plains from the Rocky Mountains during the fall of 1972 (Williams 1973; Kleen and Bush 1973). There are two additional records: 1, shot, 12 Oct 1907, near Louisiana, Pike Co. (Widmann 1908); 1, shot, 28 Oct 1894, Kansas City (Bryant 1895). The deposition of these specimens is unknown.

Winter: Only two records, besides the birds that lingered in 1972: 1, 14 Feb 1962, near Nevada, Vernon Co. (T. Pucci-**BB** 29[1]:20); 1, at suet feeder, 10 Jan–31 Mar 1969, Pleasant Hill, Cass Co. (H. Williams, L. Warner-**BB** 38[1]:2).

Fig. 24. This Clark's Nutcracker is one of three birds that spent November and December of 1972 at a feeder in North Kansas City, Clay Co. During that year there was a major movement of nutcrackers across the Great Plains from the Rocky Mountains. The photograph was taken on 6 November by David Easterla.

Black-billed Magpie (*Pica pica*)

Status: Accidental fall transient and winter resident; former accidental summer resident.

Documentation: Specimen: female, 5 Dec 1925, Jackson Co. (KU 39167).

Habitat: Open country with scattered trees, especially along watercourses.

Records:

Summer: After the extraordinary winter invasion of 1921–22, at least two pairs bred within 3 miles of Corning, Holt Co., in the summer of 1922 (C. Dankers-**BL** 24:353).

Fall Migration: There are three additional, old records, besides the one above: 4, 1 Nov 1890, Saline Co. (L. Corder; Widmann 1907); 1, 12 Nov 1927 (J. Peeler; Bennitt 1932); 1, taken in the late fall of 1913 in Holt Co. (Harris 1919b).

Winter: Formerly more frequently reported. The largest number reported was the 50+ that were recorded in the winter of 1921–22 in Holt Co. This was part of a major invasion of this species in the upper midwest during the fall of 1921 (Dinsmore et al. 1984). Other old records are as follows: 1, killed in trap, early Jan 1937, Lewis Co., and another shot within two days of the preceding bird at LaBelle, Lewis Co. (Musselman 1937); 1, Jan 1938, Fairfax, Atchison Co. (A. White-**BB** 5[3]:30). There is another unpublished record with no specific date. A bird was seen near Cosby, Andrew Co., along the Platte R. in the 1930s (G. Kephart; fide FL,DAE).

Only two modern observations that may have been escapees: 1, 26 Dec 1980, Gray Summit, Franklin Co. (L. Brenner-**NN** 53:15); 1, 17 Jan 1975, Busch WA (K. Boldt-**NN** 47:22).

Comments: Recent observations (1985–89; n=3) in the southwestern corner of the state were apparently the result of deliberate release(s) of birds in that area (PM-**BB** 56[3]:91).

American Crow (*Corvus brachyrhynchos*)

Status: Common permanent resident.

Documentation: Specimen: female, 3 Dec 1967, 10 miles west of Maryville (NWMSU, LCW 769).

Habitat: Seen in a wide range of habitats when foraging, but nests and roosts primarily in open forest and woodland.

Records:

Spring Migration: The large winter roosts have mostly dissipated by the beginning of Mar. Migrating flocks are seen through Mar, with mainly pairs or very small groups seen thereafter. High count: 600, 25 Feb 1977, St. Joseph (FL).

Summer: This species is significantly more common in the Ozarks than in any other region. It is least common in the western half of the Glaciated Plains and the Mississippi Lowlands (Map 34).

Fall Migration: Obvious migrant groups are seen in Sept, with an increase in flock number and size during Oct. By Nov, impressive roosts may be seen (primarily along the larger rivers). High count: 5,000, 25 Nov 1965, St. Joseph (FL).

Winter: This corvid is most common in the Mississippi Lowlands (average 235.6 birds/pa hr) and least common in the Ozark Border (average 2.5 birds/pa hr). There are several traditional sites where thousands of crows roost annually, e.g., along the Missouri R. at St. Joseph, along the Mississippi R. at several sites, and Mingo. Widmann (1907) states that "hundreds of thousands" congregated along the Mississippi R. at St. Louis (principally during Nov and Dec) when the city dumped refuse into the river. High counts: 13,361, 20 Dec 1980, St. Joseph CBC; 11,735, 27 Dec 1969, Big Oak Tree SP CBC; 10,045, 27 Dec 1969, Kansas City CBC.

Fish Crow (*Corvus ossifragus*)

Status: Rare permanent resident in Mississippi Lowlands; rare summer resident along Mississippi R. possibly as far north as Pike Co.

Map 34.
Relative breeding abundance of the American Crow among the natural divisions. Based on BBS data expressed as birds/route.

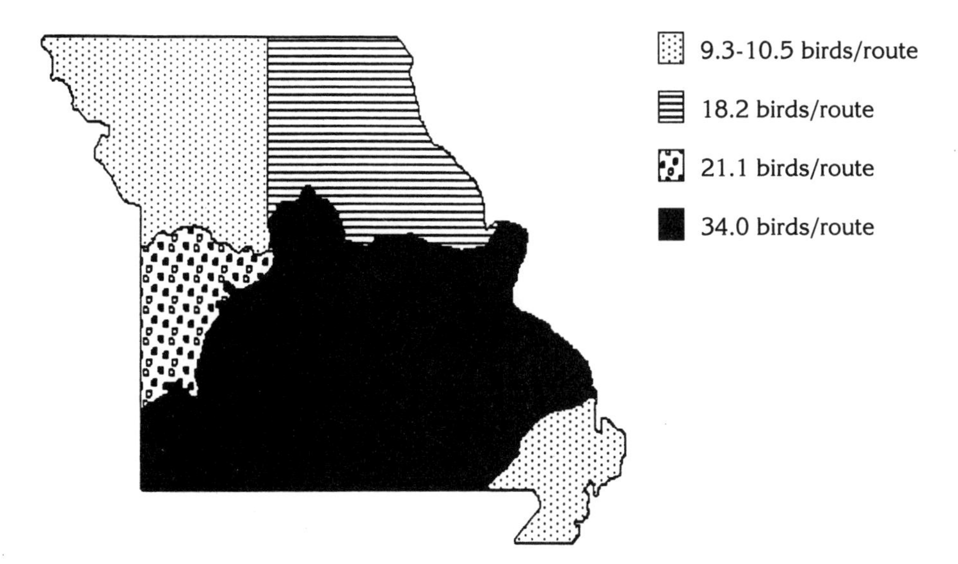

Documentation: Specimen: male, 9 June 1965, Big Oak Tree SP area (Easterla 1965a; NWMSU, DAE 902).

Habitat: Principally found in riparian areas, but also seen foraging in cultivated fields.

Records:

Spring Migration: There is an influx of birds into the state during Mar and early Apr. In early Apr of 1987 it was found at Ted Shanks, extending the known range ca. 100 km farther north (KM-**BB** 54[3]:14). Prior to this discovery it was known only as far north as the St. Louis area, where it now appears to be increasing. Surprisingly, it has not been found along the Missouri R. (away from the St. Louis area), nor along the rivers in the extreme south central section of the state that drain into Arkansas. The farthest west it has been recorded is L. Wappapelo, Wayne Co. and along the Black R. in the southeastern corner of Ripley Co. on 5 Aug 1989 (LG-**BB** 57[1]:49). Eventually it should appear along Table Rock and Bull Shoals lakes, since it is found all along the White R., in Arkansas (James and Neal 1986).

An unpublished observation of 1, 22 Apr 1954, along the Mississippi R. in Pemiscot Co. (BK) predates the earliest published report (Easterla 1965a) by ten years. High counts: 34, 16 Mar 1969, Mississippi Co. (RAR-**BB** 38[1]:3); 25, 10 Apr 1983, St. Louis (DJ-**BB** 50[3]:13).

Summer: This crow apparently breeds throughout the Mississippi Lowlands. In addition, breeding occurs along the Mississippi R. to at least

St. Louis and possibly as far north as Pike Co.

Fall Migration: Virtually no information exists on its movements during this period.

Winter: With the exception of one record, all observations for winter come from the Mississippi Lowlands. Northernmost record: 1, 30 Dec 1977, St. Charles Co. (K. Boldt-**NN** 50:8). High count: 4, 30 Dec 1972, Big Oak Tree SP CBC.

Common Raven (*Corvus corax*)

Status: EXTIRPATED; former permanent resident along primarily rivers.

Documentation: Specimen: clutch of 5 eggs collected, which cannot be located now, 5 Apr 1901, Hahatonka, Camden Co. (P. Smith; Widmann 1907).

Habitat: Primarily found along river cliffs.

Comments: Apparently this species was uncommon in Missouri until the 1870s, but it disappeared from most of the state shortly thereafter (Widmann 1907; Harris 1919b). The above record is the last nesting report for Missouri. Smith (in Widmann 1907) stated that about 6 pairs were in the vicinity, and he also mentioned a colony that nested "a few years earlier" along the Gasconade R., near Vienna. There are no substantiated records of this species since that time.

Although the raven may never breed in the state again, it may reappear as an accidental transient or winter visitant, since it has been recorded relatively recently during the fall and the winter in Iowa (Dinsmore et al. 1984). All raven observations should include details on behavior and morphology (e.g., details on tail wear) that might indicate that the bird in question was an escapee.

Family Paridae: Titmice

Black-capped Chickadee (*Parus atricapillus*)

Status: Common permanent resident in Glaciated and Osage plains; accidental in winter in the Ozarks and Mississippi Lowlands.

Documentation: Specimen: male, 14 Apr 1979, 11 miles north of Maryville (LSUMNS 99735).

Habitat: Found in virtually every habitat with trees, but most common in riparian areas.

Records:

Summer: Hybridization occurs at the range interface of this species and the Carolina Chickadee (Robbins et al. 1986b; Braun and Robbins 1986). The approximate location of the contact zone is depicted on Map 35. Fur-

Map 35.
Approximate zone of contact between the Black-capped (above shaded area) and Carolina (below shaded area) chickadees.

ther fieldwork is needed to clarify the precise location of the zone east of Henry Co. The relative abundance of the Black-capped is fairly uniform throughout the Glaciated Plains and the northern Osage Plains (ca. 5.0 birds/route); however, this figure should be considered conservative, as this species vocalizes very infrequently and is generally inconspicuous at the time of year the BBS routes are conducted.

Fall Migration and Winter: It makes irregular movements in fall and winter south of the breeding range shown in Map 35. Birds that exhibited Black-capped morphology have been recorded well south of the breeding range: 2, at a feeder, 29 Nov 1987, Farmington (SD-**BB** 55[1]:12); 2, 31 Dec 1973, Big Oak Tree SP CBC; 3 Feb 1973, near Springfield (NF et al.-**AB** 27:624). An average of 3.7 birds/pa hr has been recorded on Glaciated Plains CBCs. High counts, both on the Kansas City Southeast CBC: 845, 20 Dec 1980; 819, 19 Dec 1981.

Comments: Vocalizations are not reliable indicators of a particular bird's morphology because some individuals sing both Black-capped and Carolina song types (a characteristic that is apparently learned).

A mist-netted bird on 5 Aug 1971, at Brickyard Hill WA, originally purported to be a Boreal Chickadee, proved to be an abnormally pigmented Black-capped Chickadee (KU 67051).

Carolina Chickadee (*Parus carolinensis*)

Status: Common permanent resident in Ozarks, Ozark Border, and Mississippi Lowlands.

Documentation: Specimen: male, 10 Mar 1975, Goodman Lookout Tower, Newton Co. (LSUMNS 99778).

Habitat: Same as Black-capped Chickadee.

Records:

Summer: See Map 35 for distribution of this species in Missouri. An average of ca. 5.0 birds/route is recorded on Ozark BBS routes, whereas only 1.2 birds/route is found in the Mississippi Lowlands.

Winter: Not known to make fall or winter movements like the Black-capped. An average of 3.6 and 2.2 birds/pa hr have been recorded on Mississippi Lowland and Ozark CBCs, respectively. High counts: 364, 1 Jan 1990, Springfield CBC; 282, 2 Jan 1988, Taney Co. CBC.

Comments: See Black-capped Chickadee.

Tufted Titmouse (*Parus bicolor*)

Status: Common permanent resident statewide.

Documentation: Specimen: male, 19 Apr 1973, Maryville (NWMSU, DS 1).

Habitat: Common in forest and woodland. Less common in open areas with scattered trees.

Records:

Summer: This parid is most common in the Ozarks and least abundant in the Mississippi Lowlands (Map 36). Although the overall mean for the Glaciated Plains is 5.7 birds/route, the titmouse is slightly less common in the western than the eastern half of this region (west, 5.3; east, 6.2 birds/route). If data were available, it would likely depict a similar pattern within the Osage Plains.

Winter: It is least common in the Glaciated Plains (average < 1.0 birds/pa hr), especially in the northwestern corner. It is most common in the Ozark Border and Ozarks where an average of 2.0 birds/pa hr are recorded on CBCs. The average number found on CBCs in the Mississippi Lowlands is similar to that of the Glaciated Plains, but during Widmann's time, prior to the deforestation of the southeast, he remarked that it was "much more numerous [in the] southeast than [the] northwest." High counts: 339, 27 Dec 1987, Columbia CBC; 305, 27 Dec 1986, Taney Co. CBC.

Map 36.
Relative breeding abundance of the Tufted Titmouse among the natural divisions. Based on BBS data expressed as birds/route.

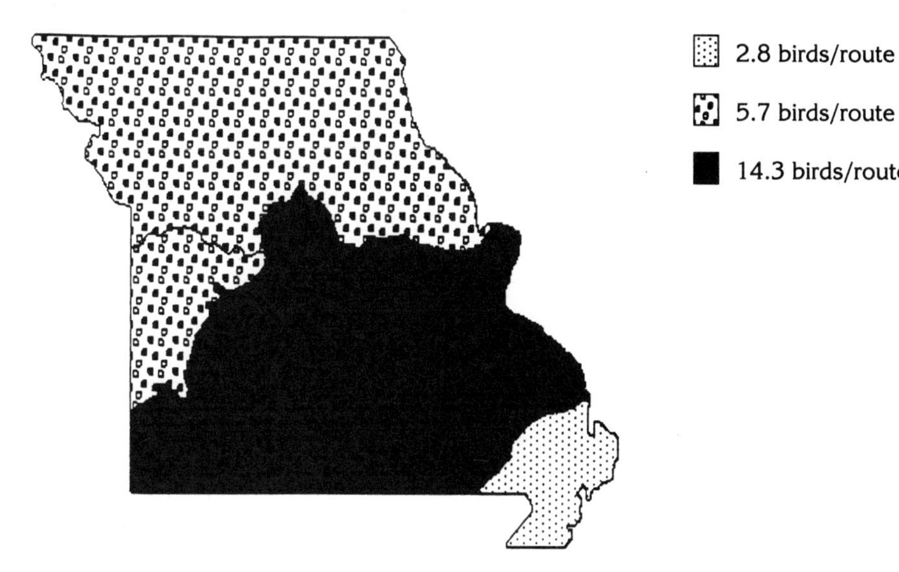

Family Sittidae: Nuthatches

Red-breasted Nuthatch (*Sitta canadensis*)

Status: Irregular, uncommon winter resident; accidental summer resident and summer visitant.

Documentation: Specimen: male, 24 Oct 1965, 10 miles west of Maryville (NWMSU, DAE 1028).

Habitat: Mostly seen in coniferous trees, but seen regularly in deciduous woods especially during migration.

Records:

Spring Migration: There is a definite movement of birds through the state during the end of Apr through mid-May, when birds are commonly seen in purely deciduous woods. This movement is especially pronounced following falls and winters when there has been a major influx of birds into the state. High count for Apr–May: 4–5, 23 Apr 1987, Carter Co. (FL, LG-**BB** 54[3]:14). Latest date: 21 May 1907, St. Louis (Widmann 1907).

Summer: A single nesting record: 2 young fed by adults, 2–12 June 1955, Kansas City (Schoen 1955). There are two additional observations involving single birds: 1, 8 June 1983, St. Louis (MP-**BB** 50[4]:34); 1, 21–22 June 1958, St. Joseph (BB-**BB** 25[8]:4).

Fall Migration: During most years this species is not detected until the second or third week of Sept, with a gradual increase in sightings and numbers as the fall progresses. The arrival of birds in Aug or early Sept portends a major influx of birds later in the fall and winter. The invasion during the fall of 1968 was one of the most spectacular recorded. The first individual was noted on the extraordinarily early date of 23 Aug 1968 in St. Louis (RA-**BB** 36[4]:12), and 4 were noted in Maryville on 31 Aug (MBR). Some of the largest birds/day totals were recorded during the invasion years of 1963 and 1975, e.g., 20, 19 Oct 1963, Kansas City (DAE-**BB** 30[4]:19); 20+, 1 Dec 1963, Missouri Botanical Arboretum (E. Hath-**NN** 35:77); 60+, 6 Nov 1975, St. Louis area (fide JEC-**BB** 44[1]:22).

Winter: Numbers vary considerably from year to year, with an apparent correlation of massive southward movements with low pine cone crops in the northern boreal forest (Bock and Lepthien 1972). Although it can be locally common in any region of the state during some winters, it is perhaps most common and more consistently seen in the Shortleaf Pine region of the state (see Map 3), e.g., 17, 16 Dec 1986, in Carter, Oregon, Ripley, and Shannon counties (MBR-**BB** 54[2]:14). High counts, both on the Kansas City Southeast CBC; 58, 19 Dec 1981; 21, 20 Dec 1980. Table 7 gives years of invasions since 1960.

Table 7. Winters of major movements of Red-breasted Nuthatches into Missouri since 1960–61.

1961–62	1968–69	1980–81	1983–84	1989–90
1963–64	1969–70	1981–82	1985–86	
1965–66	1975–76	1982–83	1986–87	

White-breasted Nuthatch (*Sitta carolinensis*)

Status: Common permanent resident.

Documentation: Specimen: male, 24 Aug 1920, Taney Co. (KU 39193).

Habitat: Most common in forest and woodland, but found in relatively sparse woodland.

Records:

Summer: This nuthatch is most common in the Ozarks and least abundant in the Mississippi Lowlands (Map 37). In addition, it is less common in the extreme western portions of the Glaciated and Osage plains than it is in the eastern sections of these areas.

Winter: Most common in Ozarks and Ozark Border where an average of 6.0–7.0 birds/pa hr are recorded. An average of 3.0–4.0 are observed/pa hr

Map 37.
Relative breeding abundance of the White-breasted Nuthatch among the natural divisions. Based on BBS data expressed as birds/route.

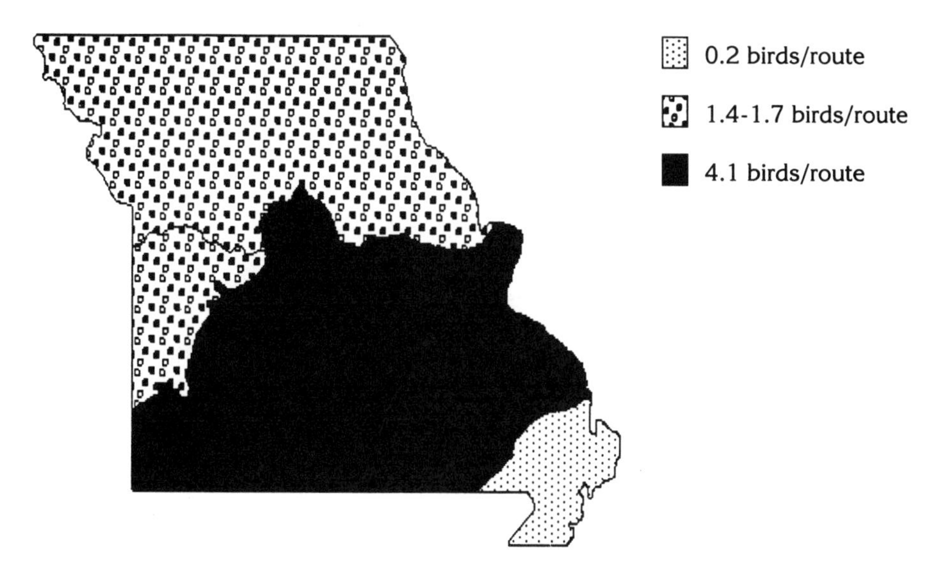

in the remainder of the state. High counts: 115, 2 Jan 1988, Taney Co. CBC; 113, 19 Dec 1982, Columbia CBC.

Brown-headed Nuthatch (*Sitta pusilla*)

Status: EXTIRPATED; Former accidental transient and summer resident.

Documentation: Specimen: pair, 19 Mar 1907, near Ink, Shannon Co. (Woodruff 1908; AMNH 230089–230090).

Habitat: Formerly in mature stands of Shortleaf Pine.

Comments: The past status of this species is perhaps the most perplexing of any bird recorded in the state. Oddly, there are only two records, and one of these involved the pair mentioned above, and the other was of a single bird in St. Louis on 6 May 1878 (Widmann 1907). Presumably this species was more common and widespread (although always local?) in the Shortleaf Pine region prior to the devastation of this habitat by the lumber industry between the 1880s and early 1900s. Perhaps the bird observed by Widmann at St. Louis was indicative of a greater nuthatch population in the pine area of the state at that time. It seems odd that the Red-cockaded Woodpecker, a species with more specific pine habitat requirements than the nuthatch, persisted until the mid–1940s and the nuthatch did not. In fact, seemingly suitable nuthatch habitat is available even today.

Family Certhiidae: Creepers

Brown Creeper (*Certhia americana*)

Status: Common transient; uncommon winter resident; casual summer resident in the Mississippi Lowlands.

Documentation: Specimen: male, 18 Dec 1921, Jackson Co. (KU 39201).

Habitat: Forest and woodland during migration and winter; recorded breeding in cypress swamps and river bottoms.

Records:

Spring Migration: Migrants are detected by mid-Mar and peak in late Mar or early Apr. Numbers drop off considerably after mid-Apr, with a few birds seen through the first few days of May. High counts: 3–5/mixed species flock, which consisted of chickadees, titmice, and Golden-crowned Kinglets, 31 Mar–4 Apr 1980, St. Clair and Bates counties (MBR, MB); 6, 21 Apr 1983, St. Louis (RK). Latest migrant date: 1, 27 May 1945, St. Louis (JEC-**BB** 12[8]:47).

Summer: Formerly "a regular inhabitant" of the Mississippi Lowlands in cypress swamps (Widmann 1907). A pair also was found breeding in the Missouri R. floodplain at St. Louis in 1909 (Betts 1909). However, there was a hiatus in records from then until May 1985, when three territorial birds (one nest) were located at Big Oak Tree SP (Robbins 1986). A singing bird was heard at Big L., Mississippi Co., on 18 May 1987 (JH), and another was seen on 6 June 1987 along the Meramec R. near Pacific (E. Bever-**BB** 54[4]:28). The gap in breeding records for the above period is undoubtedly the result of the lack of coverage in the southeastern corner during the breeding season and the difficulty of detecting this species in general. Nevertheless, it certainly is less common and more spotty in distribution than at the turn of the century, since all but a handful of cypress swamps have been drained and cleared. It should be looked for in the Missouri and Mississippi river floodplains, since creepers have been found breeding in floodplain forest in southeastern Nebraska (Cortelyon 1975) and southern Illinois (Bohlen and Zimmerman 1989).

Fall Migration: Usually not detected until the last week of Sept, but relatively large numbers (3–5/day) are seen during the first week of Oct; peak is about a week later. By the end of Oct most migrants have passed through the state. Earliest dates: 1, 1 Sept 1975, Montgomery City, Montgomery Co. (RW-**BB** 44[1]:22); 1, 11 Sept 1970, St. Joseph (FL).

Winter: Most common in the Ozarks, Ozark Border, and floodplain forest of the larger rivers. CBC data are unreliable for distinguishing differences in the relative abundance of this species among natural divisions because of the great variability among observers in detecting it. High counts: 34, 2 Jan 1989, Mingo CBC; 24, 16 Dec 1978, Kansas City Southeast CBC; 24, 22 Dec 1979, Squaw Creek CBC.

Family Troglodytidae: Wrens

Rock Wren (*Salpinctes obsoletus*)

Status: Accidental transient, summer and winter visitant.
Documentation: Specimen: immature male, tower kill, 5 Oct 1972, Maryville (Easterla and Ball 1973; NWMSU, DAE 2691).
Habitat: Rock quarries, rock and log piles.
Records:
Summer: Single record: 1 singing, 16 July 1964, St. Joseph (JHA; Easterla and Ball 1973).
Fall Migration: Only three additional records besides the above: 1, 7 Oct 1984, near Owl's Bend, Shannon Co. (J. Greenberg-**BB** 52[1]:32); 1, 4 Nov 1950, present for several weeks at an abandoned rock house foundation, at Gentle Slopes Resort, south of Gravois Mills, L. of the Ozarks (DAE; Easterla and Ball 1973); 1, 4 Nov–2 Dec 1990, Maryville Water Plant, Nodaway Co. (DAE et al.; Fig. 25).
Winter: A bird seen on 23 Jan 1966 on Bull Shoals L. near Cedar Creek, Taney Co., is the only observation for this period (NF et al.-**BB** 33[1]:15).

Carolina Wren (*Thryothorus ludovicianus*)

Status: Uncommon to common permanent resident in the Mississippi Lowlands, Ozarks, and Ozark Border; rare in the Glaciated and Osage plains.
Documentation: Specimen: male, 29 Oct 1967, near Arkoe, Nodaway Co. (NWMSU, LCW 688).
Habitat: Most common in relatively dense undergrowth of riparian forest, but occurs in upland forest and infrequently in residential areas.
Comments: Status and abundance fluctuate considerably with the severity of winter; for example, this species virtually disappeared from the northern areas of the state, and it was greatly reduced in number in the south after the severe winters of 1960–62. Gradually the bird increased and

Fig. 25. The Rock Wren is an accidental western stray to Missouri. This bird was photographed by David Easterla at the Maryville Water Plant, Nodaway Co., on 4 November 1990. Ironically, exactly 40 years earlier the photographer discovered the first Rock Wren in Missouri near Gravois Mills, Morgan Co.

recolonized the northern areas, until the three consecutive harsh winters of 1976–78, when again the Missouri population was decimated. The crash and the gradual recovery is best documented for the latter cycle (Graph 9).

BBS data exhibit that this bird is most common in the Mississippi Lowlands, Ozarks, and Ozark Border (Map 38). It is much less common in the Glaciated and Osage plains. Within the Glaciated Plains, this species is rarest in the extreme north (ave 0.2 north; 0.9 south) and the west (0.5 west; 0.9 east). High winter counts: 112, 2 Jan 1989, Mingo CBC; 103, 2 Jan 1972, Weldon Spring CBC.

Bewick's Wren (*Thryomanes bewickii*)

Status: Uncommon summer and rare winter resident in the Ozarks and Ozark Border; rare and more local summer resident in Osage Plains; casual summer resident and rare transient in Glaciated Plains.

Documentation: Specimen: male, 26 Apr 1978, Maryville (DAE; NWMSU, MF 106).

Habitat: Most common around abandoned farm equipment, sheds, etc., and in open, brushy areas at the edge of woods.

Records:

Spring Migration: The winter population is supplemented with migrants as early as the beginning of Mar, with the bulk of the birds arriving at the end of the month and in early Apr. The majority of the northern Missouri observations are from mid-Mar through mid-Apr.

Summer: This wren is most common in the Ozarks and Ozark Border. It is a rare summer resident in the extreme southern and eastern sections of the Glaciated and Osage plains, respectively. It is not known to breed in the Mississippi Lowlands.

Apparently the Bewick's Wren was formerly more common and uniform in distribution across the Ozarks and Ozark Border. It was described as being common by Widmann (1907) and "very numerous" in the late 1930s in the Steelville area, Crawford Co. (JEC-**BB** 6[8]:67). A number of observers have implied that the House Wren replaced the Bewick's in several areas. However, the decline of the Bewick's Wren and the increase in the House Wren is probably coincidental and unrelated to one species "out competing" the other for the same resources. The decrease of the Bewick's is probably related to the clearing of brushy habitats, as well as the "cleaning" of areas around farmsteads. Cowbird parasitism may also play a role in this wren's decline. The House Wren readily adapts to human habitation, and it clearly has benefited from the availability of artificial nest sites. Nevertheless, the Bewick's Wren remains the more common bird in nonurban areas of the Ozarks. Over the past 23 years, BBS data indicate that it was most common for the period of 1968–70, with a significant drop in the early 1970s (Graph 10). The population now appears stable.

Graph 9.
Missouri BBS data from 1967 through 1989 for the Carolina Wren. See p. 23 for data analysis.

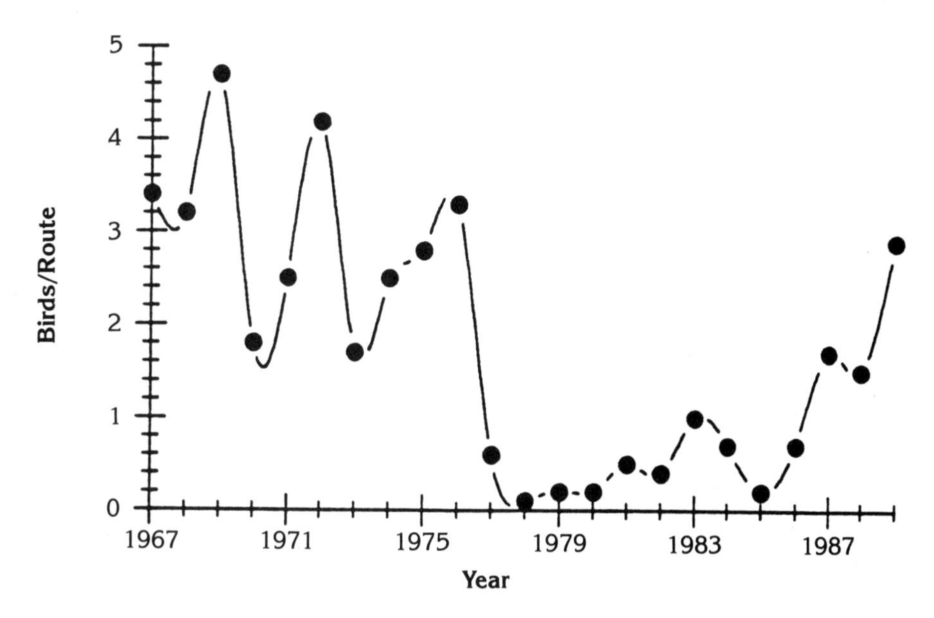

Map 38.
Relative breeding abundance of the Carolina Wren among the natural divisions. Based on BBS data expressed as birds/route.

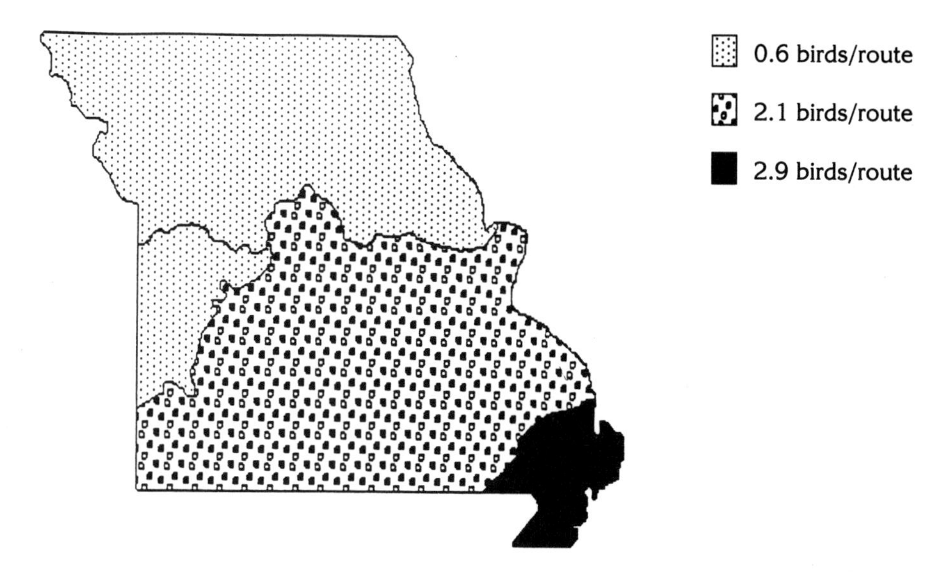

Graph 10.
Missouri BBS data from 1967 through 1989 for the Bewick's Wren. See p. 23 for data analysis.

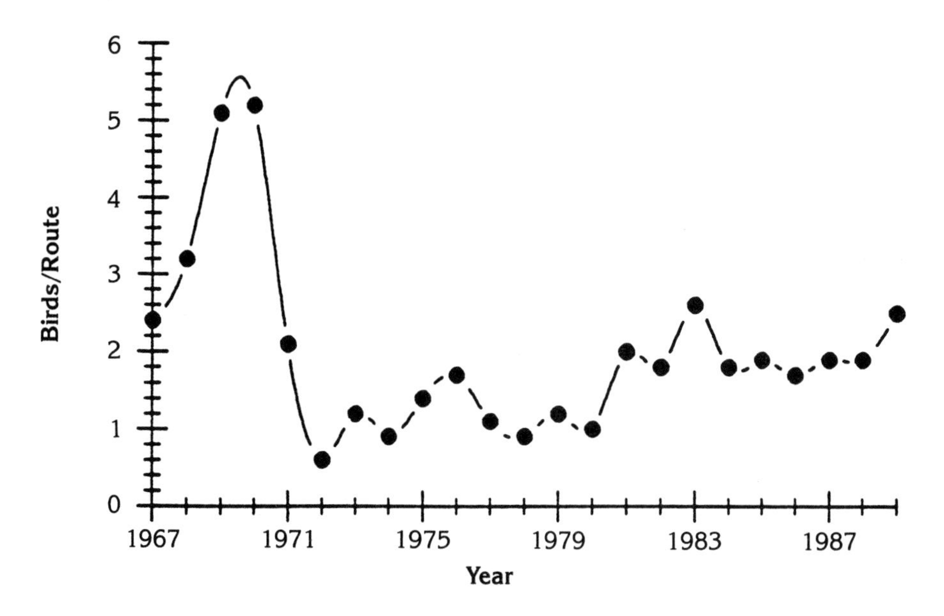

Fall Migration: There is virtually no information on when the majority of individuals leave the state in the fall. Presumably most leave by early Oct.

Winter: The Bewick's is a rare but annual winter resident in the Ozarks and Ozark Border. There are no winter records for north of 39° north (roughly Kansas City, Columbia, and St. Louis). High counts, both on Springfield CBC: 3, 26 Dec 1968; 3, 21 Dec 1974.

House Wren (*Troglodytes aedon*)

Status: Common summer resident in all regions except the south central Ozarks and the Mississippi Lowlands where it is uncommon and local. Casual winter visitant in south, accidental in north.

Documentation: Specimen: male, 17 Sept 1976, Maryville (NWMSU, LWT 100).

Habitat: Most common in urban areas, but also at forest and woodland edge.

Records:

Spring Migration: Normally the first migrants are seen by mid-Apr, with peak in early May. Earliest dates: 1, 28 Feb 1962 (winter resident?), near St. Louis (JEC-**AFN** 16:344); 1, 6 Mar 1958, St. Louis (E. Hath-**BB** 25[4]:3); 1, 16 Mar 1970, St. Louis (C. Hath-**BB** 38[3]:9).

Map 39.
Relative breeding abundance of the House Wren among the natural divisions. Based on BBS data expressed as birds/route.

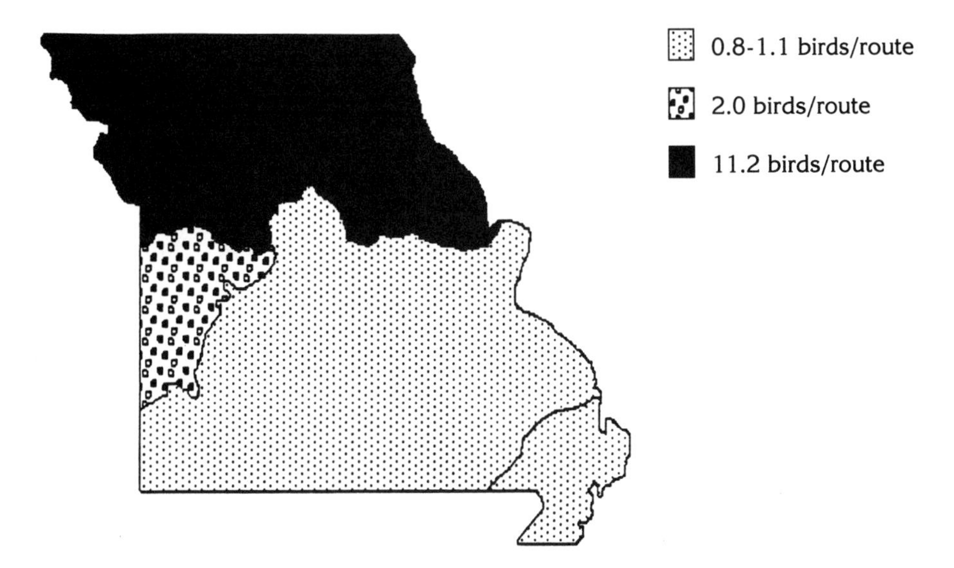

Summer: This wren is decidedly more abundant in the Glaciated Plains than in any other region (Map 39). It apparently has increased and expanded its range southward since the 1960s. Widmann (1907) stated that it was common in the north, extending southward along the Mississippi R. counties to St. Genevieve, and in the western half of the state to Jasper Co. At that time, it apparently was absent or rare in the Ozarks and Mississippi Lowlands.

This pretty much remained the situation until at least the 1960s, since in the late 1930s the House Wren had not been recorded breeding in the Steelville, Crawford Co., or the Lebanon, Laclede Co., areas (JEC-**BB** 6[8]:67; G. Moore-**BB** 4[8]:86). However, observers began noting House Wrens in residential areas of the Ozarks in the 1960s, with a dramatic increase during the 1970s. This same pattern also commenced at the same time in the Arkansas Ozarks (James and Neal 1986). It is least common and certainly most local in the south central section of the Ozarks.

In the Mississippi Lowlands, it was not recorded on the BBS route 1 in Mississippi Co. until 1976; however, just to the north, where the Ozark Border and the Mississippi Lowlands interdigitate, it was observed at the outset of BBS route 7 in 1967. It primarily remains an uncommon and local summer resident in the northern section of the Mississippi Lowlands.

Fall Migration: It is quite common through the third week of Sept, with smaller numbers (max. 3–4 birds/day) being seen through mid-Oct. During the last two weeks of Oct it is rare, with only widely scattered individuals observed. It is casually encountered during Nov. High count: 11, tower kill, 6 Oct 1962, Cape Girardeau (PH-**BB** 30[1]:7).

Winter: At least seven records, all of single birds. All but one of the records are from mid-Dec through early Jan. Only records north of Missouri R.: 1, 2 Jan 1982, Orchard Farm CBC; 1, 4 Jan 1989, near Craig, Holt Co. (LG-**BB** 56[2]:55). Only midwinter observation: 1, 29 Jan 1966, Taney Co. (SH-**BB** 33[1]:14).

Winter Wren (*Troglodytes troglodytes*)

Status: Uncommon transient; uncommon winter resident in Mississippi Lowlands and Ozarks, rare elsewhere.

Documentation: Specimen: female, 12 Oct 1960, Cape Girardeau (SEMO 119).

Habitat: Primarily found along logs and in brush piles, especially along wooded and forested streams.

Records:

Spring Migration: Migrants are apparent by mid-Mar, and their numbers gradually increase to peak in early to mid-Apr. A few birds are encountered until the first of May. High counts: 10, 11 Apr 1987, St. Louis (SRU-**BB** 54[3]:15); 6, 17 Apr 1983, Jefferson Co. (MP-**NN** 55:50). Latest date: 1, 3 May 1975, St. Louis (RA-**AB** 29:861).

Fall Migration: Usually appears by the last few days of Sept, with peak in mid-Oct. Earliest dates: 1, 15 Sept 1989, Tower Grove Park (JVB-**NN** 61:74); 1, 17 Sept 1981, St. Louis (C. Roberts-**BB** 49[1]:14). High counts: 6+, 18 Nov 1978, Mingo (CB et al.-**BB** 46[1]:26); 6, 23 Nov 1969, Brickyard Hill WA (DAE, FL-**BB** 38[3]:6).

Winter: The Winter Wren is easily overlooked unless one is familiar with its call note. It is most common in the Mississippi Lowlands, where an average of 2.2 birds/10 pa hrs is recorded on the sole CBC. It is rare in the Glaciated and Osage plains and Ozark Border, where only 0.1–0.2 birds/10 pa hrs are recorded. High counts: 24, 2 Jan 1989, Mingo CBC; 20+, 12–13 Feb 1986, Donaldson Point, New Madrid Co. (JW-**BB** 53[2]:9).

Sedge Wren (*Cistothorus platensis*)

Status: Uncommon to locally common summer resident in Glaciated and Osage plains, rare elsewhere; casual winter resident.

Documentation: Specimen: o?, 28 Dec 1965, near Puxico, Stoddard Co. (Easterla 1966b; NWMSU, DAE 1053).

Habitat: Sedge marshes, wet meadows, fields, and prairie.

Records:

Spring Migration: Decidedly less common at this season than in late summer and early fall. The first migrants are normally detected the last week of Apr. Earliest date: 1, 16 Apr 1977, Squaw Creek (TB-**BB** 44[4]:27). High counts: 4–5, 3 May 1977, Bigelow Marsh (MBR); 4, 8 May 1982, Marais Temps Clair (RK-**BB** 49[3]:49).

Summer: Recorded in relatively low numbers (3–4/site) through June and the first half of July, but during the last half of July birds suddenly appear statewide (primarily in non-Ozark regions) and begin breeding in areas where they previously were absent. It is unclear where these birds are coming from.

Fall Migration: Numbers of singing birds continue to increase until mid-Aug. In some areas this species becomes locally abundant, with well over 20 birds recorded in a single field. Typically singing ceases in early Sept, and few birds are found until late Sept when obvious migrants appear (birds at nonbreeding sites; tower kills in late Sept and early Oct). Small numbers (max. 3–4 birds/locality) are recorded through Oct. Thereafter it is only casually recorded. High counts: 51 singing males (15 [13 dummies, 2 active] nests), 16 Aug 1961, on 154.5 acres of Tucker Prairie, Callaway Co. (Easterla 1962a); 50–70, 16 Aug 1979, Marais Temps Clair (HB, V. Bucholtz-**BB** 47[1]:25); 50, 15 Aug 1965, Squaw Creek (FL-**BB** 32[3]:14). Latest date: 1, 27 Nov 1952, Scott Co. (R. Myklebust-**BB** 20[5]:2).

Winter: At least ten records, all involving single birds between mid-Dec and early Jan. Only three north of the Missouri R. as follows: 21 Dec 1941, Parkville, Platte Co. (HH-**BB** 9[2]:11); 21 Dec 1980, Trimble CBC; 1 Jan 1963, Columbia CBC.

Marsh Wren (*Cistothorus palustris*)

Status: Uncommon transient; rare summer resident in north; casual winter resident.

Documentation: Specimen: male, 26 Sept 1962, Cape Girardeau (SEMO 162).

Habitat: Primarily cattail marshes, but also seen in brush piles and wet fields during migration.

Records:

Spring Migration: In years when there is an unusually warm, "early" spring, the initial migrants are seen as early as the last week of Mar. But normally the first arrivals are not encountered until mid-Apr. Peak occurs during the first week of May. Earliest dates: 2, 16 Mar 1990, Duck Creek (BRE-**BB** 57[3]:139); 1, 25 Mar 1980, Marais Temps Clair (F. Hallet-**NN** 52:35). High count: 10, 7 May 1989, Duck Creek (RB, BRE-**BB** 56[3]:93).

Summer: Only known to breed in marshes (primarily cattail) north of the Missouri R. In very dry years (low water levels) birds may be very scarce or completely absent. High count: 7 territorial birds, 7 June 1976, Bigelow Marsh; additional birds at Squaw Creek (MBR, DAE-**BB** 44[2]:21).

Fall Migration: Migrants begin to reappear by mid-Sept and peak in late Sept and early Oct. Birds are regularly seen through early Nov, with scattered individuals casually encountered until the end of the period. High counts: 25, 24 Sept 1939, eastern Jackson Co. (WC-**BB** 6[10]:85); 19, 25 Sept 1988, Big L. SP (MBR- **BB** 56[1]:15).

Winter: In those years when the fall and early winter are mild, birds are casually recorded during the final two weeks of Dec and in early Jan. High counts, both on the Squaw Creek CBC: 11, 19 Dec 1978; 5, 18 Dec 1977.

Family Muscicapidae: Muscicapids

Golden-crowned Kinglet (*Regulus satrapa*)

Status: Common transient; uncommon winter resident.

Documentation: Specimen: male, 7 Nov 1921, Jackson Co. (KU 39256).

Habitat: Primarily seen in evergreens; also found in strictly deciduous woods, especially during migration.

Records:

Spring Migration: Migrants begin supplementing the winter population by mid-Mar and peak during the first week of Apr. Virtually all have left the state by the end of Apr. High count: 5–10/mixed species flock (with chickadees, titmice, and creepers), 31 Mar–4 Apr 1980, near Collins, St. Clair Co., and Butler, Bates Co. (MBR, MB). Latest dates: 1, 17 May 1983, Van Meter SP, Saline Co. (CH, KH-**AB** 37:877); 1, 5 May 1939, Webster Groves, St. Louis Co. (JEC-**BB** 6[6]:51).

Fall Migration: Birds begin to reappear by the last week of Sept and peak during the second and third weeks of Oct. Numbers remain relatively high through early Nov but drop off markedly thereafter. Earliest date: 1, 16 Sept 1965, St. Louis (H. Hill-**NN** 37:5). High count: 200+, 20 Oct 1985, St. Louis (RA-**NN** 57:85).

Winter: It is most common in the Mississippi Lowlands and least abundant in the Glaciated and Osage plains (Map 40). High counts: 75, 15 Dec 1974, Weldon Spring CBC; 54, 2 Jan 1989, Mingo CBC.

Ruby-crowned Kinglet (*Regulus calendula*)

Status: Common transient; uncommon winter resident in Mississippi Lowlands, rare in Ozarks and Ozark Border, casual in Glaciated and Osage plains.

Documentation: Specimen: female, 11 Sept 1932, Jackson Co. (KU 19359).

Habitat: Woodland and forest edge, brushy areas.

Records:

Spring Migration: Arrives and peaks a little later than the preceding spe-

Map 40.
Early winter relative abundance of the Golden-crowned Kinglet among the natural divisions. Based on CBC data expressed as birds/10 pa hrs.

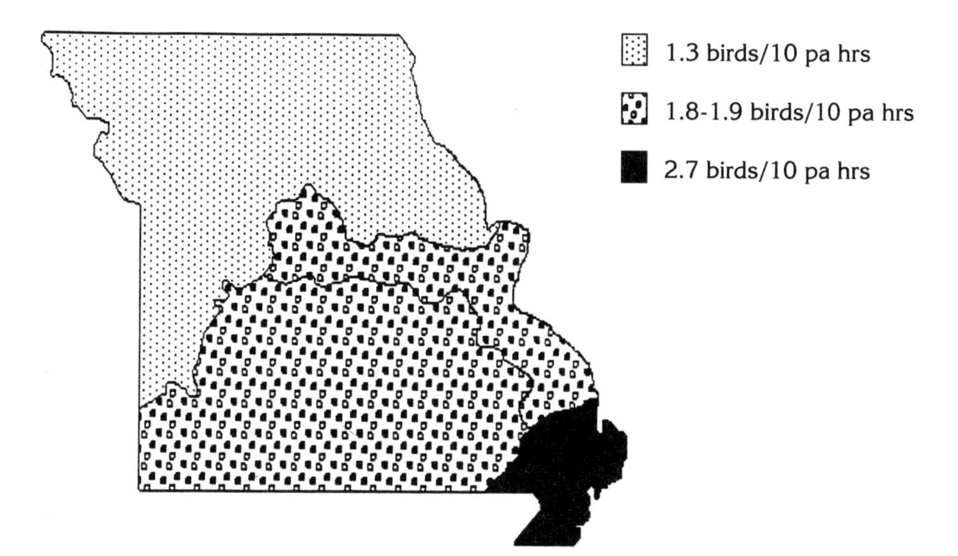

cies. The first birds appear at the beginning of the last week in Mar in the south and about a week later in the north. Peak is usually at the end of Apr. Only a few individuals are still encountered at the end of the second week in May, and virtually all have left a few days later. High count: 42, 18 Apr 1988, St. Louis (SRU). Latest dates: 1, 28 May 1986, Knob Noster SP, Johnson Co. (MBR); 3, 23 May 1989, Big L. SP (MBR-**BB** 56[3]:92).

Fall Migration: A few begin to trickle back into the state by the second week of Sept. Peak is in early Oct. Numbers are greatly reduced by the beginning of Nov, but small groups (< 10 birds) occasionally may be encountered in coniferous stands in the north until mid-Nov. Earliest dates: 1, 26 Aug 1968, 10 miles north of Maryville, Nodaway Co. (MBR, DT); 1, 30 Aug 1957, near Eureka, St. Louis (SHA-**BB** 24[8–9]:4). High counts: 50, 7 Oct 1979, L. of Ozarks SP (FL et al.); 44, 6 Oct 1987, Brickyard Hill WA (MBR-**BB** 55[1]:12).

Winter: CBC data for the sole count in the Mississippi Lowlands are somewhat surprising because they indicate that the Ruby-crowned (average 4.2 birds/10 pa hrs) outnumbers the Golden-crowned (average 2.7 birds/10 pa hrs) in that region. Outside this region, the Rudy-crowned is considerably rarer, with 0.0–0.4 birds/10 pa hrs. Northernmost record: male, 20 Dec 1969, west of Maryville (NWMSU, DAE 2302). High counts: 22, 28 Dec 1965, Mingo CBC; 14, 27 Dec 1974, Big Oak Tree SP CBC.

Map 41.
Relative breeding abundance of the Blue-gray Gnatcatcher among the natural divisions. Based on BBS data expressed as birds/route.

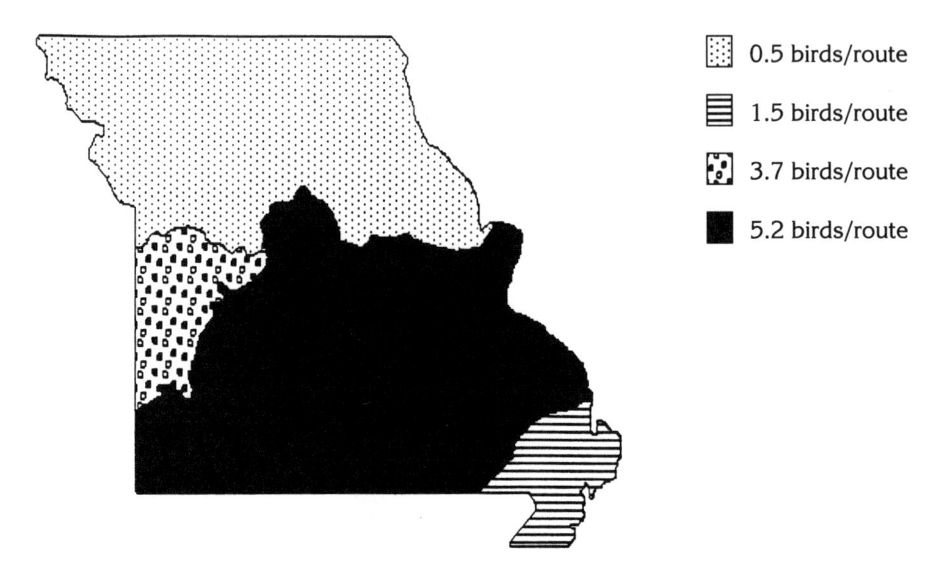

Blue-gray Gnatcatcher (*Polioptila caerulea*)

Status: Common summer resident.
Documentation: Specimen: female, 18 Apr 1965, near Salem, Dent Co. (NWMSU, DAE 874).
Habitat: Deciduous forest and woodland edge.
Spring Migration: The first individuals are seen at the beginning of Apr in the south and about a week later in the north. Peak is in late Apr in the south and during the first week of May in the north. Earliest dates: 1, 23 Mar 1978, Forsyth, Taney Co. (PM-**BB** 55[1]:14); 2, 27 Mar 1968, Big Oak Tree SP (JH). High counts for early in season: 25, 5 Apr 1974, Big Oak Tree SP (JH); 60, 18 Apr 1987, along 18 mile stretch of Current R., Shannon Co. (RK).
Summer: It is now most abundant in the Ozarks and Ozark Border and is least common in the Glaciated Plains (Map 41). This species was perhaps most abundant in the Mississippi Lowlands prior to the region being denuded at the turn of the century.
Although the number of hectares of floodplain forest remained nearly the same, the conversion of the upland forest to parkland apparently resulted in a significant decrease of this species at Swope Park, Jackson Co., between 1916 (n=31 males; A. Shirling 1920) and 1973 (n=1 male; Branan and Burdick 1981).

Fall Migration: The majority of the summer residents leave and migrants pass through the state during the last two weeks of Aug. Numbers are greatly reduced in the north even by early Sept. It is rarely encountered in the state after mid-Sept. There are no Oct observations, although there is one Nov record: 1, 12 Nov 1980, Squaw Creek (FL-**BB** 48[1]:9).

Eastern Bluebird (*Sialia sialis*)

Status: Common summer resident; uncommon in the winter in all regions of the state, except the Glaciated Plains where it is rare.

Documentation: Specimen: male, 22 Feb 1922, Jackson Co. (KU 39236).

Habitat: Relatively open areas; nests and roosts in dead trees and old wood fence posts adjacent to woodland and forest.

Records:

Spring Migration: The initial migrants begin to appear at nonwinter sites as early as late Feb. Flocks of migrants are most conspicuous from mid-Mar through early Apr (especially at the end of the third and the beginning of the fourth weeks of Mar). High count: 40, 24 Mar 1972, St. Joseph (FL).

Summer: It is most common in the Ozarks and Ozark Border and least prevalent in the Mississippi Lowlands (Map 42). In the Glaciated Plains it is more common in the eastern than the western portion. The utilization of artificial nest boxes by this species has dramatically increased the number of young produced in the state. The most successful example of this is at Pickering Ranch near Foristell, St. Charles Co., where as many as 280 young have fledged in a single season (1975) (**BB** 43[1]:24).

Graph 11 depicts how the summer abundance of this species is negatively affected following severe winters, such as the three consecutive hard winters of 1976–78.

Fall Migration: Small migrant flocks may be detected as early as mid-Sept, but the larger-sized flocks (> 40 birds/flock) do not appear until mid-Oct. These relatively large-sized groups are regularly seen until mid-Nov. Thereafter, principally wintering birds are observed. High counts: thousands, 9–10 Oct 1922, Blue R., Jackson Co. (H. Harris-**BL** 24:354); 200, 20 Oct 1984, Kansas City (**BB** 52[1]:32).

Winter: CBC data reveal that this bluebird is most common in the Ozarks and Ozark Border and least common in the Glaciated and Osage plains (Map 43). High counts, both of the Taney Co. CBC: 455, 27 Dec 1986; 348, 30 Dec 1989.

Comments: The Eastern Bluebird was designated the official state bird in 1927.

Map 42.
Relative breeding abundance of the Eastern Bluebird among the natural divisions. Based on BBS data expressed as birds/route.

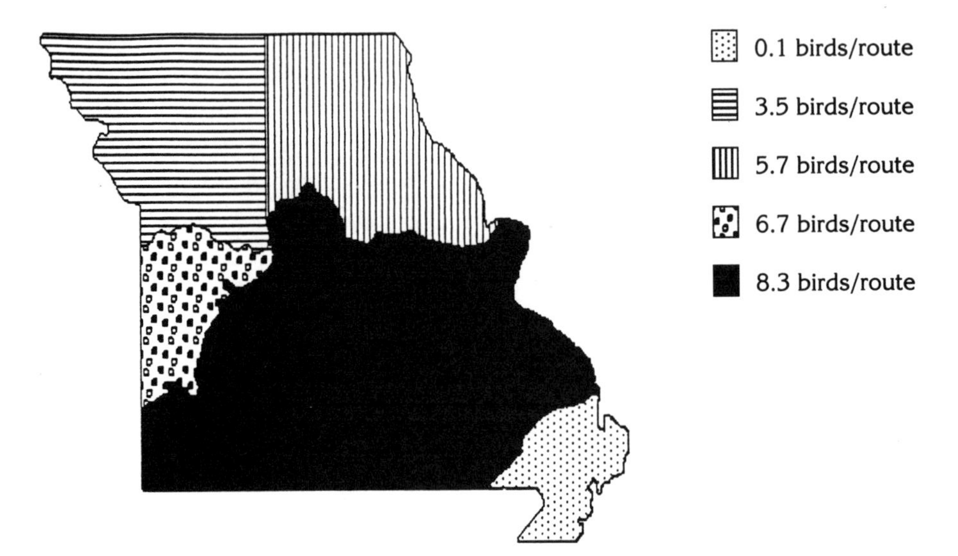

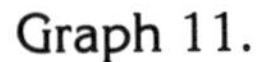

Graph 11.
Missouri BBS data from 1967 through 1989 for the Eastern Bluebird.
See p. 23 for data analysis.

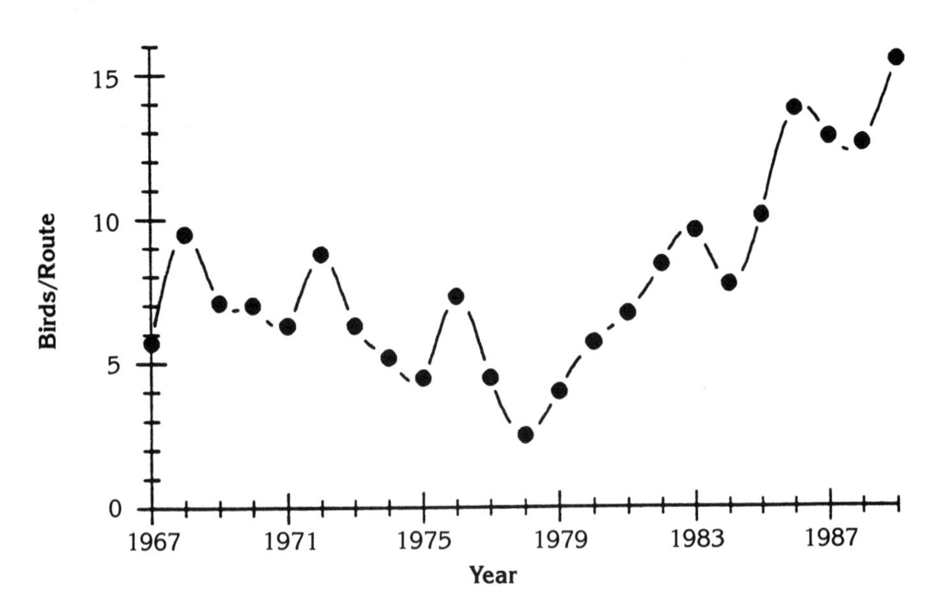

Map 43.
Early winter relative abundance of the Eastern Bluebird among the natural divisions. Based on CBC data expressed as birds/10 pa hrs.

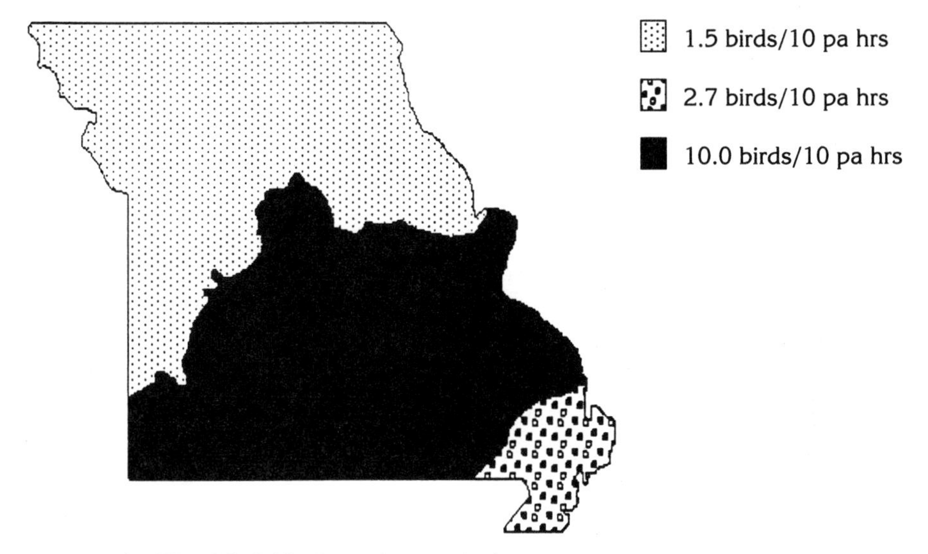

Mountain Bluebird (*Sialia currucoides*)

Status: Accidental transient and winter resident.

Documentation: Photograph: male, 10–14 Apr 1990, near Cedar Springs, Cedar Co. (JG, E. Johnson, I. and R. Rhorer-**BB** 57[3]:139; Pl. 8d).

Habitat: Open areas with scattered bushes and trees.

Records:

Spring Migration: Two records besides the above documented observation: male, 23–24 Mar 1974, Squaw Creek (FL et al.-**BB** 41[2]:3; photo, DAE, VIREO x08/15/001); female, 2 Apr 1990, Prairie State Park, Barton Co. (J. Loomis et al.-**BB** 57[3]:139).

Winter: Three records: pair, 18 Dec–13 Jan 1988–89, SW of Peculiar, Cass Co. (JJ et al.; Garrett 1989); male, initially found on 20 Dec 1986, Jefferson City CBC, and remained until at least 29 Dec (m.ob.); 1, 23 Dec 1950, near Independence (HH; Rising et al. 1978).

Comments: Rising et al. (1978) state that a bird was found dead along Shoal Creek, Clay Co., by Dr. Monahan (no additional information).

Townsend's Solitaire (*Myadestes townsendi*)

Status: A rare, irregular winter resident in the northwestern corner.

Documentation: Specimen: male, found dead, 14 Jan 1967, St. Joseph (NWMSU, DAE 1172).

Map 44.
Counties where the Townsend's Solitaire has been recorded.

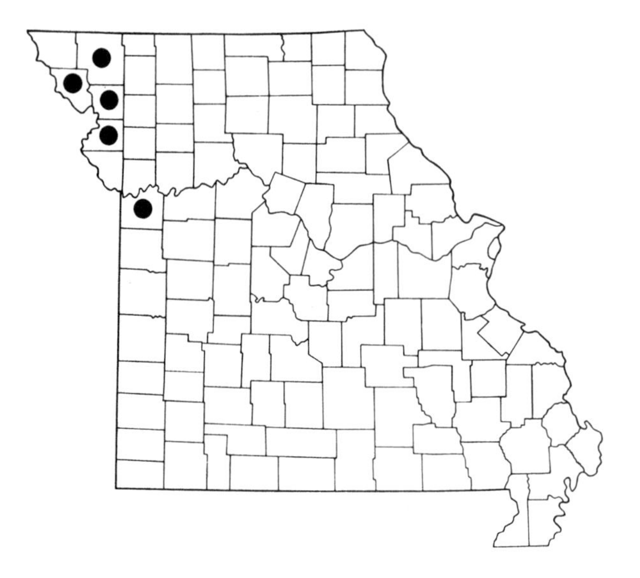

Habitat: All birds have been associated with cedar stands.

Records:

Spring Migration: Occasionally one of the birds that has overwintered lingers until late Feb or early Mar: 1, until 6 Mar 1977, St. Joseph (LG et al.-**BB** 46[3]:25).

Fall Migration and Winter: All of the ca. 15+ records for this period are concentrated in the northwestern counties (Map 44). With only a couple of exceptions, all records have been between mid-Dec and mid-Feb. Earliest fall date: adult male, 1 Dec 1963, Mt. Washington Cemetery, Jackson Co. (Rising 1965; KU 43407). The two seen on 10 Dec 1972, Reed WA (CH, KH-**BB** 40[2]:8), is the only record involving more than one bird.

Veery (*Catharus fuscescens*)

Status: Uncommon transient in east, rare in west.

Documentation: Specimen: male, 11 May 1969, near Maryville (NWMSU, DAE 2469).

Habitat: Moist deciduous woodland and forest, especially along watercourses.

Records:

Spring Migration: Normally not encountered until the last week of Apr, with peak during mid-May. A few are still seen passing through during the final week of May. Earliest date: 1, 16 Apr 1965, Big L., Mississippi Co. (JH).

An average of 0.1 birds/hr was recorded between 25 Apr–20 May 1979–90 at Forest Park (RK; n=421 hrs). High count in the east: 11, 23 May 1985, Hayti, Pemiscot Co. (JW-**BB** 52[3]:15); west: 3, 18 May 1969, Maryville (DAE).

Summer: Only two records: 1, 3 June 1984, Wallace SP, Clinton Co. (CH et al.-**BB** 51[4]:17). The other is of a band recovery of a presumed early fall migrant; the bird was initially banded in St. Louis on 5 May 1929 and was found dead there on 30 July 1931 (Cooke 1937).

Fall Migration: Much less conspicuous in the fall than in the spring; few published records for this season. Observations span from late Aug through early Oct, with no discernible peak. Earliest date: 1, 27 Aug 1968, Big Oak Tree SP (JH). Latest date: 14 Oct (no year given), Kansas City (Harris 1919b).

Comments: The Veery has recently been recorded breeding as far south as central Iowa and central Illinois (Dinsmore et al. 1984; Bohlen and Zimmerman 1989); thus, it might occasionally breed in the northeastern corner of Missouri.

Gray-cheeked Thrush (*Catharus minimus*)

Status: Uncommon transient.

Documentation: Specimen: male, 7 May 1967, near Maryville (NWMSU, DAE 1188).

Habitat: Deciduous woodland and forest.

Records:

Spring Migration: First appears during the final days of Apr and peaks in mid-May. Like the other migrant thrushes it is seen through May. An average of 0.2 birds/hr was recorded at Forest Park between 25 Apr–20 May 1979–90 (RK; n=421 hrs). Earliest dates: 1, 22 Apr 1990, Bee Creek WA, Taney Co. (CL, PM-**BB** 57[3]:139); 3, 23 Apr 1990, Forest Park (RK). High count: 15, 14 May 1967, St. Joseph/Squaw Creek (FL).

Summer: One record with details, although Widmann (1907) mentions that it lingers into June: 1, 1 June 1983, Busch WA (SRU-**BB** 50[4]:34).

Fall Migration: Most of the information for this period comes from tower kill data. Peak is during the third week of Sept, with a few birds seen into the first week of Oct. Earliest date: 1, 5 Sept 1979, St. Joseph (FL). High count: 22, 19 Sept 1966, Columbia (Elder and Hansen 1967). Latest date: 1, 24 Oct 1975, Busch WA (JC-**BB** 44[1]:22).

Swainson's Thrush (*Catharus ustulatus*)

Status: Common transient.

Documentation: Specimen: o?, 24 Apr 1970, Cape Girardeau (SEMO 317).

Habitat: Found in virtually any size woodland and forest.

Records:

Spring Migration: The most common *Catharus* during migration. Like most other members of this genus, the first individuals appear during the last week of Apr. Peak is during the first week of May in the south and at the end of the second week in the north. Migrants are readily seen into June. At Forest Park, Korotev has recorded an average of 2.8 birds/hr in 421 hrs of observation (25 Apr–20 May 1979–90). Earliest date: 1, 14 Apr 1983, Forest Park (m.ob.-**NN** 55:50). High counts: 50, 14 May 1967, St. Joseph/Squaw Creek (FL); 45+, 6 May 1990, Barry Co. (JH, PM-**BB** 57 [3]:139).

Summer: Late migrants are regularly seen through the first week of June. Latest date: 1, 9 June 1945, Hannibal, Marion Co. (WC-**AM** 47:38).

Fall Migration: Although there are no published observations for late Aug, this species undoubtedly begins to reappear then. Peak is at the end of the third or at the beginning of the fourth week of Sept. Smaller numbers are seen during the first two weeks of Oct, with an occasional bird encountered until the end of Oct. Earliest date: 1, 5 Sept 1965, St. Joseph (FL). High counts, both from tower kills at Columbia: 53, 19 Sept 1966; 53, 22 Sept 1965 (Elder and Hansen 1967). Latest date: 1, 27 Oct 1975, Busch WA (JC-**BB** 44[1]:22).

Hermit Thrush (*Catharus guttatus*)

Status: Uncommon transient; uncommon winter resident in the extreme south, rare in Ozarks and Ozark Border, casual in the Glaciated and Osage plains.

Documentation: Specimen: female, 30 Jan 1966, Taney Co. (CMSU 212).

Habitat: Most common in bottomland woodland and forest.

Records:

Spring Migration: This thrush is the earliest of the *Catharus* to arrive and pass through the state, with the first migrants appearing in late Mar. Peak is in mid-Apr, with an occasional bird encountered as late as early May. High count: 10, 2 Apr 1989, St. Louis (B. Boesch-**BB** 56[3]:92). Latest dates: 1, 12 May 1990, Tower Grove Park (BRO et al.-**NN** 62:53); two records for 7 May.

Fall Migration: Normally not observed until the beginning of Oct, with the bulk of the birds passing through during the second and third weeks of Oct. After the first of Nov, it is only rarely encountered in the north. Earliest dates: 1, 28 Aug 1985, St. Louis (B. Hely-**NN** 57:69); 1, 10 Sept 1981, St. Louis (E. Hath-**NN** 53:60).

Winter: Outside of the Mississippi Lowlands (average 3.8 birds/10 pa hrs), and the extreme southern tier of Ozark counties it is a rare bird (average 0.1–0.2 birds/10 pa hrs in the Ozark Border and Ozarks), with very few observations from the Glaciated and Osage plains. During severe winters it can be quite scarce even at the southern border. High counts, both on the Taney Co. CBC: 16, 2 Jan 1988; 11, 27 Dec 1986.

Wood Thrush (*Hylocichla mustelina*)

Status: Common summer resident.

Documentation: Specimen: female, 26 Apr 1971, Cape Girardeau (SEMO 328).

Habitat: Most common in moist, shady woodland and forest.

Records:

Spring Migration: The initial migrants usually arrive by mid-Apr in the south, but not until the last week of Apr in the north. It is common statewide after the first week of May. Earliest dates: 1, 17 Mar 1982, St. Louis (CP-**BB** 49[3]:21); 1, 3 Apr (no year given), Dunklin Co. (Widmann 1907).

Summer: Although it is found throughout the state anywhere there are relatively large tracts of moist woodland and forest, it is most common in the Ozarks where it is three times as prevalent than in any other region (Map 45). BBS data indicate that Missouri's Wood Thrush population appears to be relatively stable (Graph 12).

Fall Migration: Most leave the state during Sept, becoming less common towards the latter half of the month. Small numbers are seen until mid-Oct, but thereafter it is only accidentally encountered. High counts, tower kill data: 5, 20 Sept 1963, Columbia (George 1963); 4, 6 Oct 1962, Cape Girardeau (Heye 1963; SEMO 205). Latest date: 1, 29 Oct 1963, St. Joseph (FL).

American Robin (*Turdus migratorius*)

Status: Common summer resident; uncommon to locally common winter resident in south, rare in north.

Documentation: Specimen: female, 7 June 1934, Campbell, Dunklin Co. (MU 101).

Habitat: Woodland, parks, residential areas, etc.

Records:

Spring Migration: The initial migrants begin supplementing the winter population as early as mid-Feb in the south and about a week later in the north. Literally overnight, during mid-Mar (in the south) to early Apr (in the north), hundreds to a few thousand birds may arrive in an area. By the end of Apr mainly breeding birds remain. High counts: 1,500, 25 Mar 1969, St. Joseph (FL); 1,000+, 28 Mar 1982, St. Joseph/Squaw Creek (FL).

Summer: This is one of the most abundant species at this season in the state. It is most common in towns and the suburbs. The influence of park-like habitat on the abundance of this species is readily seen by the 1916

Map 45.
Relative breeding abundance of the Wood Thrush among the natural divisions. Based on BBS data expressed as birds/route.

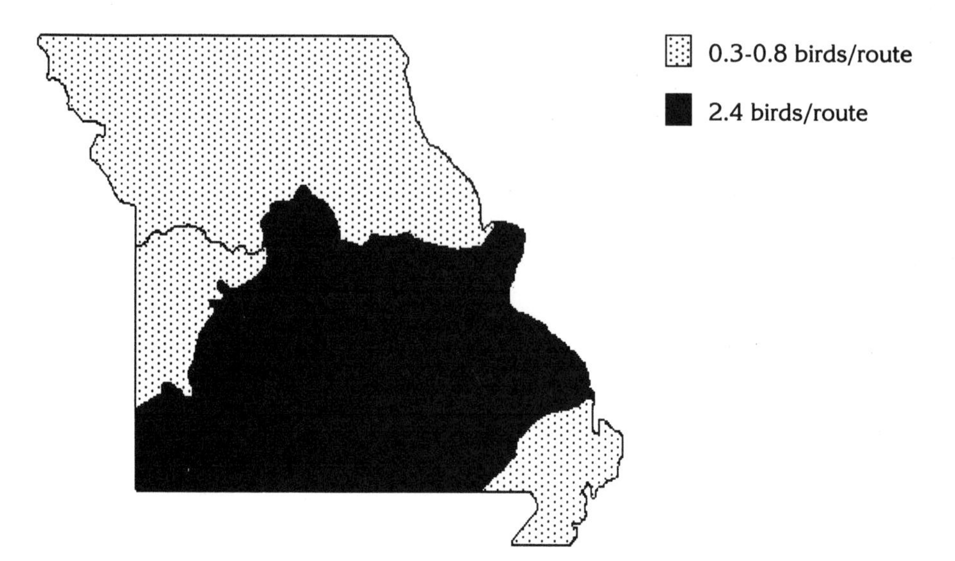

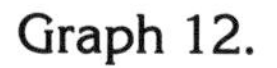
Graph 12.
Missouri BBS data from 1967 through 1989 for the Wood Thrush. See p. 23 for data analysis.

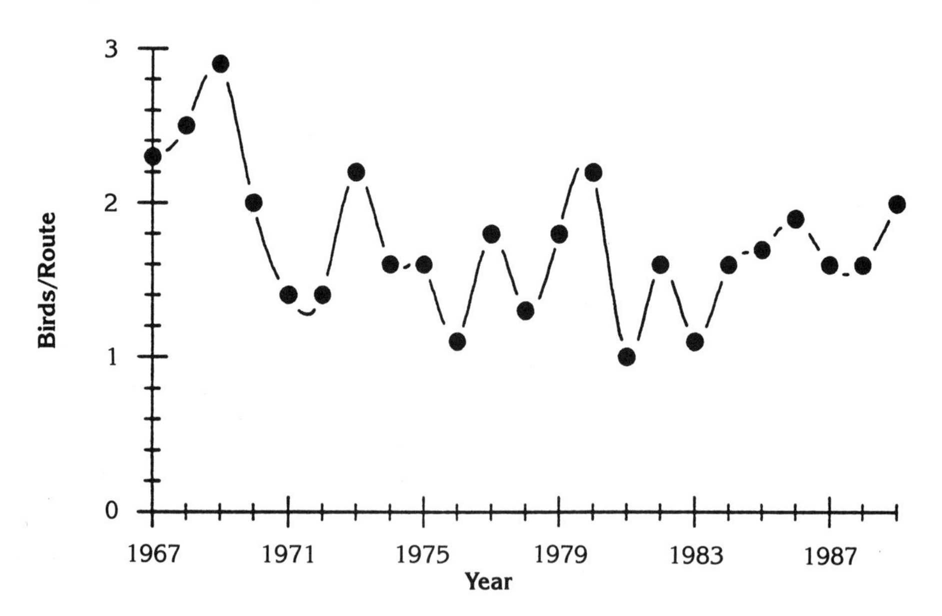

Map 46.
Relative breeding abundance of the American Robin among the natural divisions. Based on BBS data expressed as birds/route.

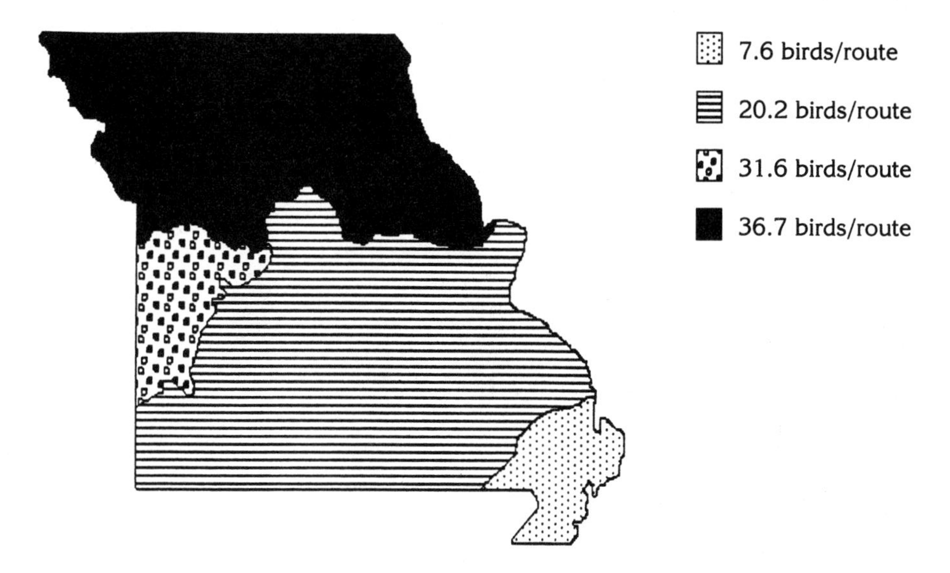

(n=31; A. Shirling 1920) vs. 1973 (n=197; Branan and Burdick 1981) Swope Park comparison. Much of the natural habitat was converted to parkland in the intervening period, which resulted in a 635% increase of this species. The highest densities occur in the Glaciated Plains and the lowest in the Mississippi Lowlands (Map 46).

Fall Migration: Although relatively large-sized flocks (in the hundreds of birds) may be seen from the latter half of Sept through the end of the period, the largest groups are usually observed during Oct when the bulk of this species passes through the state. High counts: 100,000–250,000, late Nov 1988, Taney Co. (PM, JH et al.-**BB** 56[1]:15); 2,000, 4 Nov 1962, St. Joseph (FL).

Winter: Winter roosts are quite local; most occur along river and stream bottoms and in cedar groves in the southern half of the state. The two largest robin roosts reported were as follows: one at Washington SP, Washington Co., between late Dec and late Jan 1964–65, where over 3.5 million birds were estimated during Jan (Anderson 1965); estimated 1.5 million, late Jan 1989, Taney Co. (PM, JH et al.-**BB** 56[2]:55). CBC data clearly show that this species is rarest in the Glaciated and Osage plains where an average of 0.5 birds/pa hr are recorded. The Ozarks has the highest mean, 4.6 birds/pa hr. Other high count: 6,984, 31 Dec 1977, Sullivan CBC.

Varied Thrush (*Ixoreus naevius*)

Status: Accidental winter resident.

Documentation: Photograph: adult male, 21 Jan 1985, Maryville (DAE; VIREO x05/2/001–002; Fig. 26).

Records: The above bird was present at a feeder from 20 Jan through 7 Mar 1985 (P. Robbins; Robbins and Easterla 1985). The only other record is of a bird that also appeared at a feeder from 21 Dec–19 Jan 1986–87, near Columbia (C. Cockrell, JW, B. Clark-**BB** 54[1]:37; photo in ref and VIREO x05/3/002).

This species is now recorded almost annually in the winter in Iowa (Dinsmore et al. 1984).

Family Mimidae: Mockingbirds, Thrashers, and Allies

Gray Catbird (*Dumetella carolinensis*)

Status: Common summer resident; casual winter resident.

Documentation: Specimen: o?, 13 Apr 1957, Cape Girardeau (SEMO 42).

Habitat: Thickets, hedgerows, dense understory at woodland and forest edge.

Records:

Spring Migration: Casually seen in Mar, but during most springs it is not seen until mid-Apr in the south and about a week later in the north. Peak is in early May, and it is common thereafter. Earliest date: 1, 7 Mar 1987, Busch WA (M. Botz-**BB** 54[3]:15). High count: 40, 26 Apr 1964, St. Joseph/ Squaw Creek (FL).

Fig. 26. This adult male Varied Thrush spent January through early March of 1985 at a feeder in Maryville, Nodaway Co. It was the first record for the state. Photograph taken on 25 January by David Easterla.

Map 47.
Relative breeding abundance of the Gray Catbird among the natural divisions. Based on BBS data expressed as birds/route.

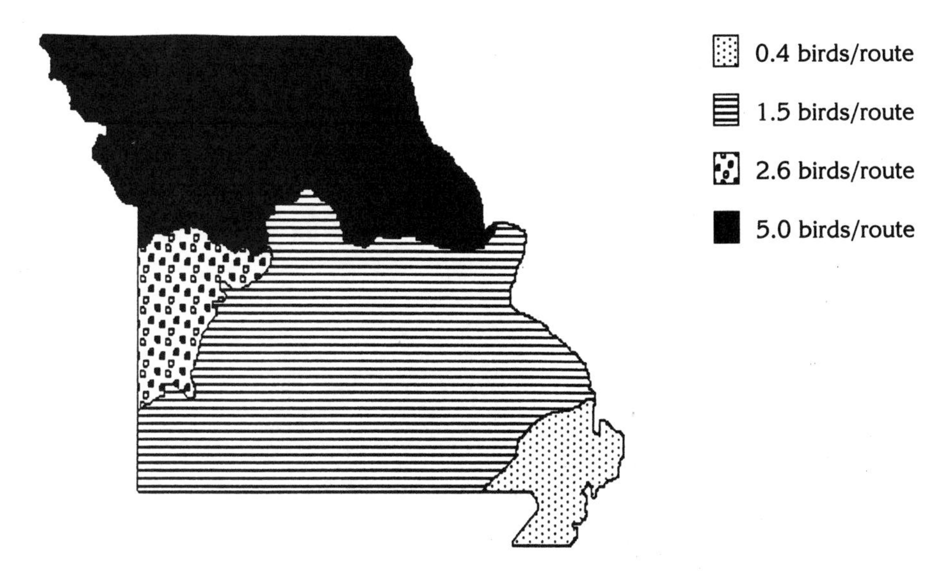

Summer: This species is most abundant in the Glaciated and Osage plains (Map 47).

Fall Migration: The bulk of the migrants pass through during the last ten days of Sept and the first few days of Oct. But by the beginning of the second week of Oct few birds are seen, and after early Nov it is only casually encountered. High counts, both tower kills: 300, 6 Oct 1954, Clinton, Henry Co. (M. James-**AFN** 4:204); 60, 22 Sept 1965, Columbia (Elder and Hansen 1967).

Winter: At least ten records involving single birds, with all but a couple of observations during the last two weeks of Dec and the first few days of Jan. Only one record north of 39° north (Kansas City): 1 until 1 Jan 1971, St. Joseph (FL).

Northern Mockingbird (*Mimus polyglottos*)

Status: Common summer resident in all regions except the extreme northern and western sections of the Glaciated Plains where it is rare; uncommon winter resident in south, rare in north.

Documentation: Specimen: female, 28 Dec 1931, Jackson Co. (KU 18503).

Habitat: Mainly in open areas with dense hedgerows, scattered scrubs, and thickets.

Map 48.
Relative breeding abundance of the Northern Mockingbird among the natural divisions. Based on BBS data expressed as birds/route.

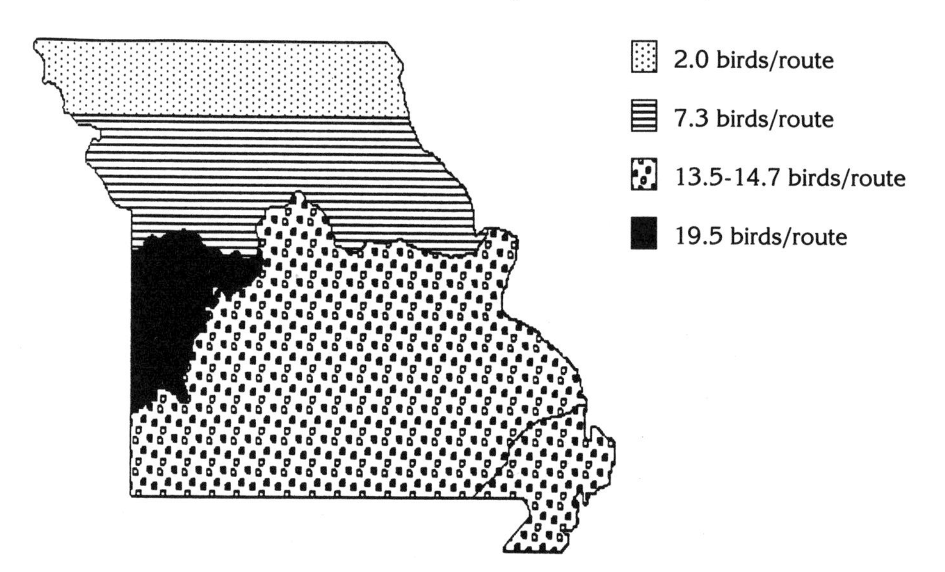

Records:

Spring Migration: Migration commences at the end of Mar and continues through Apr, with no discernible peak.

Summer: Generally it is most common across the southern half of the state where it is most abundant in the Osage Plains (Map 48). It is least common in the Glaciated Plains, especially in the northern and western sections. When BBS data for this region are subdivided the following is revealed: north (average ca. 2.0 birds/route) vs. south (ca. 7.3 birds/route); and west (4.4 birds/route) vs. east (6.4 birds/route). The breeding population, like those of Northern Bobwhite, Carolina Wren, and Eastern Bluebird, is depressed following severe winters (Graph 13).

Fall Migration: Although data are scant on movements in the fall, it appears that the bulk of the migrants pass through during late Sept and Oct; however, there clearly is some movement throughout the season.

Winter: As in the summer season, it is least prevalent in the Glaciated Plains region (average 2.0 birds/10 pa hr; it is rarest in the northern section). The CBC data reveal that it is most common in the Ozarks where an average of 7.0 birds/10 pa hr is recorded. This species is much less common during severe winters. High counts, both on the Springfield CBC: 113, 26 Dec 1972; 102, 17 Dec 1983.

Graph 13.
Missouri BBS data from 1967 through 1989 for the Northern Mockingbird. See p. 23 for data analysis.

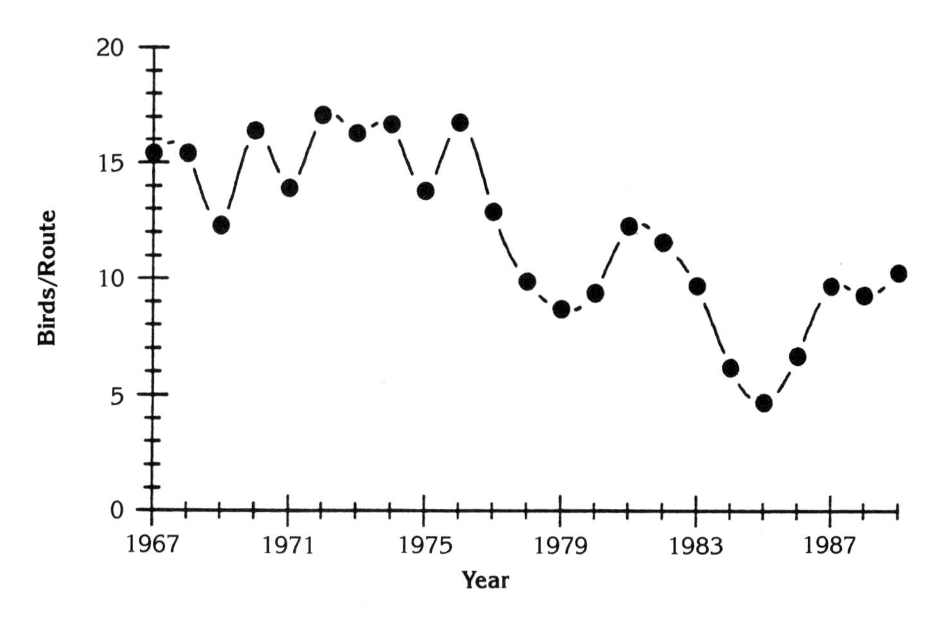

Sage Thrasher (*Oreoscoptes montanus*)

Status: Accidental winter resident.

Documentation: Photograph: 1, 28 Dec–14 Jan 1966–67, Lampe, Stone Co. (BBR-**BB** 35[1]:2; Fig. 27).

Comments: The above is the only record; the bird was frequently seen feeding on Eastern Red Cedar berries.

Brown Thrasher (*Toxostoma rufum*)

Status: Common summer resident; rare winter resident in southeast, very rare elsewhere.

Documentation: Specimen: male, 30 Apr 1968, near Maitland, Holt Co. (NWMSU, RAR 291).

Habitat: Thick brushy areas, hedgerows, and dense second growth at woodland and forest edge.

Records:

Spring Migration: The first of the nonwintering birds appear by mid-Mar in the south and at the end of Mar or early Apr in the north. Peak is during mid-Apr in the south and late Apr and early May in the north. High counts: 50, 21 Apr 1990, Tower Grove Park (JZ-**BB** 57[3]:139); 40, 26 Apr 1964, St. Joseph/Squaw Creek (FL).

Fig. 27. This photograph documents the only record of the Sage Thrasher for Missouri. The bird was photographed near Lampe, Stone Co., on 14 January 1967 by James Key.

Summer: Like the Gray Catbird, this species is most common in the Glaciated and Osage plains (Map 49). Unlike the breeding populations of the Carolina Wren, E. Bluebird, and Northern Mockingbird, the Brown Thrasher breeding population exhibited no significant declines following the severe winters of the late 1970s. Perhaps the bulk of the Missouri breeding thrasher population winters farther south than do the summer Missouri populations of the above species. Presumably the more northern wintering thrasher populations (Missouri, northern Arkansas, etc.) that were most affected by those winters breed to the north of our state.

Fall Migration: The bulk of the birds pass through during the last two weeks of Sept, but by early Oct few remain. Primarily only single birds are seen through the remainder of Oct, and thereafter it is rarely encountered. High count: 86, 17 Sept 1953, Calvary Cemetery, St. Louis (JEC-**NN** 23:42).

Winter: Most wintering birds are found in the southern part of the Ozarks and in the Mississippi Lowlands, although during most winters a few are scattered throughout the state. The Mingo and Big Oak Tree SP CBCs have recorded an average of 23.0 and 9.0 birds/100 pa hrs, respectively. Most other counts have an average of 1.0 bird/100 pa hrs. High counts, both on the Mingo CBC: 18, 28 Dec 1965; 17, 18 Dec 1982.

Family Motacillidae: Wagtails and Pipits

American Pipit (*Anthus rubescens*)

Status: Uncommon transient; casual winter resident.

Documentation: Specimen: o?, 10 May 1964, Cooley L. (NWMSU, DAE 548).

Map 49.
Relative breeding abundance of the Brown Thrasher among the natural divisions. Based on BBS data expressed as birds/route.

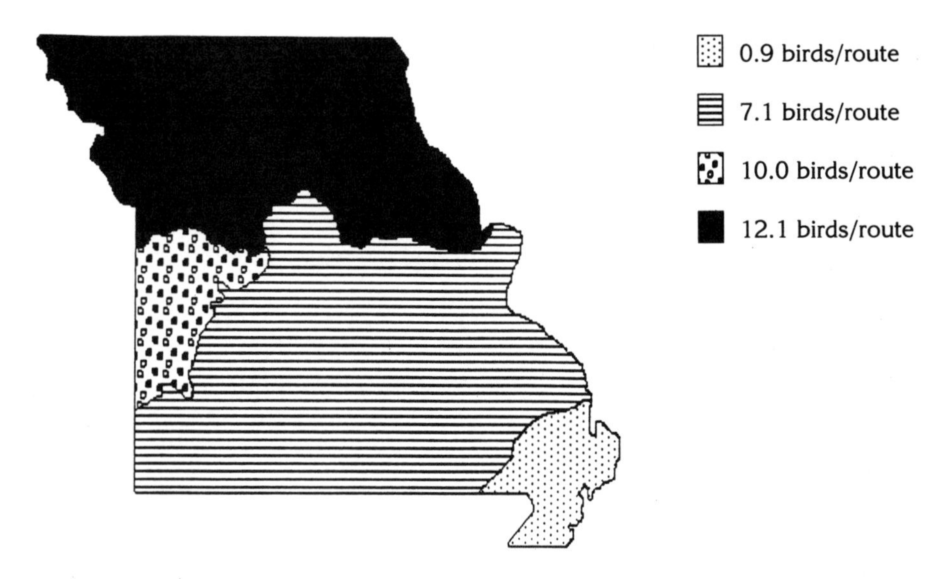

Habitat: Flooded fields, mudflats, edges of ponds and lakes.

Records:

Spring Migration: The first birds appear during the second week of Mar in the south and about a week later in the north. By late Mar, moderate-sized flocks (> 25 birds/group) are observed. The largest flocks (100+ birds) are seen in early to mid-Apr, with numbers dropping off rapidly by the end of Apr. Only individuals or small-sized groups are seen in early May, with no records after mid-May. Earliest dates: 2, 28 Feb 1985, Springfield (L. Hill-**BB** 52[2]:17); no.?, 1 Mar 1959, Sugar L. (FL). High counts: 150, 5 Apr 1984, Excelsior Springs, Clay Co. (CH-**BB** 51[3]:13); several reports of 100+ birds. Latest dates, both at Maryville SL: 1, 13 May 1976 (MBR, K. Kaufman-**BB** 44[2]:18); 2, 12 May 1975 (MBR, DAE).

Fall Migration: The first arrivals appear by mid-Sept, and most Sept observations are of solitary birds. In early Oct flocks consisting of 50+ birds are encountered, with peak in mid-Oct. Small-sized groups are seen until mid-Nov, but thereafter only single birds are observed. Earliest dates: 1, 7 Sept 1964, Mississippi Co. (JH); 1, 12 Sept 1975, Maryville SL (DAE-**BB** 44[1]:22). High counts, both by DAE at the Maryville SL: 160, 20 Oct 1972 (**BB** 40[1]:9); 110, 20 Oct 1988 (**BB** 56[1]:15). Latest date: 21, 11 Dec 1953, Creve Coeur L. (JEC et al.-**NN** 25:62).

Winter: The following are the only reliable winter observations: 1, 20 Dec 1987, Jackass Bend CBC; 2, 21 Dec 1988, Big Oak Tree SP CBC; 1, 29 Dec

1973, Kansas City North CBC; 1, 31 Dec 1971, Big Oak Tree SP CBC; 1, 4 Jan 1986, Orchard Farm CBC; 2, 27 Jan 1962, L. Springfield (F. Shumate-**BB** 25[10]:20).

Comments: Formerly known as the Water Pipit; the A.O.U. Check-list Committee (1989) recently split this species from the European bird (*A. spinoletta*). Occasionally, birds in full nuptial plumage may be encountered in the spring. A male in such plumage was observed 15 miles east of Columbia, Callaway Co., on 3 May 1961 (DAE; NWMSU, no prep. number).

Sprague's Pipit (*Anthus spragueii*)

Status: Rare transient in west, hypothetical in east; hypothetical winter resident.

Documentation: Specimen: o?, 29 Oct 1977, SW of Maryville (NWMSU, DAE 2987).

Habitat: Dry, open short-grass prairie, especially on hilltops, harvested soybean fields, and pastures.

Records:

Spring Migration: There are fewer observations for this season than in the fall; all are from the western half of the state. Birds begin to appear at the end of Mar. Most observations are in Apr, but there are a few sightings for early May. Earliest date: 1, 13 Mar 1982, Taberville Prairie (LG-**BB** 49[3]:21). High count: 2 (1 specimen, apparently discarded from freezer, JR, pers. comm.), 4 Apr 1978, near Liberal, Barton Co. (JR-**BB** 45[3]:18). Latest date: 1, singing while circling over prairie, 13 May 1986, Taberville Prairie (S. Schoech).

Fall Migration: To date there are no Sept records, although it surely does arrive by then. The earliest records are for the beginning of the second week of Oct, with peak during late Oct. There are no reliable Dec observations. Earliest date: female, 7 Oct 1987, 3 miles west of Forest City, Holt Co. (MBR-**BB** 55[1]:12; ANSP 180002). High counts: 7+, 22 Oct 1989, St. Joseph (MBR, DAE, RF, LG, K. Jackson-**BB** 57[1]:50); 6, 13 Oct 1979, Taberville Prairie (TB-**BB** 47[1]:25). Latest date: 1, 26 Nov 1979, Osage Prairie, Vernon Co. (TB).

Winter: No verified record.

Comments: Although there are no records for the eastern half of the state, it undoubtedly does pass through there on a regular basis. Bohlen and Zimmerman (1989) list it as a regular transient in Illinois.

Family Bombycillidae: Waxwings

Bohemian Waxwing (*Bombycilla garrulus*)

Status: Rare, highly irregular winter resident.
Documentation: Specimen: male, 18 Dec 1976, Maryville (NWMSU, DAE 2989).
Habitat: Usually encountered at cedars and fruiting bushes.
Records:
Spring Migration: Birds are occasionally recorded during this period after major flight years. For example, a number of birds from the 1961–62 invasion lingered until late Mar on into mid-Apr. Latest date: 3, 15 Apr 1962, Trimble (FL-**BB** 29[2]:18).
Fall Migration: There is only a single reliable observation for this period: 1, 6 Dec 1973, Busch WA (m.ob.-**NN** 46:20).
Winter: This species is observed very infrequently. In winters when it does appear, usually only individuals or small flocks (< 5 birds) are observed. However, two major invasions have been recorded. The first documented invasion was the one during the winter of 1919–20. "Flocks aggregating 600 birds" were observed in Holt Co. that winter (C. Dankers), and 175 (4 collected) were recorded in Jan 1920 in eastern Jackson Co. (H. Harris-**BL** 22:107). The other major influx occurred during the winter of 1961–62, and birds were encountered across the northern half of the state. The largest group reported for that flight was the 40 at St. Joseph on 11 Feb (FL). Nonetheless, just across the state line in Johnson Co., Kansas, up to 1,000 birds were seen in Jan 1962 (Rising et al. 1978). During the past twenty years, it has been recorded only twice on CBCs, both of single birds on the Maryville count.

Cedar Waxwing (*Bombycilla cedrorum*)

Status: Common transient; rare summer resident; uncommon, irregular winter resident.
Documentation: Specimen: female, 28 May 1883, Jackson Co. (KU 39261).
Habitat: Most readily encountered at fruiting trees in parks, residential areas, and forest edge.
Records:
Spring Migration: There are two distinct periods when birds move through the state at this season. The earlier, more inconspicuous migration (less clearly defined because of the presence of winter residents), occurs primarily in Mar. The second movement is during the last two weeks of May and early Jun. During the latter period, birds are encountered statewide in relatively large numbers (in the hundreds). It is possible that the

earlier movement consists of birds that have wintered in more northern climes (the southern United States), while the May migration is perhaps composed of birds that have wintered much farther south (possibly in Mexico, Central America). High count: 2,100, 10 Mar 1986, Taney Co. (PM-**AB** 40:480); 2,000, 22 Feb 1948, Springfield (B. Eisenmeyer-**BB** 15[4]:4).

Summer: A rare and local breeder throughout the state. The southernmost counties where it has been recorded breeding are Lawrence, Christian, and Shannon. In addition to the relatively small breeding population, migrant flocks are seen in early June (see above) and occasionally in late July.

Fall Migration: A few small-sized flocks (< 20 birds/flock) are occasionally seen as early as Aug: 23, 24 Aug 1971, Swan L. (FL, MBR, JHI). However, it is generally rare until the end of Sept and early Oct, and then flocks made up of over a hundred birds are regularly seen. Although the majority pass through in Oct, less frequent, but still impressive flocks may be encountered in Nov and Dec.

Winter: This species is dependent on the availability of small fruit; therefore it is irregular and erratic in distribution, but not surprisingly, it is most common in the Ozark and Ozark Border (average 2.0–4.0 birds/pa hr; < 0.4 bird/pa hr elsewhere) where cedars are most abundant. Infrequently, huge, roving flocks (numbering in the high hundreds or low thousands) are observed. High counts: 2,135, 31 Dec 1988, Dallas Co. CBC; 1,689, 2 Jan 1972, Weldon Spring CBC.

Family Laniidae: Shrikes

Northern Shrike (*Lanius excubitor*)

Status: Casual winter resident in the western quarter of Glaciated Plains.

Documentation: Specimen: female, 1 Feb 1977, SW of Maryville (NWMSU, DAE 2985).

Habitat: Relatively open areas, especially at edge of thickets.

Records:

Spring Migration: There are three records of wintering birds lingering into early Mar as follows: male, 1 Mar 1884, Lexington, Lafayette Co. (ISU 781); 1, 7 Mar 1971, Squaw Creek (MBR-**AFN** 25:587); 1, 9 Mar 1977, Squaw Creek (TB-**BB** 44(4):27).

Fall Migration: Three records: 1 taken (specimen apparently no longer extant) on 24 Nov 1901, Holt Co. (Dankers; Harris 1919b); 1, photographed and banded, 9 Dec 1969, extreme northern Atchison Co. (H. and F. Diggs-**BB** 36[4]:19; Fig. 28); 1, 10 Dec 1988, James A. Reed WA, Jackson Co. (MMH-**BB** 56[2]:55); the latter record represents the southernmost for the state.

Fig. 28. This photograph documents the first Northern Shrike in Missouri in over 60 years. The adult was mist-netted and photographed on 9 December 1969 in northern Atchison Co. by Fitzhugh and Hazel Diggs. This shrike is now known to occur irregularly during the winter in the northwestern corner of the state.

Winter: Prior to the 1950s, most wintering shrikes in northern Missouri were automatically (and incorrectly) referred to as this species (see Widmann 1907; Bennitt 1932; CBCs). In fact, besides two specimens taken by the reliable Dankers in Holt Co. (listed above) and the above Lafayette Co. specimen, there are no other unquestionable records prior to the 1969 photographed bird. Since the 1969 bird, there have been ten additional records (includes birds collected, photographed and banded). All except one are north of the Missouri R. in the extreme western section of the Glaciated Plains. It may occur in other sections of the Glaciated Plains, but observers have been virtually nonexistent in the north-central and northeastern sections of this region. High count: 2, 26 Dec 1981, Squaw Creek and extreme SW corner of Atchison Co. (MBR, FL, LG, J. Robinson-**BB** 49[2]:17).

Loggerhead Shrike (*Lanius ludovicianus*)

Status: Uncommon permanent resident in the western Glaciated Plains, Osage Plains, and the Mississippi Lowlands; rare elsewhere.

Documentation: Specimen: male, 14 Dec 1986, near Cole Camp, Benton Co. (ANSP 178154).

Habitat: Open areas with scattered trees, bushes, and hedgerows.

Records:

Spring Migration: Migration is largely disguised by the presence of a relatively large winter population; nevertheless, there is a discernible movement during Mar into early Apr. Most summer residents have territories established by late Mar. High count: 25, 5 Apr 1964, Trimble/Squaw Creek (ca. 175 car miles; FL).

Summer: Over the past thirty years there has been an overall decline of this species in the United States (Arbib 1977). This trend is also true for Missouri, especially in the Glaciated Plains and the Mississippi Lowlands (Kridelbaugh 1981). The decline has been attributed to pesticides and the loss of breeding habitat. Indeed, the Missouri decline is consistent with these probable factors, since the most significant decreases have been noted in areas of greatest cultivation, i.e., where pesticide use and land modification is most acute.

Although there has been a decline in shrike numbers during the past three decades, it should be noted that this species is at present more abundant in some regions of the state than it was at the turn of the century. It certainly was a much rarer bird in the Mississippi Lowlands, prior to this area being cleared, and the same can be said for much of the Ozarks (Widmann 1907). Conversely, it probably is less common today in the Glaciated and Osage plains than in 1900. The BBS data are used to depict the relative abundance of this shrike in Missouri for the past 23 years (Graph 14) and the differences in abundance among the natural divisions (Map 50).

Fall Migration: As in spring, migration is difficult to detect, but there is a definite movement during the later half of Sept, which peaks in Oct, and continues through mid-Nov. A bird, which had been banded as a juvenile in Saskastchewan on 31 July 1953, was recovered at Cross Timbers, Hickory Co. on 23 Sept of the same year (a distance of 1,953 miles; Burnside 1987).

Winter: Coincident with the decline in the breeding population, there has been a decrease in the number of wintering birds, especially in the Glaciated Plains region (Kridelbaugh 1981). This species is most common in the Osage Plains and Mississippi Lowlands (average of 2.5–3.0 birds/10 pa hrs) and least abundant in the Glaciated Plains and the Ozark Border (average of < 1.0 bird/10 pa hrs). High counts: 26, 19 Dec 1982, Trimble CBC; 24, 28 Dec 1965, Mingo CBC.

Comments: The breeding and the vast majority of the wintering populations are referable to the race *migrans* (=*mexicanus,* see Phillips 1986). However, the Great Plains race *excubitorides* has been recorded during migration and in winter (see banding recovery above).

Family Sturnidae: Starlings and Allies

European Starling (*Sturnus vulgaris*)

Status: Introduced; common permanent resident.

Documentation: Specimen: o?, 27 Jan 1930, Jackson Co. (KU 39268).

Habitat: Found in nearly every habitat, except continuous, heavily forested areas.

Comments: Following the establishment of this species in New York City in 1890–91, it first reached Missouri in 1928 and was recorded on 28 Jan at Charleston, Mississippi Co. (Cooke 1928). In Jan of 1929 it was noted at

Graph 14.
Missouri BBS data from 1967 through 1989 for the Loggerhead Shrike. See p. 23 for data analysis.

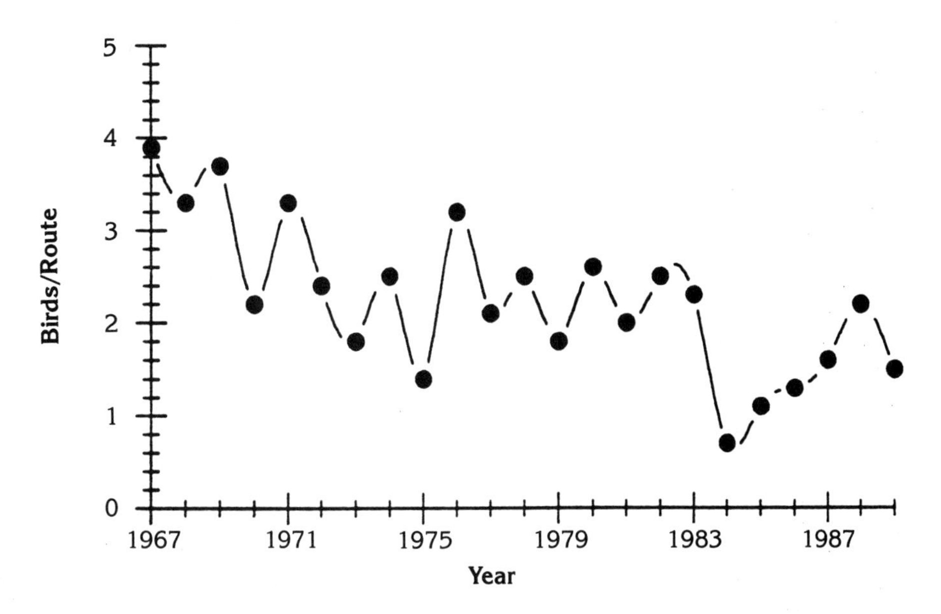

Map 50.
Relative breeding abundance of the Loggerhead Shrike among the natural divisions. Based on BBS data expressed as birds/route.

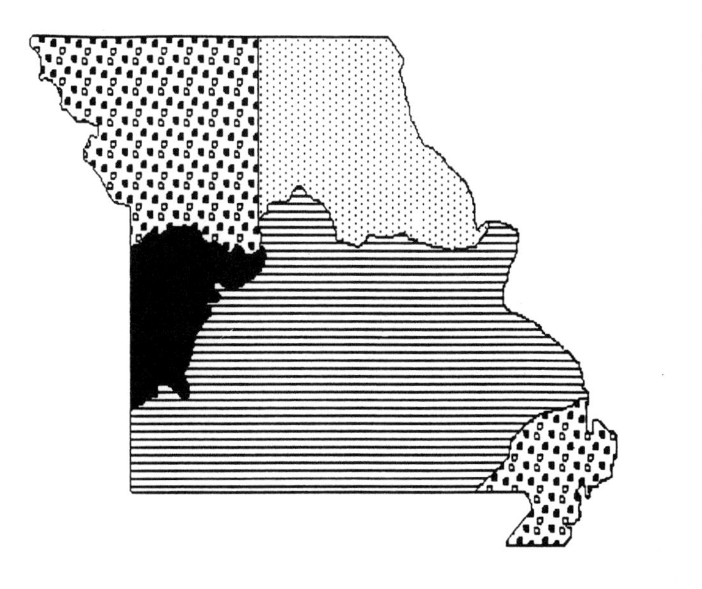

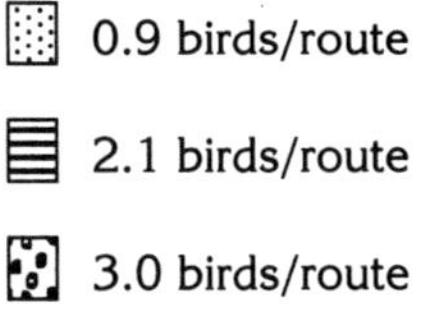

Map 51.
Relative breeding abundance of the European Starling among the natural divisions. Based on BBS data expressed as birds/route.

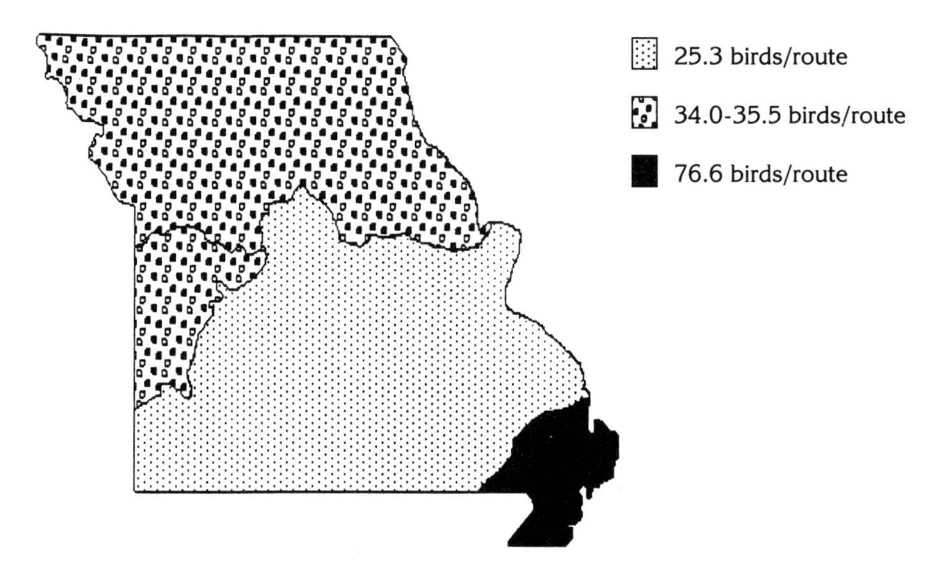

Patton, Bollinger Co. (A. Vaughn-**BB** 3[2]:11), and by Dec of that year it had reached Marionville, Lawrence Co., and West Plains, Howell Co. (Neff 1930). This species was first noted near Bethany, Harrison Co., Columbia, Boone Co., and Lebanon, Laclede Co. in 1931, and it bred at the latter two sites in 1932 and 1933, respectively. Nesting was also noted at Forest Park, St. Louis in 1932. At Maryville and Rolla it was first seen in 1932, and by the fall of 1934 flocks comprised of over 500 birds were recorded at Lebanon. Shortly thereafter it became established statewide.

An analysis of early banding data showed that birds in the midwest tended to migrate northeastward in spring and southwestward in fall (Kessel 1953). For example, starlings marked in winter at Independence, Jackson Co., primarily migrated northeastward, with a number of returns in spring in Iowa, northern Illinois, and Wisconsin. Furthermore, some birds banded in northeastern Illinois were recovered in Missouri in winter (Kessel 1953).

Map 51 shows the relative abundance of this species in the early summer across the state. CBC data show large variations in numbers, primarily the result of the presence or absence of winter roosts.

Family Vireonidae: Vireos

White-eyed Vireo (*Vireo griseus*)

Status: Locally common summer resident in Ozarks and Ozark Border, rare elsewhere.

Documentation: Specimen: male, 3 May 1968, near Maryville (NWMSU, DAE 2324).

Habitat: Brushy areas, thickets, and dense second growth at edge of forest and woodland.

Records:

Spring Migration: The initial migrants are seen during the first week of Apr in the extreme south and by the beginning of the third week of Apr in the north. Peak is during the last week of Apr in the south and the first week of May in the north. Earliest dates: 1, 29 Mar 1980, near Collins, St. Clair Co. (MBR, MB); no.?, 2 Apr 1896, Mississippi Lowlands (Widmann 1907); 1, 2 Apr 1989, south of Mincy, Taney Co. (PM, JH-**BB** 56[3]:93); 4, 3 Apr 1975, Big L., Mississippi Co. (JH).

Summer: Based on BBS data it is nearly ten times more common in the Ozarks and Ozark Border (average of 3.3 birds/route) than in any other region of the state. It is rarest in the western half of the Glaciated Plains.

Fall Migration: Summer residents and migrants are commonly seen through most of Sept, but it is rare by the first week of Oct. High count: 40+, 10 Sept 1977, Roaring R. SP (NJ, JG-**BB** 45[1]:28). Latest dates: 1, 18 Oct 1969, Columbia (BG); immature, 16 Oct 1989, near Noel, along Elk R., McDonald Co. (MBR-**BB** 57[1]:50).

Bell's Vireo (*Vireo bellii*)

Status: Uncommon summer resident in Osage and western half of Glaciated plains, rare elsewhere.

Documentation: Specimen: male, 26 May 1986, Bigelow Marsh (ANSP 178067).

Habitat: Thickets and brushy hedgerows.

Records:

Spring Migration: Usually not seen before early May, but by mid-May it is uncommon at nesting areas. Earliest dates: 1, 23 Apr 1964, St. Joseph (FL); 1, 25 Apr 1965, St. Louis (RA, KA-**NN** 37:5).

Summer: Most common in the Osage and western half of the Glaciated plains (Map 52). Rare elsewhere, except Mississippi Lowlands where it is apparently absent. This species status needs to be monitored, as it has been on a steady decline since the early 1980s BBS data.

Fall Migration: It is regularly heard singing at breeding localities into the second week of Sept, but thereafter it is rarely encountered. Latest dates:

Map 52.
Relative breeding abundance of the Bell's Vireo among the natural divisions. Based on BBS data expressed as birds/route.

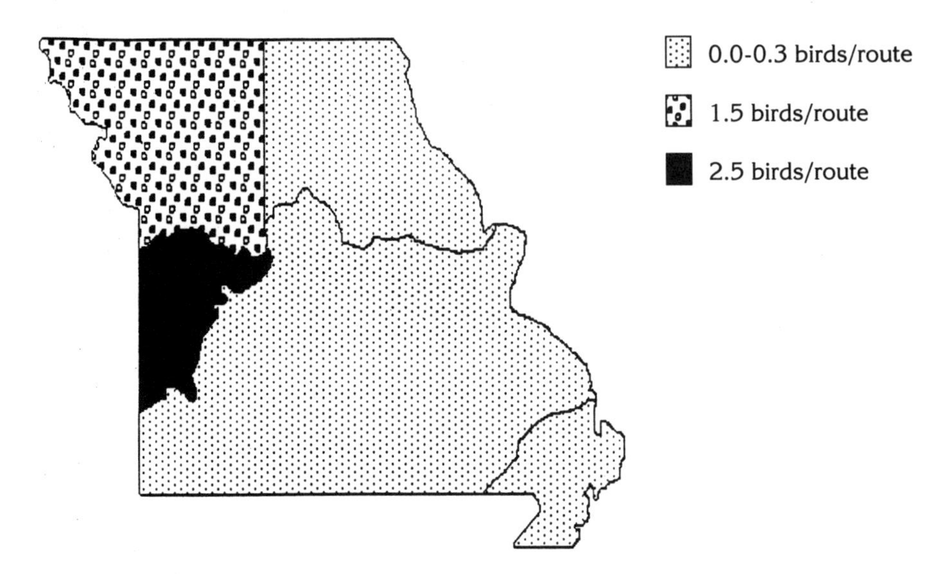

male, 1 Oct 1919, Lexington, Lafayette Co. (CMC 452); female, 26 Sept 1970, near Columbia (MU 1938).

Comments: The Bell's Vireo is one of only three species (the others being the Lark and Harris' sparrows) that were first made known to science through their discovery in Missouri. The type was taken on 6 May 1843 at St. Joseph (Cooke 1910).

Solitary Vireo (*Vireo solitarius*)

Status: Uncommon transient; accidental winter visitant.

Documentation: Specimen: male, 7 May 1967, near Maryville (NWMSU, LCW 568).

Habitat: Encountered in virtually any size or type of woodland and forest.

Records:

Spring Migration: The first individuals appear during the last week of Apr, with peak during the first week of May. It is rarely seen after mid-May. Earliest date: 1, 19 Apr 1985, St. Louis (RK). High count: 6, 28 Apr 1983, St. Louis (RK). Latest date: 1, 29 May 1983, Wallace SP, Clinton Co. (CH, MMH).

Summer: A single observation: 1, 8 June 1985, Cass Co. (JG).

Fall Migration: The first individuals appear at the beginning of Sept and peak in mid- to late Sept. It is regularly seen through early Oct, but it is only casually observed during the last two weeks of Oct. Earliest date: 1, 28 Aug

1964, St. Joseph (FL). High counts: 8, 15 Sept 1979, St. Joseph/Squaw Creek (FL); 7, 3 Oct 1988, Atchison Co. (MBR-**BB** 56[1]:15). Latest dates: 1, 26 Nov 1963, Columbia (WG-**BB** 30[4]:19); 1, at suet feeder, 25 Nov 1981, Jefferson City (JW-**BB** 49[1]:14).

Winter: One record: 1, 15 Dec 1962, St. Joseph (FL-**BB** 30[1]:32).

Comments: All Missouri records are referable to the nominate race.

Yellow-throated Vireo (*Vireo flavifrons*)

Status: Uncommon summer resident.

Documentation: Specimen: male, 31 May 1965, Gravois Mills (NWMSU, DAE 1069).

Habitat: Mature woodland and forest, especially bottomland.

Records:

Spring Migration: It arrives as early as the beginning of Apr in the extreme south and by the third week in the north. Peak is at the end of Apr or early May. Earliest dates: 30 Mar 1896, Mississippi Lowlands (Widmann 1907); 1, 3 Apr 1985, Big Oak Tree SP (JH).

Summer: Although BBS data underrepresent this species, the overall relative abundance among the regions nevertheless is accurate. It is most common in the Ozarks and Ozark Border where an average of 0.3 birds/route have been recorded. An average of ca. 0.2 birds/route have been found in the Glaciated and Osage plains. None have been recorded on the sole BBS route in the Mississippi Lowlands, but the bird does breed there.

Fall Migration: The bulk of the summer residents leave and the migrants pass through during Sept. By the end of Sept it is rarely seen, with a few lingering into early Oct. Latest dates: 17 Oct 1903, Monteer, Shannon Co. (W. Savage; Widmann 1907); 1, 14 Oct 1956, L. of Ozarks SP (FL).

Warbling Vireo (*Vireo gilvus*)

Status: Common summer resident.

Documentation: Specimen: male, tower kill, 10–11 Sept 1964, Jackson Co. (NWMSU, DAE 702).

Habitat: Most common in riparian woodland, especially in cottonwoods.

Records:

Spring Migration: In the south birds appear as early as the beginning of the second week of Apr, and in the north they arrive during the third week. Peak is in late Apr in the south and early May in the north. Earliest dates: 6 Apr 1893, Mississippi Lowlands (Widmann 1907); 1, 8 Apr 1977, Mississippi Co. (JH). High count: 40, 6 May 1962, St. Joseph/Squaw Creek (FL).

Summer: This species is very common in cottonwoods that line the major rivers in the state, especially the Missouri and Mississippi rivers. It is

Map 53.
Relative breeding abundance of the Warbling Vireo among the natural divisions. Based on BBS data expressed as birds/route.

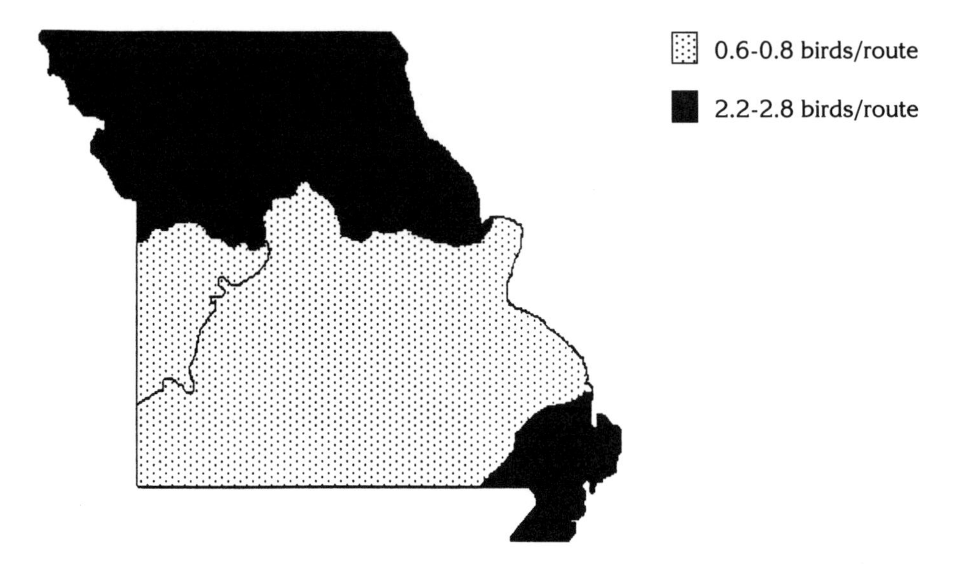

the most common vireo across the Glaciated Plains and in the Mississippi Lowlands (Map 53).

Fall Migration: Most have left the state by mid-Sept, with a few individuals regularly seen into the beginning of the fourth week of Sept. Latest dates: 1, 1 Oct 1990, Brickyard Hill WA (MBR); 27 Sept 1891, St. Louis (Widmann 1907).

Winter: An extraordinary record is of a bird that was attending a suet feeder and eventually was found dead on 20 Jan 1936, Kansas City (A. Shirling-**BB** 3[2]:12).

Philadelphia Vireo (*Vireo philadelphicus*)

Status: Rare transient.

Documentation: Specimen: female, 21 May 1966, Maryville (NWMSU, DAE 1112).

Habitat: Deciduous woodland and forest.

Records:

Spring Migration: The first arrivals are observed by the beginning of the final week of Apr. Peak is during mid-May, with a few birds occasionally seen through the final week of May. An average of 0.2 birds/hr was recorded at Forest Park between 6–25 May 1979–90 (RK; n=286 hrs). Earliest dates: 2, 18 Apr 1965, Big Oak Tree SP (JH); 1, 21 Apr 1983, Busch WA (SRU-**NN**

Graph 15.
Missouri BBS data from 1967 through 1989 for the Warbling Vireo.
See p. 23 for data analysis.

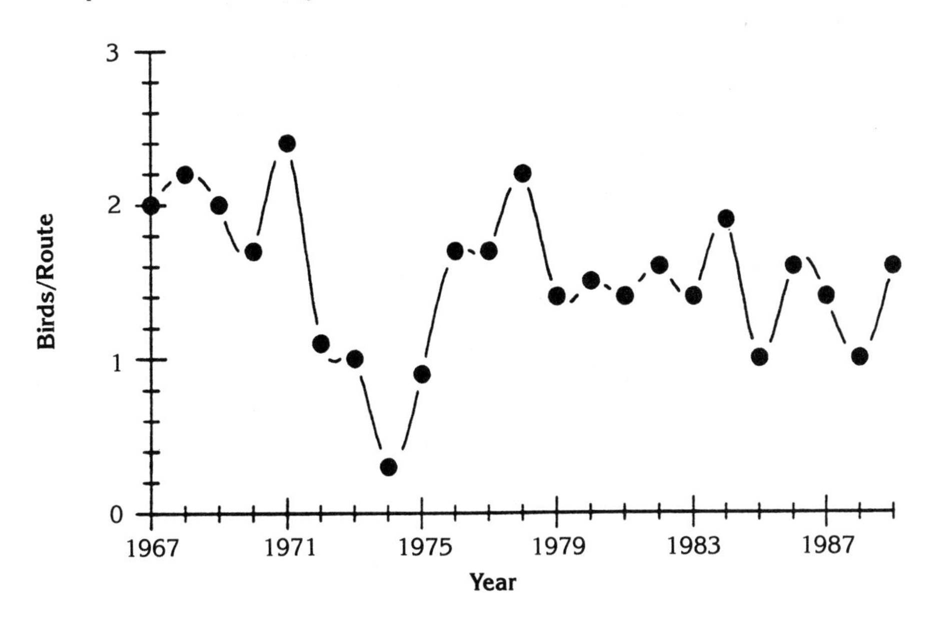

55:51). High count: 10–12, 5 May 1990, Mingo/Duck Creek (BRE, RB-**BB** 57[3]:140).

Summer: One observation of a late migrant: 1, 2 June 1988, Forest Park (RK-**BB** 55[3]:89).

Fall Migration: The first appear in early Sept, but peak is not until the end of the third or at the beginning of the fourth week of Sept. It is regularly seen through the first week of Oct but is absent thereafter. Earliest date: 1, 24 Aug 1979, Busch WA (F. Hallet-**NN** 51:63). High counts: tower kills, 15, 24 Sept 1960, Columbia (**BB** 28[1]:9); 7, 20 Sept 1963, Columbia (George 1963). Latest date: male, 24 Nov 1984, Maryville (MBR, DAE; NWMSU, MBR 1450).

Red-eyed Vireo (*Vireo olivaceus*)

Status: Common summer resident.

Documentation: Specimen: o?, 5 May 1879, Charleston, Mississippi Co. (MCZ 45105).

Habitat: Breeds in relatively large, mature tracts of woodland and forest.

Records:

Spring Migration: The first migrants appear at the southern border at the beginning of Apr, about a week later in central Missouri, but not until

Map 54.
Relative breeding abundance of the Red-eyed Vireo among the natural divisions. Based on BBS data expressed as birds/route.

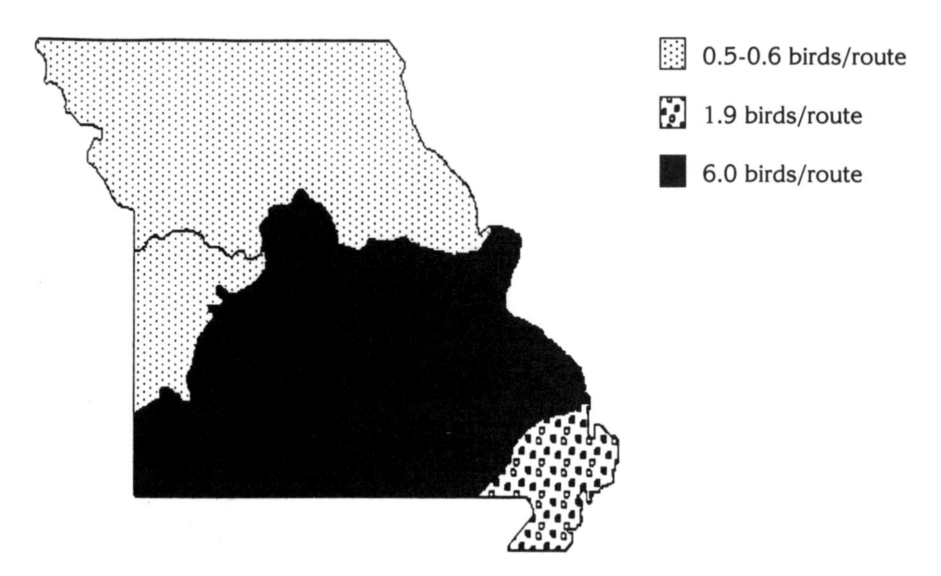

the last week in northern Missouri. Peak is not until mid-May. Earliest dates: 1, 29 Mar 1980, near Collins, St. Clair Co. (MBR, MB); 2, 3 Apr 1985, Big Oak Tree SP (JH). High count: 12, 13 May 1973, Honey Creek WA, Andrew Co. (FL).

Summer: The Red-eyed is the common vireo in the larger tracts of woodland and forest. It is most abundant in the Ozarks and least so in the Glaciated and Osage plains (Map 54).

Shirling (1920) recorded 84 males at Swope Park in 1916, while Branan and Burdick's (1981) census of the same site in 1973 revealed only 34 males. The latter workers documented over a 50% decrease in upland forest at the site between the two censuses; however, acreage of riparian forest was similar between the two censuses.

Fall Migration: The summer resident population begins to be supplemented by migrants at the end of Aug; however, peak is not until the end of the third or at the beginning of the fourth week of Sept. Birds are readily seen through the first week of Oct, e.g., 12, tower kill, 6 Oct 1962, Cape Girardeau (Heye 1963), but none are seen after mid-Oct. High counts: tower kills at Columbia; 185, 20–21 Sept 1963 (George 1963); 126, 19 Sept 1966 (Elder and Hansen 1967). Latest date: 1, 13 Oct 1962, Charleston, Mississippi Co. (JH).

Graph 16.
Missouri BBS data from 1967 through 1989 for the Red-eyed Vireo.
See p. 23 for data analysis.

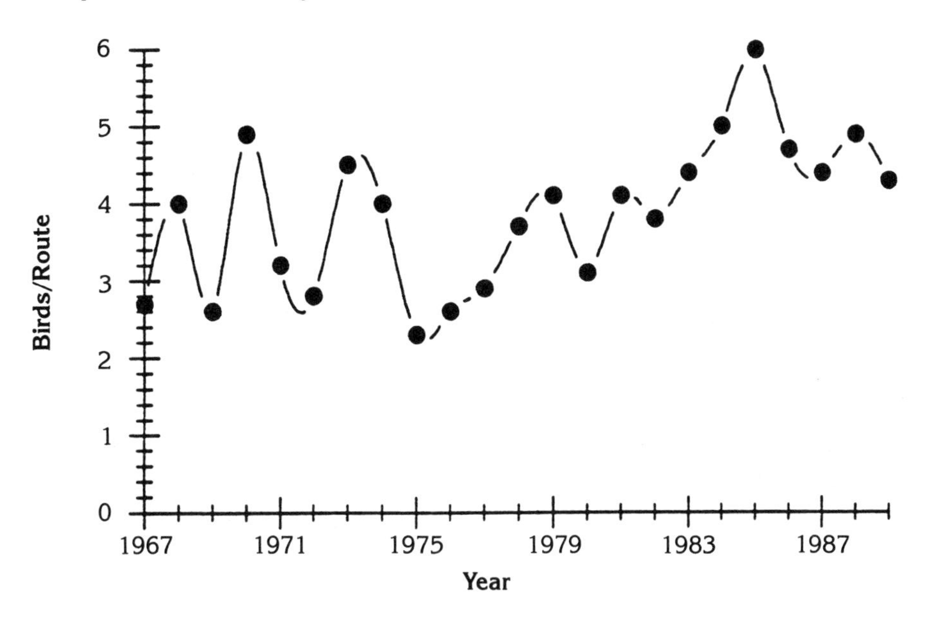

Family Emberizidae: Emberizids

Bachman's Warbler (*Vermivora bachmanii*)

Status: EXTIRPATED; possibly EXTINCT; formerly a common summer resident along the St. Francois R. drainage in Dunklin Co.; accidental vagrant in southeastern Ozarks.

Documentation: Specimens: see below.

Habitat: In dense growth in swampy woodland and forest.

Records:

Spring Migration: Widmann (1907) found singing males as early as 17 Apr 1898 in Dunklin Co. Most of his observations of this species were made in May. In fact, the nest and eggs of this species were first made known to science from Widmann's efforts during mid-May 1897 when he located a nest on 13 May in the final stages of construction. Eventually 3 eggs were laid (Widmann 1897). Howell (1911) found it "rather numerous" during 25 Apr–5 May 1909 along the St. Francois R. in Dunklin Co. He collected six males in this period (USNM 204985–204990).

Widmann apparently found this species only along the St. Francois R. in Dunklin Co.; however, Woodruff (1907) took two males during May of 1907 in the southeastern section of the Ozarks. One (AMNH 229845) was taken near Ink, Shannon Co., on 2 May, while the other (AMNH 229846) was collected near Grandin, Carter Co., on 23 May.

Apparently this species persisted in the St. Francois drainage through the late 1940s, as Cunningham found the bird west of Cardwell, Dunklin Co., in 1934 (**BB** 7[9]:64) and again in the spring of 1948 (**AFN** 2:181). The 1948 observation apparently is the last for the state; it involved a singing male "in second growth gum and cypress." Searches during the late 1950s at the former breeding sites were unsuccessful (DAE, D. Snyder).

Blue-winged Warbler (*Vermivora pinus*)

Status: Uncommon summer resident in the Ozarks and Ozark Border; rare and local in the Mississippi Lowlands and the eastern sections of the Osage and Glaciated plains.

Documentation: Specimen: male, 20 Apr 1907, Current R., Shannon Co. (AMNH 229878).

Habitat: Relatively open areas with scattered small trees and scrubs, brushy areas at woodland and forest edge.

Records:

Spring Migration: It is first detected in southern Missouri during the second week in Apr. Peak is in early May. During migration it is regularly recorded in all areas of the state, except the northwestern corner where it has been observed on only three occasions during the past 30 years. Korotev has recorded an average of 0.3 birds/hr at Forest Park between 25 Apr–8 May 1979–90 (n=246 hrs). Earliest dates: 1, 1 Apr 1988, Farmington, St. Francois Co. (BRE); 3 Apr 1896, "southern border" (Widmann 1907); 1, 9 Apr 1963, St. Louis (Hanselmann 1963).

Summer: It is most common in the Ozarks and Ozark Border, and rare and quite local in the Mississippi Lowlands and the eastern sections of the Osage and Glaciated plains. The average of 6.7 birds on BBS route 8 is the highest in the state. Formerly it was more widespread in the western Glaciated Plains. Shirling (1920) recorded as many as 17 singing males in Swope Park during June 1916, and it remained a "regular" breeder in Jackson Co. until the 1940s (WC; Rising et al. 1978). Widmann (1907) stated that it was "common" at Chillicothe, Livingston Co., on 16 May 1854.

At present it is not known to breed west of ca. 93° 00′ W, north of the Missouri R. In the Osage Plains it is found only in the extreme eastern section; it has not been recorded on any of the ten Cass County Summer Counts (1978–87; JG, pers. comm.).

Fall Migration: This is one of the earliest warblers to leave, as very few are seen after the first week of Sept. Latest dates: 1, 15 Sept 1950, south of Gravois Mills (DAE); 1, 15 Sept 1989, Mingo (BRE-**BB** 57[1]:50).

Comments: Hybrids between the Blue-winged and the Golden-winged Warbler are rarely encountered. Phenotypes of these hybrids range from birds that are largely yellow below, with a Golden-winged face pattern (referred to as "Lawrence's Warbler") to birds that are mainly white below, with a Blue-winged face pattern (called "Brewster's Warbler"). The latter is more frequently seen (RK recorded six in eleven springs, 1979–89, at Forest Park [Korotev 1990]). At least one specimen exists: male, 12 May 1907, Spring Valley, Shannon Co. (Woodruff 1908; AMNH 229871).

Golden-winged Warbler (*Vermivora chrysoptera*)

Status: Uncommon transient; former accidental summer resident.

Documentation: Specimen: male, 28 May 1968, near Maryville (NWMSU, DAE 2325).

Habitat: Brushy areas, woodland and forest edge.

Records:

Spring Migration: This species first appears at the end of Apr and peaks during the second week of May. Late migrants are seen through the end of May. Korotev recorded an average of 0.6/hr between 27 Apr–17 May 1979–90 at Forest Park (n=359 hrs). Earliest dates: 1, 24 Apr 1990, Forest Park (RK); at least three sightings for 25 Apr. High counts in east: 20, 5 May 1990, Mingo/Duck Creek (BRE, RB-**BB** 57[3]:140); 20, 13 May 1990, Tower Grove Park (JZ); in west, 8, 15 May 1982, Van Meter SP, Saline Co. (CH, KH-**BB** 49[3]:21).

Summer: There is a single, definite breeding record—quite surprisingly—from Mississippi Co. in the southeast where 2 eggs were collected from a nest on 29 May 1890 (M. Crawford; WFVZ 73393). The nest, made of bark strips lined with fine grass and hair, was located in a clump of briers and vines ca. 1.5 m above the ground (from original data sheet; WFVZ). This represents the furthest south that this species has been documented breeding in the midwestern United States. In addition, Widmann (1907) mentions that it was noted as a breeder in northeastern Missouri in Audrain Co. in 1884 by Mrs. Musick, and apparently breeding was recorded during this same period in Lee Co., Iowa (opposite Clark Co., Missouri), and in Richland Co., Illinois (opposite St. Louis). These records coincide with an increase and range expansion of Golden-wings in the northeastern United States (Connecticut, New York) in the late 1880s through the turn of the century (Gill 1980). The Missouri records are consistent with the argument that the above expansion was the result of deforestation, which temporily provided prime breeding habitat, i.e., early stages of old field succession, for Golden-wings (Gill 1980).

There have been the following recent observations of single birds (presumably of late spring and early fall migrants): 1, 8 June 1980, Bluffwoods SF (FL-**AB** 34:901); 1, singing, mid-June 1984, Tyson Valley RC (DC-**NN** 56:63); 1, 30 July 1977, Roaring R. SP (JG-**BB** 44[4]:31).

Fall Migration: Southbound birds begin to appear by the last week of

Aug. Peak is during mid-Sept, and birds are regularly seen until the end of Sept. However, there is only a single Oct observation. Earliest date: 1, 18 Aug 1985, Busch WA (JZ-**NN** 57:68). High count: 7, 17 Sept 1979, St. Louis (TB-**NN** 51:68). Latest date: 1, 23 Oct 1969, Springfield (NF-**BB** 38[3]:6).

Comments: See Blue-winged Warbler.

Tennessee Warbler (*Vermivora peregrina*)

Status: Common transient.

Documentation: Specimen: female, 16 Oct 1962, Cape Girardeau (SEMO 144).

Habitat: Woodland and forest.

Records:

Spring Migration: Normally the first arrivals appear at the end of the third week of Apr in the south and about a week later in the north. In most springs, usually during the second week of May, it is the most conspicuous of the migrant warblers, as it is heard singing in residential, as well as rural areas. A few stragglers are seen through the final week of May. An average of 5.8 birds/hr was recorded at Forest Park between 20 Apr–25 May 1979–90 (RK; n=521 hrs). Earliest dates: 1, 9 Apr 1967, St. Joseph (JHA); 1, 9 Apr 1989, Kennett, Dunklin Co. (H. Schanda). High counts: 50, 18 May 1969, St. Joseph (FL); 43, 7 May 1982, Forest Park (RK).

Summer: Late migrants are seen rarely during the first week of June and casually thereafter. Latest dates: 1 singing, 28 June 1980, Providence, Boone Co. (BG-**BB** 47[4]:13); 1 singing, 20 (not 21) June 1956, south of Gravois Mills (DAE-**BB** 23[8]:4).

Fall Migration: The first arrivals appear at the beginning of Sept and peak during the third week of Sept. Birds are regularly encountered through the first week of Oct, but they are only accidentally seen after mid-Oct. Earliest date: 1, 9 or 10 Aug 1985, St. Louis (M. Wiese-**NN** 57:67). High counts, tower kills at Columbia: 92, 19–20 Sept 1966 (Elder and Hansen 1967); 66, 20–21 Sept 1963 (George 1963). Latest dates (only Nov observations): 1, 27 Nov–4 Dec 1988, St. Louis (L. Hepler, RG et al.-**BB** 56[1]:15; photos in MRBRC files); 1, 24 Nov 1985, St. Louis (RA-**NN** 57:90); 1, 11 Nov 1983, Fayette, Howard Co. (C. Royall-**BB** 51[1]:45).

Orange-crowned Warbler (*Vermivora celata*)

Status: Common transient in west, uncommon in east; casual winter visitant.

Documentation: Specimen: male, 9 Oct 1987, Thurnau WA, near Craig, Holt Co. (ANSP 180005).

Habitat: Brushy areas, weedy fields, forest and woodland edge.

Records:

Spring Migration: Less conspicuous at this season than in the fall. More common during both spring and fall in the western half of the state. Usually not seen before the third week of Apr. The bulk passes through during the final days of Apr and the first few days of May. It is absent after mid-May. An average of 0.3 birds/hr was recorded at Forest Park between 20 Apr–10 May 1979–90 (RK; n=331 hrs). Earliest dates: 1, at feeder (winter resident ?), 27 Feb–31 Mar 1980, St. Louis (KA-**BB** 47[3]:13); 1, 9 Apr 1982, Big Oak Tree SP (JH). High counts, both at St. Joseph by FL: 15, 30 Apr 1961; 12, 7 May 1967. Latest date: 1, 27 May 1982, Missouri R., Atchison Co. (MBR).

Fall Migration: Usually not seen until mid-Sept and does not become common until the last week of Sept. Peak is during the first week of Oct. It remains fairly common through mid-Oct but is seen in only small numbers (1–3 birds/day) during the final two weeks of Oct. After the first week of Nov it is only casually recorded, although there are records for the final week of the period. Earliest date: 1, 8 Sept 1987, St. Louis (EL-**BB** 55[1]:12). High counts: 37+, 28 Sept 1974, east of Maryville (MBR); 29, 6 Oct 1987, Brickyard Hill WA (MBR-**BB** 55[1]:12); 25–30, 17 Sept 1981, St. Louis (TB-**NN** 51:68).

Winter: There are at least four reliable records: 1, 19 Dec 1981, Maryville CBC (DAE, TE); 1, at feeder, Dec 1980, St. Louis (KA-**NN** 53:15); 2, 28 Dec 1978, Mingo CBC; 1, 11 Jan 1942, Missouri R. bottoms, Jackson Co. (WC et al.-**BB** 9[4]:22).

Comments: Only the nominate race has been recorded.

Nashville Warbler (*Vermivora ruficapilla*)

Status: Common transient; accidental winter visitant.

Documentation: Specimen: male, 29 Oct 1967, near Arkoe, Nodaway Co. (NWMSU, LCW 770).

Habitat: Brushy areas, woodland, and forest.

Records:

Spring Migration: Usually not seen until the beginning of the third week of Apr in the south and about a week later in the north. Peak is at the end of the first or beginning of the second week of May. It is rarely seen during the final week of May. At Forest Park, an average of 4.5 birds/hr was recorded between 20 Apr and 15 May 1979–90 (RK; n=419 hrs). Earliest dates: 1, 8 Apr 1986, St. Louis (SRU-**NN** 58:48); 1, 11 Apr 1987, Taney Co. (PM, TB-**BB** 55[1]:14). High counts: 50, 5 May 1963, St. Louis (Hanselmann 1963); 35, 4 and 6 May 1983, Forest Park (RK). Latest date: 1, singing, 30 May 1986, northeastern corner of Clark Co. (MBR).

Summer: Single observation, presumably of a very early fall migrant: 1, 30 July 1977, Roaring R. SP (JG-**BB** 44[4]:31).

Fall Migration: Migrants begin to reappear at the end of Aug. It does not become common until the second week of Sept, with peak at the end of the third or the early part of the fourth week of Sept. It remains common through the first week of Oct, and birds are encountered in smaller numbers (1–3/day) through the third week of Oct. Earliest dates: 1, 26 Aug 1971, Maryville (MBR); 3, 27 Aug 1967, St. Joseph (FL). High counts: 24, tower kill, 20–21 Sept 1963, Columbia (George 1963); 23, 25 Sept 1968, St. Louis (JEC-**NN** 40:92). Latest dates: 1, 11 Nov 1974, Busch WA (JC-**BB** 42[2]:7); two sightings for 30 Oct.

Winter: One observation: 1, 28 Dec 1975, Weldon Spring CBC.

Northern Parula (*Parula americana*)

Status: Common summer resident.

Documentation: Specimen: male, 13 June 1987, McCormack L., Oregon Co. (ANSP 178215).

Habitat: Most common in bottomland woodland and forest.

Records:

Spring Migration: It arrives in the southern section of the state during the first few days of Apr. Usually, in the north, it is not encountered until mid-Apr. Peak is during late Apr in the south and in early May in the north. An average of 0.4 birds/hr was recorded at Forest Park between 20 Apr–10 May 1979–90 (RK; n=331 hrs). Earliest dates: 28 Mar 1896, Poplar Bluff, Butler Co. (Widmann 1907); 1, 29 Mar 1980, near Collins, St. Clair Co. (MBR, MB). High counts: 89, 18 Apr 1987, along 18 mile stretch of Current R., Shannon Co. (RK-**BB** 54[3]:15); 12, 4 May 1963, St. Louis (Hanselmann 1963).

Summer: BBS routes grossly underestimate the abundance of this species. Surveys along river and stream courses are needed to adequately assess the summering population. It is most common in the Ozarks and Ozark Border riparian areas. The average of 7.4 birds on BBS route 8 is the highest in the state. Although none have been recorded on the single BBS route in the Mississippi Lowlands, it does breed there; however, it is decidedly less common there now than it was prior to the region being cleared in the early 1900s.

A significant decline (32 vs. 11 males) was noted at Swope Park between Shirling's (1920) 1916 survey and Branan and Burdick's (1981) 1973 census. Missouri BBS data indicate that this warbler is increasing in the state (Graph 17).

Fall Migration: Summer residents become quiet and inconspicuous by the latter part of Aug and presumably many leave at that time. Migrants are noted through Sept into early Oct: 2, tower kill, 6 Oct 1962, Cape Girardeau (Heye 1963; SEMO 167). Latest date: 1, 22 Oct 1983, Duck Creek (SD-**BB** 51[1]:45).

Graph 17.
Missouri BBS data from 1967 through 1989 for the Northern Parula.
See p. 23 for data analysis.

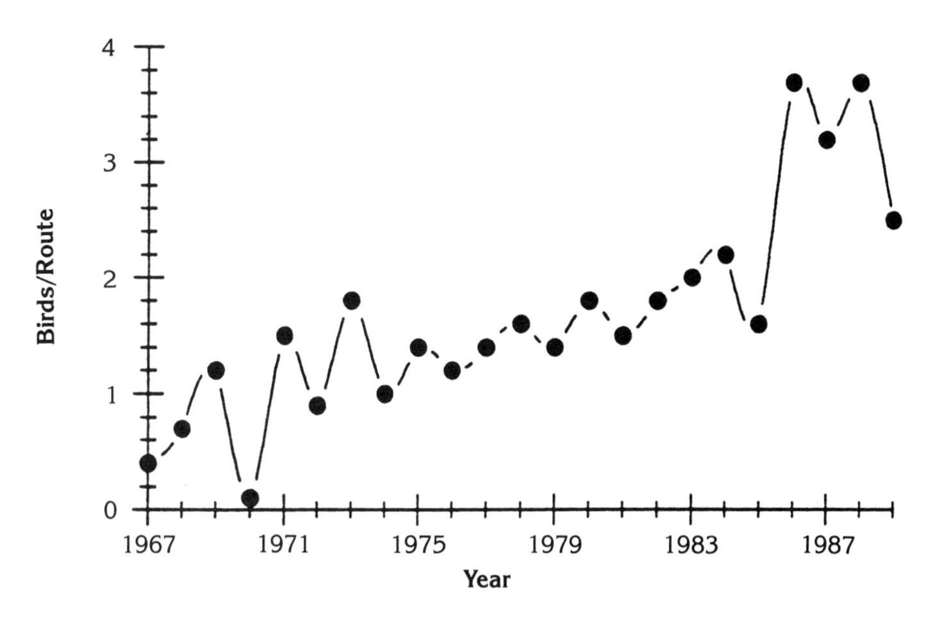

Yellow Warbler (*Dendroica petechia*)

Status: Common transient; locally common summer resident across the Glaciated Plains and along the Missouri and Mississippi river drainages; rare to uncommon elsewhere.

Documentation: Specimen: female, 11 Sept 1932, Jackson Co. (KU 19378).

Habitat: Most common in second-growth woodland and thickets near water, especially in willows.

Records:

Spring Migration: The first birds arrive during mid-Apr in the south and by the beginning of the final week of Apr in the north. Peak is during the first week of May. Earliest dates: male, 11 Apr 1923, Lexington, Lafayette Co. (CMC 359); 13 Apr 1893, Hornersville, Dunklin Co. (Widmann 1907). High count: 50, 5 May 1963, St. Joseph/Squaw Creek (FL).

Summer: As stated above it is most common in the Glaciated Plains and along the Missouri and Mississippi rivers. It is more abundant and widespread in the western than the eastern half of the Glaciated Plains. On BBS route 30 a state high average of 2.1 birds has been recorded. The Yellow Warbler is rare to uncommon along streams and rivers of the Ozarks, Ozark Border, and Osage Plains.

Fall Migration: Most have left the state by late Aug, as only small numbers (< 5 birds/day) are seen in early Sept. Usually only single birds are infrequently encountered thereafter. Three were killed at towers at Columbia on each of the following dates: 19–20 Sept 1966 (Elder and Hansen 1967); 20–21 Sept 1963 (George 1963). There are no Oct sightings, but there is one Nov observation: 1, 8 Nov 1969, Maryville (DAE-**BB** 36[4]:12).

Chestnut-sided Warbler (*Dendroica pensylvanica*)

Status: Common transient; casual summer resident in east.

Documentation: Specimen: female, 2 Oct 1964, Cape Girardeau (SEMO 220).

Habitat: During migration in woodland and forest; breeds in brushy second growth at forest edge.

Records:

Spring Migration: The first individuals appear at the end of Apr and peak during the second week of May. Migrants are seen throughout May. Korotev recorded an average of 1.9 birds/hr between 2–23 May 1979–90 at Forest Park (n=349 hrs). Earliest dates: 1, 20 Apr 1990, Farmington, St. Francois (BL); 23 Apr 1885, Mt. Carmel, Audrian Co. (Mrs. Musick; Widmann 1907). High counts: 40, 18 May 1963, St. Louis (Hanselmann 1963); 26, 8 May 1983, Forest Park (RK).

Summer: Widmann (1907) states that it "has repeatedly been found breeding in eastern Missouri in places grown with hazel, blackberry and scrub-oak." He gave definite records for the following counties: St. Louis (2 nests/eggs in 1893; USNM 265520, 265530), Iron, and Audrain.

Recently, it has been found or suspected of breeding in Dent and Oregon counties. In Dent Co., Evans (1980) found two territorial males along a powerline right-of-way in May–June 1978, and he located two nests (one which is deposited at NWMSU) in this same area in 1979. Another territorial bird was present near the Eleven Point R., Oregon Co. in 1984 (JR-**BB** 51[4]:17). This species undoubtedly is a more regular breeder in the eastern section of the Ozarks and Ozark Border than the few records indicate.

Fall Migration: A few migrants reappear at the very end of Aug, and numbers increase to peak in mid-Sept. Birds are regularly seen, although in much smaller numbers, through the first week of Oct, e.g., 5, tower kill, 6 Oct 1962, Cape Girardeau (Heye 1963). However, only single birds are casually seen thereafter. Earliest date: 16 Aug 1964, Kansas City (EC-**BB** 31[4]:19). High counts: 30, 12 Sept 1964, St. Louis (SHA-**BB** 31[4]:18); 15, tower kill, 20–21 Sept 1963, Columbia (George 1963); 15, tower kill, 24 Sept 1960, Columbia (**BB** 28[1]:9). Latest date and only Nov observation: 1, 16–28 Nov 1980, St. Louis (MP-**NN** 53:4).

Magnolia Warbler (*Dendroica magnolia*)

Status: Common transient.
Documentation: Specimen: male, 14 June 1961, St. Louis (SEMO 108).
Habitat: Primarily found in woodland and forest.
Records:
Spring Migration: Less common than the preceding species. Not seen until early May, with peak at the beginning of the third week of May. A few are seen during the final days of May. Korotev recorded an average of 1.6 birds/hr at Forest Park between 4–25 May 1979–90 (n=327 hrs). Earliest dates: 1, 13 Apr 1986, Taney Co. (JD-**AB** 40:480); 1, 21 Apr 1989, Schell-Osage (JG, E. Johnson-**BB** 56[3]:93). High counts: 18, 19–20 May 1983, Forest Park (RK).
Summer: Besides the above specimen record there are two additional observations of late migrants: 2, 3 June 1975, Springfield (NF-**BB** 43[1]:24); 1, 6 June 1988, near Farmington (BRE-**BB** 55[4]:126).
Fall Migration: Birds begin to arrive at the end of Aug and peak at the end of the second or third week of Sept. It is rarely seen in early Oct, e.g., 8, tower kill, 6 Oct 1962, Cape Girardeau (Heye 1963), and there is only one record after mid-Oct. Earliest dates: 1, 23 Aug 1986, St. Louis (JZ-**NN** 58:84); 1, 23 Aug 1987, St. Joseph (FL-**BB** 55[1]:12). High count: 11, tower kill, 24 Sept 1960, Columbia (**BB** 28[1]:9). Latest dates: 1, 1 Nov 1975, Columbia (BG-**BB** 44[1]:22); 1, 20 Oct 1985, St. Louis (RG).

Cape May Warbler (*Dendroica tigrina*)

Status: Rare transient, more regular and frequently observed in the east.
Documentation: Specimen: male, 13 May 1976, near Arkoe, Nodaway Co. (MBR, DAE, K. Kaufman; NWMSU, DAE 2976).
Habitat: Woodland and forest.
Records:
Spring Migration: Rare, but regular in the eastern half of the state where numbers vary considerably from year to year. Apparently Cape May populations are positively correlated with spruce budworm "outbreaks" on the breeding grounds. Thus, springs following productive breeding seasons are when this species is observed in relatively high numbers. During especially "good" years, such as 1966, 1968, 1982, as many as twenty birds may be observed during May in the St. Louis area; however, in most years, a total of < 5 birds may be seen in spring. Korotev has recorded a total of 73 birds, between 28 Apr–11 May 1979–90 (n=252 hrs), at Forest Park. In contrast, in northwestern Missouri, a total of < 10 have been observed over the past twenty years.

Normally this species first appears in early May and peaks during the second week of May. There are only two records during the final week of May. Earliest dates: 1, 21 Apr 1989, Schell-Osage (JG, E. Johnson-**BB** 56[3]:93); at least three observations for 26 Apr. High counts: 6, 4 May 1988, Forest Park (RK). Latest date: 28 May 1907, St. Louis (Widmann 1907).

Fall Migration: Although it still is more frequently seen in the eastern half of the state, more are observed in western Missouri at this season than in the spring. The earliest arrivals are usually detected in early Sept. With the exception of a couple of observations, all sightings are of single birds, with most records clustered in Sept. It is casually seen in Oct. Earliest date: 1, 23 Aug 1986, St. Joseph, Andrew Co. (LG-**BB** 54[1]:34). High count: 2, adult feeding immature, 15 Sept 1968, Maryville (MBR-**BB** 36[3]:4). Latest dates: 1, 1–12 Dec 1963, St. Louis (SHA-**BB** 31[1]:34); immature female, 31 Oct–11 Dec 1970, Maryville (MBR et al.-**BB** 38[4]:10); immature female, 24 Nov 1984, Maryville (MBR, DAE; NWMSU, MBR 1451).

Black-throated Blue Warbler (*Dendroica caerulescens*)

Status: Rare transient, more regular and frequently observed in the east.

Documentation: Specimen: male, 12 Oct 1974, near Brickyard Hill WA (NWMSU, DAE 2917).

Habitat: Woodland and forest.

Records:

Spring Migration: A regular, but rare transient in east (average 1–2/year; max. 4 birds/spring [1972] at St. Louis), casual in west (10–15 records). The first birds begin to appear at the beginning of May. Most sightings are concentrated in the first two weeks of May. There is only a single observation involving more than one bird: male and female, 10 May 1986, St. Louis (CS et al.-**NN** 58:65). Earliest dates: 27 Apr 1904, Iberia, Miller Co. (Widmann 1907); 1, 27 Apr 1974, Sullivan, Franklin Co. (J. Irving-**BB** 41[3]:5). Latest date: male, 29 May 1972, LaBenite Park, Jackson Co. (SP-**BB** 39[3]:7).

Fall Migration: More frequently encountered but still very rare at this season than in spring (the majority of the western Missouri records are at this season). Observations range from early Sept through Oct, with most during the last two weeks of Sept and the first few days of Oct. All sightings involve solitary birds. Earliest date: 1 Sept 1887, St. Louis (Widmann 1907). Latest dates: female, 5 Nov 1978, Busch WA (G. and T. Barker-**BB** 46[1]:27); male, window kill, 1 Nov 1980, Maryville (DAE; NWMSU, MBR 1426).

Yellow-rumped Warbler (*Dendroica coronata*)

Status: Common transient; locally uncommon winter resident in south, very rare and local in north.

Documentation: Specimen: female, 28 Nov 1948, McDonald Co. (KU 29227).

Habitat: Seen in a wide array of habitats during migration; primarily in woodland and forest edge in winter.

Records:

Spring Migration: This is the most common migrant warbler. Birds begin appearing at nonwinter locations by mid-Mar, but the bulk does not pass through until the final week of Apr. It is still common during the first week of May, but thereafter numbers drastically drop off. It is only casually seen after the second week of May. An average of 8.2 birds/hr was recorded at Forest Park between 8 Apr–14 May 1979–90 (RK; n=449 hrs). High counts: 78, 26 Apr 1982, Forest Park (RK); 60, 24 Apr 1963, St. Louis (Hanselmann 1963); 60, 7 May 1967, St. Joseph (FL). Latest date: 21 May 1907, St. Louis (Widmann 1907).

Summer: Only one old record: singing male, 21 June 1897, St. Louis (Widmann 1907).

Fall Migration: Normally not encountered until the third week of Sept, and it does not become common until the last few days of Sept or in early Oct. Most pass through during the first half of Oct. By mid-Nov very few are seen in the north, but migrants are still regularly seen in the south. Earliest date: 15 Aug 1964, Kansas City (EC-**BB** 31[3]:19); 1, 27 Aug 1975, Montgomery City, Montgomery Co. (RW-**BB** 44[1]:22). High counts: 150, 3 Oct 1988, St. Francois Co. (HF-**BB** 56[1]:15); 90+, 6 Oct 1987, Atchison Co. (MBR).

Winter: CBC data indicate that this warbler is most common in the Ozark Border, followed by the Mississippi Lowlands (Map 55). However, counts recently established in the White River section of the southwestern Ozarks suggest that the Yellow-rumped is most common in this region of the state (average of 18.5 birds/pa hr; n=7 years, 1982–88). Numbers fluctuate considerably depending on the availability of small-sized fruit. High counts: 217, 24 Dec 1988, Eagle Rock CBC; 165, 27 Dec 1986, Taney Co. CBC.

Comments: The western race, *D. c. auduboni,* has been recorded in the state on at least four occasions: adult, 8 Apr 1966, Duck Creek (DAE-**BB** 33[2]:10); 1, 19 Apr 1959, Kansas City (J. Isenberger; Rising et al. 1978); 1, 11 May 1980, western Macon Co. (S. Hein-**BB** 47[3]:13); 1, 14 Oct 1973, St. Louis (RA-**BB** 41[1]:4).

Hermit Warbler (*Dendroica occidentalis*)

Status: Accidental winter visitant.

Documentation: Specimen: immature male, 20 Dec 1969, 10 miles west of Maryville (Easterla 1970b; NWMSU, DAE 2290).

Comments: The above bird was obtained from a mixed-species flock in a large, isolated pine grove. It represents the only state record and is one of

Map 55.
Early winter relative abundance of the Yellow-rumped Warbler among the natural divisions. Based on CBC data expressed as birds/10 pa hrs.

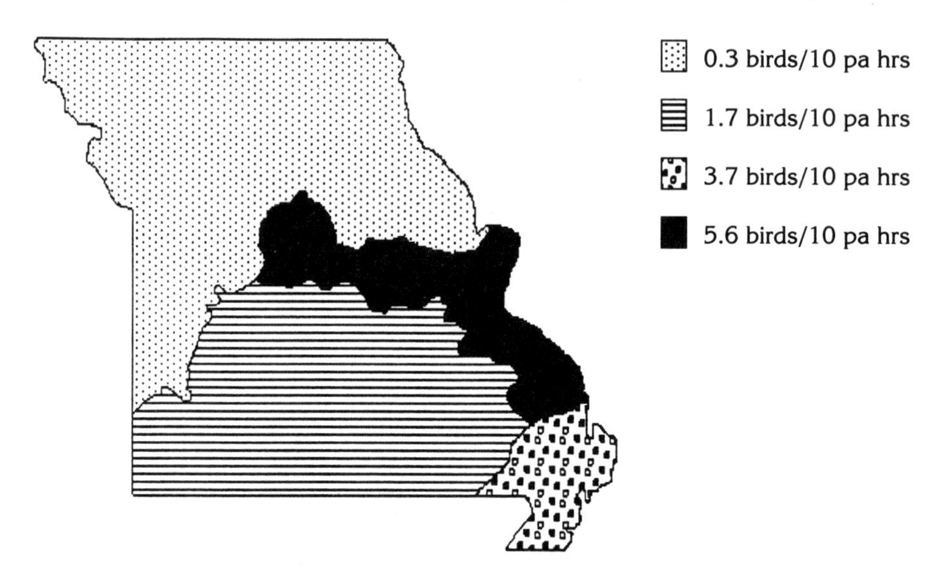

the few United States winter records outside coastal California.

Black-throated Green Warbler (*Dendroica virens*)

Status: Common transient.
Documentation: Specimen: male, 6 Oct 1954, Jackson Co. (KU 31983).
Habitat: Woodland and forest.
Records:
Spring Migration: A few appear as early as the first week of Apr in the extreme south, but normally it is not recorded in the south until mid-Apr, with the first individuals reaching the north at the end of Apr. Peak is during the first ten days of May, with a few stragglers occasionally seen up until the final days of the month. Between 20 Apr and 20 May 1979–90 (n=474 hrs), an average of 1.1 birds/hr was recorded at Forest Park (RK). Earliest date: 1, 31 Mar 1981, Springfield (CB et al.-**AB** 35:831). High count: 13, 6 May 1983, Forest Park (RK). Latest date: 1, 31 May 1975, Big Oak Tree SP (DAE-**BB** 43[1]:21).
Summer: Two June and one July observations: 1, mid-Jun 1874, Warrensburg, Johnson Co. (Scott 1879); male, 24 June 1945, Busch WA (JEC-**BB** 12[7]:41); singing male, 4 July 1980, Columbia (BG-**BB** 47[4]:13).
Fall Migration: Birds begin reappearing by late Aug, with peak during mid- to late Sept. Relatively high numbers are still passing through in early

Oct, e.g., 9, tower kill, 6 Oct 1962, Cape Girardeau (Heye 1963). A few solitary birds are regularly seen through the third week of Oct; thereafter, it is only casually observed. Earliest date: 1, 20 Aug 1980, St. Louis (MP- **NN** 52:51). High count: 20, 12 Sept 1964, St. Louis (SHA-**BB** 31[4]:18). Latest date: 1, 16 Nov 1972, Maryville (DAE-**BB** 40[1]:9).

Blackburnian Warbler (*Dendroica fusca*)

Status: Uncommon transient.

Documentation: Specimen: male, 30 Apr 1970, Cape Girardeau (SEMO 318).

Habitat: Woodland and forest.

Records:

Spring Migration: Normally not seen until the end of Apr in south, early May in north. Peak occurs during the middle of the month, but a few are seen up until the last few days of May. An average of 0.7 birds/hr was recorded between 3–18 May 1979–90 at Forest Park (RK; n=276 hrs). Earliest dates: male, 9 Apr 1989, Roaring R. SP (LG-**BB** 56[3]:93); 1, 24 Apr 1986 and 24 Apr 1990, Forest Park (RK). High counts: 20+, 13 May 1990, Tower Grove Park (JZ-**BB** 57[3]:140); 13, 7 May 1983, Forest Park (RK). Latest dates: 1, 31 May 1984, St. Louis (RK); 30 May 1904, Kansas City (Widmann 1907).

Summer: One observation: singing male, 8 June 1983, Eleven Point R., Oregon Co. (JW-**BB** 50[4]:34).

Fall Migration: Birds reappear as early as the beginning of the final week of Aug, with peak usually during the second or third week of Sept. It is casually seen after the first of Oct. Earliest dates: 1, 21 Aug 1981, Forest Park (MP-**NN** 53:56); 2, 22 Aug 1965, St. Joseph (FL). High counts: "an amazing flight" of 100 were seen on 3 Sept 1938 in St. Charles Co. (WS-**BB** 5[10]:97); 8, 11 Sept 1966, Knob Noster SP, Johnson Co. (SH-**BB** 34[1]:9). Latest dates: 2, 11 Oct 1977, Pleasant Hill, Cass Co. (JG); 1, tower kill, 6 Oct 1962, Cape Girardeau (Heye 1963).

Yellow-throated Warbler (*Dendroica dominica*)

Status: Locally common summer resident in the Ozarks and Ozark Border; more local and uncommon in the northcentral and northeastern sections of Glaciated Plains; very rare or absent in northwestern corner of Glaciated Plains; rare in Osage Plains.

Documentation: Specimen: male, 4 Apr 1907, Spring Valley, Shannon Co. (AMNH 229962).

Habitat: Breeds in two totally different habitats: most common along rivers with sycamores, but not uncommon in upland stands of Shortleaf Pine.

Records:

Spring Migration: One of the earliest warblers to arrive, with a few appearing as early as the end of Mar. Most are at breeding sites by the final week of Apr. Earliest date: 21 Mar 1894, Hornersville, Dunklin Co. (Widmann 1907). High counts: 27, migrating north along stream, 1 Apr 1990, Bee Creek WA, Taney Co. (JH, PM-**BB** 57[3]:140); 20+, 20 May 1978, 13 mile stretch of Niangua R., Dallas Co. (MBR, FL-**BB** 45[3]:19).

Summer: It is most common in the Ozarks where it is found in two distinct habitats, as mentioned above. However, it formerly was most abundant in the Mississippi Lowlands where it commonly bred in cypress swamps (Widmann 1907). The BBS route 10 has the highest average of Yellow-throateds, with 1.9 birds. Presently, it is rarest in the Osage Plains (where it is principally found along the Osage and Grand rivers and their tributaries) and the extreme northwestern section of the Glaciated Plains. In the latter region it is not known to breed north of Buchanan Co. (Bluffwoods SF) or west of Grundy Co. (Crowder SP), although at least one pair bred at Honey Creek WA, Andrew Co. until the early 1970s. In the central and eastern sections of the Glaciated Plains it is found along nearly all the major rivers to the Iowa border, although it is less common and more local than in the Ozarks.

Fall Migration: Rarely seen after the first of Sept, with only single birds recorded. Latest dates: 11 Oct 1885, St. Louis (Widmann 1907); 1, tower kill, 6 Oct 1962, Cape Girardeau (Heye 1963; SEMO 166).

Pine Warbler (*Dendroica pinus*)

Status: Locally common summer resident in Shortleaf Pine in the Ozarks; away from the Ozarks pine country it is a rare transient in the east, casual in the west; very rare winter resident in Ozark pinelands.

Documentation: Specimen: male, 15 Dec 1986, near Ellsinore, Carter Co. (ANSP 178157).

Habitat: Primarily Shortleaf Pine.

Records:

Spring Migration: This is the earliest of the breeding warblers to appear. A few arrive at the end of Feb, with numbers gradually increasing through Mar. Presumed northbound migrants (birds breeding in the northern United States and southern Canada) are detected through the northern half of Missouri during the last two weeks of Apr and the first few days of May. At that time it is a rare but regular transient through the St. Louis area. It is only casually encountered during this same period in western Missouri. High counts for early in the season: 10, 25 Feb 1981, Forsyth, Taney Co. (D. Jones-**AB** 35:306); 10, 11 Mar 1986, Hawn SP, St. Genevieve Co. (BRE).

Summer: Although this species is local, it is fairly common; its breeding range corresponds with the distribution of Shortleaf Pine (see Map 3). Thus the heart of the range of this species is in the southeastern section of the Ozarks. There are, however, a few isolated pockets of birds in the extreme southwestern counties of the state. BBS route 11 has the highest yearly average, with 3.4 birds.

Fall Migration: Migrants are seen primarily from mid-Sept through early Oct, e.g., 3, 14 Sept 1978, St. Louis (m.ob.-**BB** 46[1]:27–28); 4, 26 Sept 1978, Springfield (CB et al.). Presumably the bulk of the breeding population leaves in Sept, but data are scant. Widmann (1907) lists three Aug records for St. Louis, the earliest on 20 Aug 1905. Late date for northern Missouri: immature female, 16 Nov 1975, Maryville (DAE, DT-**BB** 44[1]:22; NWMSU, DAE 2974).

Winter: A regular but very rare winter resident in the heart of the Shortleaf Pine range. It is accidental elsewhere: 1, 18 Dec 1982, Kansas City Southeast CBC; 1, 21 Jan 1967, Missouri Botanical Arboretum (JEC et al.-**BB** 34[1]:23). High count and first winter record: 4, 27 Dec 1965, near Winona, Shannon Co. (DAE, SH-Easterla 1966b; NWMSU, DAE 1052).

Comments: Like other species that are restricted to Shortleaf Pine in Missouri (Table 1), the Pine Warbler is now less common and widely distributed than it was before the pine was largely lumbered at the turn of the century.

Kirtland's Warbler (*Dendroica kirtlandii*)

Status: Accidental transient.

Documentation: Specimen: male, 8 May 1885, St. Louis (Widmann 1885; now cannot be located).

Comments: The above record was obtained when the Kirtland's Warbler population apparently was much greater, as there were a number of what are now considered extralimital records during this same period (Walkinshaw 1983).

There is an additional reliable observation of a single bird at Busch WA on 30 Sept 1950 (JEC, JC-**NN** 22:46). This bird was observed foraging 3–4 m above the ground in a semi-open area. The observers noted that it "continually wagged its tail" as it foraged.

Prairie Warbler (*Dendroica discolor*)

Status: Uncommon summer resident in Ozarks and Ozark Border; accidental transient in Glaciated Plains.

Documentation: Specimen: male, 26 Apr 1907, Spring Valley, Shannon Co. (Woodruff 1908; AMNH 229991).

Habitat: Most common in fields with scattered cedars; also found in young, brushy successional growth.

Records:

Spring Migration: The initial migrants appear during the third week of Apr. By early May it is uncommon at breeding sites. It is accidental as a migrant in the Glaciated Plains. There is a single observation at this season for northwestern Missouri: male, 18 May 1975, Pigeon Hill WA, Buchanan Co. (D. Reynolds, FL-**AB** 29:861; SJM 76.1.191). Dierker (1979) did not record this species at Hannibal during the ten year period of 1966–75. Earliest date: 1, 14 Apr 1990, Poplar Bluff, Butler Co. (S. Hudson).

Summer: Local but not rare in the Ozarks and Ozark Border where an average of 1.1 birds/route have been recorded. BBS routes 11 and 8 have had the highest densities, with an average of 4.5 (n=18 years) and 5.2 birds/route (n=11 years), respectively. This species is perhaps most common in the Cedar Glades of the White River section of the Ozarks (see Map 3). It is virtually absent from the Glaciated Plains, Mississippi Lowlands, and all but the extreme eastern portion of the Osage Plains.

Fall Migration: Birds leave the state during the latter part of Aug and early Sept. There are very few observations after early Sept. The bird seen at St. Joseph, Andrew Co., on 23 Aug 1986 was quite unusual (LG-**BB** 54[1]:34). Latest dates: 1, 23 Sept 1890, St. Louis (Widmann 1907); 1, 20 Sept 1952, south of Gravois Mills (DAE).

Palm Warbler (*Dendroica palmarum*)

Status: Common transient in east, uncommon in west; accidental winter resident in south.

Documentation: Specimen: male, 27 Apr 1963, Platte Co. (KU 41455).

Habitat: Usually seen in open situations: woodland edge, brushy fields, and second growth.

Records:

Spring Migration: In most years, the first arrivals appear at the southern border during the second week of Apr and about a week later in the north. Peak is during the last few days of Apr or the first few days of May. Virtually none are seen after 15 May. It is considerably more common in the eastern half of the state. Between 20 Apr–10 May 1979–90 at Forest Park an average of 1.8 birds/hr was recorded (RK; n=331 hrs). Earliest date: 1, 7 Apr 1978, Charleston, Mississippi Co. (JH). High count east: 30, 27 Apr 1963, St. Louis (Hanselmann 1963); High count west: 10+, 8 May 1978, Independence (RF-**BB** 45[3]:19). Latest dates: 15–18 May 1902, Jasper, Jasper Co. (W. Savage; Widmann 1907); several additional observations for 15 May.

Fall Migration: Only accidentally seen prior to the second week of Sept; it does not become common until the last week of Sept. Peak occurs at the end of Sept and early Oct. Birds are regularly seen throughout Oct, with an occasional bird lingering into Nov. It is accidental from mid-Nov through the end of the period. As in spring, it is more abundant in eastern Missouri. Earliest dates: 1, 12 Aug 1977, Trimble WA (SP, M. Myers-**BB** 45[1]:28); 1, 21 Aug 1977, Independence (CH). High count: 10, tower kill, 24–25 Sept 1960, Columbia (**BB** 28[1]:9). Latest date: 1, 5–9 Dec 1974, Busch WA (JC-**BB** 42[3]:14).

Winter: At least four reliable records: 1, 15 Dec 1973, Kansas City Southeast CBC; immature male, specimen of the race *hypochrysea,* 21 Dec 1965, south of Clinton, Henry Co. (DAE, SH; Easterla 1966b; NWMSU, DAE 1050); 1, until 26 Dec 1982, Busch WA (T. Barker-**AB** 37:308); and 1, 27 Dec 1979, Big Oak Tree SP CBC.

Comments: The nominate race is the common migrant, while the Yellow Palm (*D. p. hypochrysea*) is casually recorded.

Bay-breasted Warbler (*Dendroica castanea*)

Status: Uncommon transient.

Documentation: Specimen: female, 21 Oct 1956, Campbell, Dunklin Co. (SEMO 36).

Habitat: Woodland and forest.

Records:

Spring Migration: Usually first appears in early May and peaks in mid-May. A few are seen passing through up until the last few days of the month. It is more common in the eastern half of the state. Korotev recorded an average of 1.2 birds/hr at Forest Park between 4–25 May 1979–90 (n=327 hrs). Earliest dates: both at St. Louis on 26 Apr 1979 and 1984 (**NN** 51:46, 56:50). High counts east: 18, 13 May 1981, St. Louis (RK); 17, 19 May 1983 (RK). High counts west: 5, 6 May 1962, St. Clair Co. (SH-**BB** 29[2]:19); 5–6, 15 May 1979, near Maryville (MBR, DAE, TB).

Summer: Two June observations: 2 June 1907, St. Louis (Widmann 1907); 1 (same bird?), 10 and 23 June 1984, St. Louis (R. Edwards-**NN** 56:63).

Fall Migration: More common statewide in the fall. Birds are first detected in early Sept and peak between the end of the second and the beginning of the fourth weeks of Sept. It is rarely seen during the first half of Oct. Earliest date: 4 Sept 1905, St. Louis (Widmann 1907). High counts: 40, 12 Sept 1964, St. Louis (SHA-**BB** 31[4]:18); 17, tower kill, 20–21 Sept 1963, Columbia (George 1963). Latest dates: 1, 1 Nov 1975, Columbia (BG-**BB** 44[1]:22); 1, 22 Oct 1955, Creve Coeur L. (JEC-**BB** 22[11]:2).

Blackpoll Warbler (*Dendroica striata*)

Status: Common transient in spring; in fall, rare in east and casual in west.

Documentation: Specimen: male, 16 May 1968, Maryville (NWMSU, PER 46).

Habitat: Woodland and forest.

Records:

Spring Migration: Like the preceding species, it is a late migrant and rarely seen before the final days of Apr in the south, and usually not until the end of the first week in May in the north. During peak (mid-May) it is one of the most common migrant warblers, rivaling the Tennessee in numbers. However, unlike that species, it goes largely unnoticed because its song is relatively high pitched. It remains common through the beginning of the final week of May, and it is regularly seen and heard during the final days of May. An average of 2.6 birds/hr was recorded at Forest Park between 30 Apr–26 May 1979–90 (RK; n=395 hrs). Earliest dates, both at St. Louis: 1, 19 Apr 1982 (RK-**BB** 49[3]:22); 1, 21 Apr 1987 (**BB** 54[3]:15). High counts: 60, 11 May 1975, St. Joseph/Squaw Creek (FL); 30, 11 May 1969, St. Joseph/Squaw Creek (FL).

Summer: Late migrants are occasionally seen into the first week of June. Latest dates: 1, 9 June 1945, Hannibal, Marion Co. (WC-**AM** 47:38); male, 6 June 1964, west of Cardwell, Dunklin Co. (DAE; NWMSU, DAE 586).

Fall Migration: A very rare bird in fall, with few reliable records. Many purported observations pertain to the Bay-breasted Warbler. Examination of tower kill data across the state reveal no records. Lawhon had only three sightings in 35 years in northwestern Missouri. Although listed as uncommon for St. Louis in Anderson and Bauer (1968), there are very few published records for that region. Nevertheless, a few additional unpublished records do exist for St. Louis, with most during the last two weeks of Sept (PS, pers. comm.).

Given the fall migratory route of this species (Murray 1989), it would be expected to be most frequently encountered in the northeastern corner of the state; however, data are lacking. Earliest date: 1, 30 Aug 1984, St. Joseph (FL). High count: 5, 5 Sept 1984, St. Joseph (FL).

Comments: A melanistic male was recorded near Arkoe, Nodaway Co., on 9 May 1975 (NWMSU, DAE 2945).

Cerulean Warbler (*Dendroica cerulea*)

Status: Uncommon summer resident in the Ozarks, Ozark Border, and Mississippi Lowlands; rare in the Glaciated and Osage plains.

Documentation: Specimen: male, 3 May 1969, near Maryville (NWMSU, DAE 2323).

Habitat: Primarily in mature bottomland woodland and forest.

Records:

Spring Migration: At the southern border birds begin to appear during the second week of Apr, but they do not arrive in the north until the final week of Apr. Peak in the south is in early May. Earliest dates: 1, 6 Apr 1958, south of Gravois Mills (DAE); 1, 6 Apr 1986, Gray Summit, Franklin Co.

(RG). High count: 50, 9 May 1946, Big Oak Tree SP (WC-**AM** 48:114).

Summer: Today the Cerulean is less common than it was prior to the early 1900s in the Mississippi Lowlands, the Missouri R. valley, and in at least some areas of the Glaciated Plains. Perhaps the most dramatic decline has occurred in the Mississippi Lowlands where most of the bottomland forest had been cleared by the early 1900s. Widmann (1907) states that it was "especially numerous in the southeast." He also provides earlier accounts on the abundance of it in the bottomland forests of the Missouri R. valley as far north as Platte Co.; for example, at the latter locality, it was described as being common in May 1891. At Swope Park, Shirling (1920) recorded 34 singing males in June 1916, but it had virtually disappeared as a breeder from this site by about 1940 (WC; Rising et al. 1978).

Fall Migration: Breeding birds are rarely encountered, presumably they have departed, after mid-Aug. Migrants, birds seen away from breeding locales, are seen during the later half of Aug into early Sept. Latest dates: 28 Sept 1897, Dunklin Co. (Widmann 1907); 1, 26 Sept 1968, Busch WA (JEC-**NN** 40:92).

Black-and-white Warbler (*Mniotilta varia*)

Status: Common transient; uncommon summer resident in Ozarks and Ozark Border, rare elsewhere.

Documentation: Specimen: male, 13 May 1879, Charleston, Mississippi Co. (MCZ 45171).

Habitat: During migration in woodland and forest; in the breeding season primarily found in mature bottomland forest.

Records:

Spring Migration: Males begin to appear by the last week of Mar at the southern border but not until mid-Apr in the north. Peak is during the last week of Apr in the south and in early May in the north. Earliest dates: 10 Mar 1894, Dunklin Co. (Widmann 1907); 23 Mar 1907, Shannon Co. (Woodruff 1908). High counts: 20, 25 Apr 1963, St. Louis (Hanselmann 1963); 9, 8 May 1983, Forest Park (RK).

Summer: Most common in the Ozarks and Ozark Border. The highest yearly average is recorded on BBS route 8 where an average of 1.4 birds is recorded. Not surprisingly, due to its requirement of relatively mature forest, it is rare in the Glaciated and Osage plains. This species undoubtedly has declined since the early 1900s in the same areas and for the same reasons that beset the Cerulean Warbler. High count: 21, 18–19 June 1982, Barry Co. (MMH-**AB** 36:984).

Fall Migration: Migrants are detected as early as the latter part of Aug, with numbers gradually increasing to peak in mid-Sept. Although less common, birds are readily encountered through the first week of Oct, and casually thereafter until the beginning of Nov. High counts, both at tower kills at Columbia: 22, 24 Sept 1960 (**BB** 28[1]:9); 18, 19 Sept 1966 (Elder and Hansen 1967). Latest dates: 1, 1–2 Nov 1969, Maryville (MBR-**BB** 38[3]:6); 1, 2 Nov 1981, Busch WA (M. Richardson-**NN** 54:6).

American Redstart (*Setophaga ruticilla*)

Status: Common transient; common summer resident in Glaciated Plains and along the Missouri and Mississippi rivers; uncommon and more local in other regions.

Documentation: Specimen: female, 6 Oct 1962, Cape Girardeau (SEMO 200).

Habitat: During migration found in virtually any woodland and forest, but principally breeds in riparian woodland and forest.

Records:

Spring Migration: In the extreme south the first birds appear by mid-Apr, but they usually are not encountered until the final days of Apr in the north. The second week of May is normally peak. Earliest date south: 11 Apr 1893, Mississippi Lowlands (Widmann 1907). Earliest date north: 21 Apr 1922, Kansas City (H. Harris-**BL** 24:223). High count: 17, 8 and 15 May 1983, Forest Park (RK).

Summer: Quite common in moist woodland and bottomland forest throughout the Glaciated Plains and along the Missouri and Mississippi river drainages where suitable forest still remains. Uncommon and more local elsewhere. Being a riparian species, it is particularly underrepresented on BBS routes.

Fall Migration: Relatively large numbers of migrants, presumably a combination of summer residents and more northern breeders, are seen during the latter half of Aug and early Sept. It is decidedly less common by mid-Sept, and only small numbers (usually 1–2 birds/day) are observed during the final two weeks of Sept and the first week of Oct. It is accidental after the first week of Oct. High counts: 50, 30 Aug 1978, St. Louis (S. Hosler-**NN** 50:65); 8, tower kill, 24 Sept 1960, Columbia (**BB** 28[1]:9). Latest dates: 1, 13 Oct 1931, St. Louis (Barger-**BL** 33:413); 10 Oct 1904, Shannon Co. (Widmann 1907).

Prothonotary Warbler (*Protonotaria citrea*)

Status: Common summer resident in Ozarks, Ozark Border, and in the remaining tracts of wet woodland in the Mississippi Lowlands; more local and rarer in Glaciated and Osage plains.

Documentation: Specimen: male, 12 June 1926, Buchanan Co. (KU 39292).

Habitat: Common in wet, bottomland woodland and forest, especially river oxbows; also common in woodland and forest bordering Ozark streams and rivers.

Records:

Spring Migration: In the southeast birds appear at the onset of Apr and peak there in late Apr. Birds do not appear in northern Missouri until the end of Apr or in early May. Earliest dates: no.?, 31 Mar 1896, Dunklin Co. (Widmann 1907); 1, 2 Apr 1983, St. Louis (RG-**NN** 55:49). High counts: 100+, 9 May 1946, Big Oak Tree SP (WC-**AM** 48:114); 30+, territorial birds, 20 May 1978, along 13 mile stretch of Niangua R., Dallas Co. (MBR, FL-**BB** 45[3]:18).

Summer: The Prothonotary is quite common along Ozark and Ozark Border streams and rivers. Likewise it is still abundant in the remaining tracts of swampy forest in the Mississippi Lowlands. It is less common and more local in similar habitat in the Glaciated and Osage plains. Censuses along watercourses are needed to accurately assess the abundance of this species, since it is vastly underrepresented on BBS routes.

Fall Migration: Most have left the state by late Aug, although they are commonly seen up until that time, e.g., 14, 26–27 Aug 1976, Bennett Springs SP, Dallas Co. (JG-**BB** 44[3]:22). Very few are seen in Sept, and virtually none are encountered during the final two weeks of the month. Latest dates: 1, 7 Oct 1958, St. Joseph (FL); 1, 27 Sept 1987, Busch WA (BRO, TP, RK, SRU-**BB** 55[1]:12).

Worm-eating Warbler (*Helmitheros vermivorus*)

Status: Locally uncommon summer resident in Ozarks and Ozark Border, rare elsewhere.

Documentation: Specimen: male, 25 Apr 1907, Spring Valley, Shannon Co. (Woodruff 1907; AMNH 229842).

Habitat: Forested ravines bordered by steep slopes.

Records:

Spring Migration: The first individuals arrive during the final week of Apr in the south, and in early May in the north. Peak is during mid-May. An average of 0.1 birds/hr was recorded at Forest Park between 20 Apr–7 May 1979–90 (RK; n=277 hrs). Earliest dates: 1, 17 Apr 1981, Indian Creek, Washington Co. (SD); 1, 18 Apr 1982, Forest Park (RK).

Summer: It was formerly more common in at least some areas of the Glaciated Plains; for example, Shirling (1920) recorded 21 singing males at Swope Park in June 1916, but WC noted that it had "mostly vanished" as a breeder there by the late 1920s (Rising et al. 1978). Presently it is known to breed in very small numbers along the forested Missouri R. bluffs at least as far north as Bluffwoods SF (FL). It is quite local across the remainder of the Glaciated Plains and found almost exclusively in steep, wooded ravines bordering streams and rivers. Not unexpectedly, given its very flat physiogeography, the Worm-eating Warbler is virtually absent from the Osage Plains region.

The Worm-eating is poorly represented on BBS routes, probably, in part, because its insectlike song is easily overlooked. The average for Ozark and Ozark Border BBS routes is 0.2 birds/route. Route 8 has the highest average, with 1.9 birds/year.

Fall Migration: Few remain in the state after late Aug. It is only casually recorded after mid-Sept, with only one reliable observation after the beginning of the final week of Sept. Latest dates: 1, 2 Oct 1955, south of Gravois Mills (DAE); 1, 23 Sept 1971, Busch WA (JEC et al.-**NN** 43:119); 1, 23 Sept 1983, Babler SP, St. Louis (R. Laffey-**NN** 55:82).

Swainson's Warbler (*Limnothlypis swainsonii*)

Status: Locally rare summer resident along river floodplains in the southern two tiers of counties; accidental transient elsewhere.

Documentation: Specimen: male, 5 June 1964, near Big Oak Tree SP (NWMSU, DAE 584).

Habitat: Primarily found in cane (*Arundinaria gigantea*) along southern Ozark streams and rivers.

Records:

Spring Migration: The first birds are heard at the end of the third week of Apr. Peak appears to be in mid-May. Earliest date: 2, 16 Apr 1986, below Greer Springs, Eleven Point R., Oregon Co. (JW); 1, 18 Apr 1974, Big Oak Tree SP (JH). High count: 8, 3 May 1970, Big Oak Tree SP (RAR-**BB** 38[3]:9). There are several extralimital sightings, all south of the Missouri R. in the eastern edge of the Ozarks or Ozark Border. The northernmost of these are as follows: 1, 22 Apr 1950, St. Louis (M. Tuttle-**BB** 18[1]:1); 1, 23 Apr 1973, St. Louis (PS-**NN** 45:67); 1, 3 May 1981, south of Westphalia, Osage Co. (JG, J. Warner-**BB** 48[3]:20); 1, 22 May 1977, Gray Summit, Franklin Co. (RG-**BB** 44[4]:27). Rising et al. (1978) list three May records for Johnson Co., Kansas, which is across the state line from southern Kansas City, Missouri.

Summer: Recent surveys (commencing in the early 1980s) by the MDC have revealed that this species is more common and widespread, although still rare and local, than originally believed. Birds have been found from as far west as Taney Co. (there is a single observation from Barry Co.) and east across the southern tier of counties to Butler Co., with most breeding birds concentrated along the Eleven Point and Current rivers in stands of cane. The importance of cane to this species was recognized by Widmann (1895a); he first found the bird breeding in the state in Dunklin Co. in 1894. It has not been recorded breeding at Big Oak Tree SP (5 pairs, 17 June 1972, JH-**BB** 39[4]:10), since the early 1970s when much of the cane was degraded by mowing.

Fall Migration: There is very little information available for this period; presumably birds leave in Aug. Haw noted three birds at Big Oak on 3 Sept 1968. Latest dates: 1, 16–17 Sept 1972, Busch WA (J. and N. Strickling-**BB** 40[1]:9); 1, 12 Sept 1989, western Taney Co. (BJ-**BB** 57[1]:50).

Ovenbird (*Seiurus aurocapillus*)

Status: Common transient; uncommon summer resident in Ozarks and Ozark Border, rarer elsewhere.

Documentation: Specimen: male, 13 May 1969, Maryville (NWMSU, CLT 130).

Habitat: Breeds in relatively large tracts of upland woodland and forest.

Records:

Spring Migration: Birds begin arriving during the second week of Apr in the southeast but not until the final week of Apr in the north. Peak is in early May in the south and mid-May in the north. An average of 1.0 birds/hr was recorded between 30 Apr–22 May 1979–90 at Forest Park (RK; n=369 hrs). Earliest dates: Widmann (1907) gives two observations at St. Louis for 12 Apr. High counts: 20, 13 May 1990, Tower Grove Park (JZ); 20+, 14 May 1976, Maryville area (MBR).

Summer: Data from BBS routes underestimate the abundance of this species in the state. It is widespread and uncommon throughout the Ozarks and Ozark Border. As the result of fewer, large tracts of mature forest, it is less widespread and scarcer in the Mississippi Lowlands, Glaciated and Osage plains; nevertheless, where appropriate habitat exists, it is uncommon in these areas. The Ovenbird undoubtedly is less common today than it was before most of the state's forest was cleared or subdivided into smaller parcels. The average of 3.5 birds/route (n=9 years) on Route 9 is the highest in the state.

Fall Migration: Migrants begin to reappear by the end of Aug, and numbers gradually build to a peak between the 15–25 Sept. It is rare (max. 1–2 birds/day) but regular during the first half of Oct and only casually observed thereafter. High counts, tower kills: 197, 20–21 Sept 1963, Columbia (George 1963); 87, 24 Sept 1960, Columbia (**BB** 28[1]:9). Latest dates, both at St. Louis: 1, 8 Nov 1975 (JE-**BB** 44[1]:22); 1, 24 Oct 1978 (H. Brammeier-**NN** 50:77).

Northern Waterthrush (*Seiurus noveboracensis*)

Status: Common transient.

Documentation: Specimen: male, 28 Apr 1967, near Maryville (NWMSU, LCW 361).

Habitat: Primarily seen in wet thickets, edge of flooded woodland and forest, and along streams.

Records:

Spring Migration: The first individuals arrive about two weeks later than the Louisiana Waterthrush, usually during the second week of Apr in the south and not until the final week of Apr in the north. There are a couple of Mar sightings, but given the difficulty in distinguishing this species from the Louisiana, they are suspect. Numbers gradually increase to peak during the first week of May in the south and the second week in the north. It remains common through the third week of May, with a few individuals occasionally seen until the end of the month. An average of 0.7 birds/hr was recorded at Forest Park between 23 Apr–15 May 1979–90 (RK; n=390 hrs). Earliest date: 1, 8 Apr 1987, St. Louis (SRU-**BB** 54[3]:15). High counts: 15, 9 May 1946, Big Oak Tree SP (WC-**AM** 48:114); 7, 2 and 4 May 1981, Forest Park (RK). Latest date: 31 May 1897, St. Louis (Widmann 1907).

Summer: One sighting of a late migrant: 1, 3 June 1984, Clinton Co. (CH-**BB** 51[4]:17).

Fall Migration: Birds begin to reappear by mid-Aug, but they do not become prevalent until the end of the month. Peak is usually during the second or third week of Sept, with many still present up until the final days of Sept. A few birds (1–2/day) are regularly seen through the first week of Oct, but thereafter it is only accidentally encountered. Earliest dates: 1, window kill, 8 Aug 1972, Kansas City (SP-**BB** 39[4]:10); two records for 12 Aug. High counts: tower kills at Columbia; 93, 20–21 Sept 1963 (George 1963); 9, 19 Sept 1966 (Elder and Hansen 1967). Latest dates: 17 Oct 1885 (Widmann 1907); 1, tower kill, 6 Oct 1962, Cape Girardeau (Heye 1963; SEMO 178).

Louisiana Waterthrush (*Seiurus motacilla*)

Status: Common summer resident in Ozarks and Ozark Border, uncommon elsewhere.

Documentation: Specimen: male, 30 May 1964, Gravois Mills (NWMSU, DAE 800).

Habitat: During the breeding season it is restricted to streams and relatively small rivers that are bordered by woodland and forest.

Records:

Spring Migration: This is one of the earliest warblers to arrive, and the first individuals appear in the Mississippi Lowlands at the beginning of the third week of Mar. Usually it is not detected until the end of Mar in the north. Peak is during mid-Apr in the south and about a week later in the north. Earliest dates: 12 Mar 1894, Mississippi Lowlands (Widmann 1907); 1, 17 Mar 1986, Mingo (BRE-**AB** 40:480). High count: 60, 18 Apr 1987, along 18 mile stretch of Current R., Shannon Co. (RK-**BB** 54[3]:15).

Summer: It is common and widespread along Ozark and Ozark Border streams and rivers. It is uncommon and more local in the Mississippi

Lowlands, Glaciated and Osage plains. Like other riparian breeders this species is poorly represented on BBS routes.

Fall Migration: Most have left the state by early Sept, but tower kill data show that birds are still passing through until early Oct. Widmann (1907) states that it was "still common and in song in early Oct 1896" in Dunklin Co. in the extreme southeastern corner of the state, although this observation is presumably an unusual occurrence. Tower kill data: 20, 20–21 Sept 1963, Columbia (George 1963); 5, 24 Sept 1960, Columbia (**BB** 28[1]:9); 2, 6 Oct 1962, Cape Girardeau (Heye 1963).

Kentucky Warbler (*Oporornis formosus*)

Status: Uncommon summer resident in Ozarks and Ozark Border, less common elsewhere.

Documentation: Specimen: female, 28–29 Apr 1970, Cape Girardeau (SEMO 320).

Habitat: Prefers relatively large, mature stands of woodland and forest.

Records:

Spring Migration: It first appears in the Mississippi Lowlands during the second week of Apr but not until the final week of Apr in the north. Peak is in early to mid-May in the south and north, respectively. An average of 0.2 birds/hr was recorded at Forest Park between 23 Apr–15 May 1979–90 (RK; n=390 hrs). Earliest date: 1, 30 Mar 1975, Charleston, Mississippi Co. (JH-**AB** 29:699). High counts: several of 4 birds/day.

Summer: This species undoubtedly is less common than it was at the turn of the century, primarily the result of the clearing of forest, especially in the Mississippi Lowlands and the Ozarks. At Swope Park, in the Glaciated Plains, Shirling (1920) counted 74 singing males there in June 1916, but a relatively recent survey (1973) by Branan and Burdick (1981) found only a single male. Over 50% of the upland forest at Swope Park had been removed in the interval between these two censuses.

Today it is uncommon but widespread throughout the Ozarks and Ozark Border (average 1.1 birds/route) and least widespread and abundant in the relatively unforested Glaciated Plains (average 0.2 birds/route), Mississippi Lowlands (average 0.4 birds/route), and the Osage Plains (0.7 birds/route). The 9.5 yearly average on BBS route 8 is the highest in the state.

Fall Migration: Like the preceding species most have left the state by the latter part of Aug. A few individuals are encountered during the first half of Sept. Latest dates: adult male, 18 Oct 1977, Busch WA (PS-**BB** 45[1]:28); 1, tower kill, 8–9 Oct 1967, Cape Girardeau (PH).

Connecticut Warbler (*Oporornis agilis*)

Status: Rare transient.

Graph 18.
Missouri BBS data from 1967 through 1989 for the Kentucky Warbler. See p. 23 for data analysis.

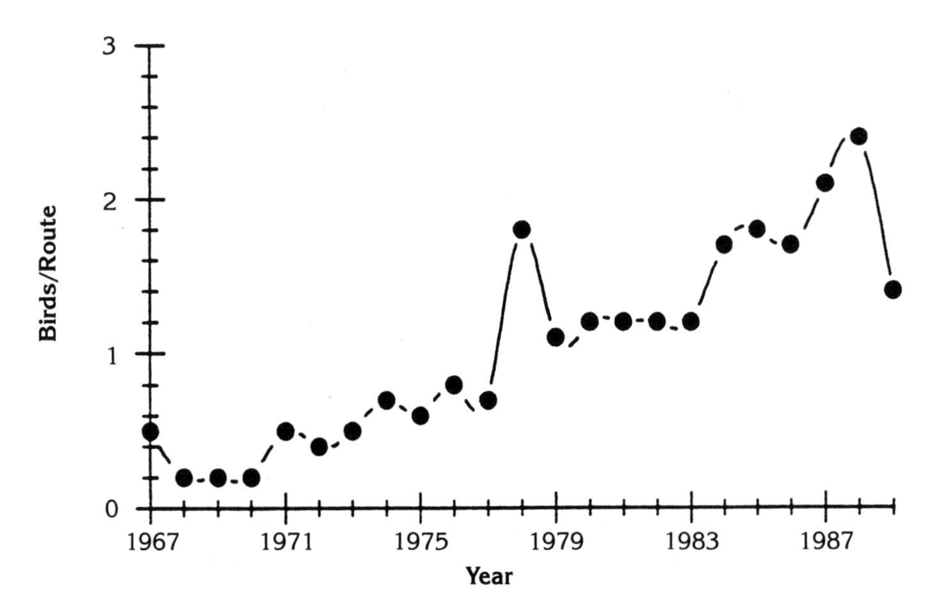

Documentation: Specimen: male, 18 May 1974, near Maryville (NWMSU, DAE 2918).

Habitat: Woodland and forest understory, especially in moist areas.

Records:

Spring Migration: One of the later migrants. Although there are a few records for late Apr, it usually is not seen until the second week of May in the south and not before the beginning of the third week in the north. Most observations are of single birds, but a number of them involve two birds at a single site.

Since the migratory route of this species moves southeast to northwest in the spring, this species is probably most common in the northeastern quarter of the state. It is regularly seen even along the entire western border. Earliest dates: 1, 24 Apr 1987, Farmington (BL-**BB** 55[1]:14); 1, 25 Apr 1979, St. Louis (CP-**NN** 51:45).

Summer: Late migrants are infrequently seen during the first few days of June.

Fall Migration: Most records are for the first three weeks of Sept, but there are at least three records for the first half of Oct. It is perhaps more widespread during this season than in the spring. This warbler is largely overlooked in the fall because it is not vocal and few observers work its habitat in late Aug or early Sept, when most should pass through the state.

Earliest dates: 1, 14 or 15? Aug 1974, St. Louis (JE-**NN** 46:105); 1, 1 Sept 1969, Maryville (MBR). High count: 2, tower kill, 10–11 Sept 1964, Kansas City (DAE-**BB** 31[4]:18). Latest dates: 1, 13 Oct 1983, Marais Temps Clair (M. Scudder, F. Ruegsegger-**NN** 55:91); 1, 9 Oct 1938, St. Charles Co. (WS-**BB** 5[11]:103).

Mourning Warbler (*Oporornis philadelphia*)

Status: Uncommon transient.

Documentation: Specimen: o?, 14 Sept 1918, Lexington, Lafayette Co. (CMC 361).

Habitat: Dense thickets, especially at woodland and forest edge.

Records:

Spring Migration: Like the preceding species this is a late migrant. The first birds appear by the second week of May and peak during the last ten days of the month. An average of 0.2 birds was recorded at Forest Park between 8–25 May 1979–90 (RK; n=243 hrs). Earliest dates: 1, 27 Apr 1981, St. Louis (B. Hely-**NN** 53:39); 1, 28 Apr 1935, Hahatonka, Camden Co. (IA et al.-**BB** 2[4]:22). High counts: 15, 30 May 1963, St. Louis (Hanselmann 1963); at least three counts of 6 birds from both sides of the state.

Summer: There are a number of observations of late migrants for the first week of June. Latest dates: 1, 9 June 1945, Hannibal, Marion Co. (WC-**AM** 47:38); 1, 9 June 1981, St. Louis (A. Roth-**BB** 48[3]:25). In addition, there is a single July record, presumably of a very early fall migrant: 1, 30 July 1977, Roaring R. SP (JG-**BB** 44[4]:31).

Fall Migration: The first arrivals are detected at the end of Aug. Apparently, peak is during mid-Sept, with an occasional bird observed during the first few days of Oct. Earliest dates: 1, 16 Aug 1979, Big Oak Tree SP (JH); 1, 25 Aug 1970, Maryville (MBR). High counts, tower kills: 9, 20–21 Sept 1963, Columbia (George 1963); 5, 10–11 Sept 1964, Kansas City (**BB** 31[4]:18). Latest dates: 1, 7 Oct 1980, Springfield (CB-**BB** 48[1]:9); 1, tower kill, 6 Oct 1962, Cape Girardeau (Heye 1963).

Comments: Purported hybrids between this species and the MacGillivray's Warbler have proved to represent extreme plumage variants of pure birds (Pitocchelli 1990). Most of these have been Mourning Warblers that resemble MacGillivray's. Spring males encountered in Missouri that do not possess eye-arcs *and* dark lores are Mourning Warblers. Song is diagnostic, and wing minus tail measurements separate all but the extreme plumage variants (Pitocchelli 1990).

MacGillivray's Warbler (*Oporornis tolmiei*)

Status: Accidental spring transient.

Documentation: Photograph: male, netted, 1 May 1974, extreme northern Atchison Co. (F. and H. Diggs; VIREO x05/1/020; Fig. 29).

Habitat: Same as that of the Mourning Warbler.

Comments: The identification of the above photographed bird was verified by experts (J. Pitocchelli, G. Hall) familiar with the morphological variation in Mourning and MacGillivray's warblers. See comments under Mourning Warbler.

Common Yellowthroat (*Geothlypis trichas*)

Status: Common summer resident; very rare winter resident.

Documentation: Specimen: male, 4 May 1974, Maryville (NWMSU, JWG 17).

Habitat: Marshes, wet, tall meadows, and thickets bordering water.

Records:

Spring Migration: In the southeast the initial migrants arrive during the second week of Apr but not until the beginning of the fourth week in the north. Peak is during the first and second weeks of May in the south and north, respectively. Earliest dates: 1, 12 Mar 1990 (winter resident?), Mingo (BRE); 1, 13 Mar 1987 (winter resident?), Duck Creek (BRE). High count: 15, 18 May 1963, St. Louis (Hanselmann 1963).

Fig. 29. This male MacGillivray's Warbler was netted and photographed by Fitzhugh and Hazel Diggs on 1 May 1974 in northern Atchison Co. It represents the only record for the state.

Map 56.
Relative breeding abundance of the Common Yellowthroat among the natural divisions. Based on BBS data expressed as birds/route.

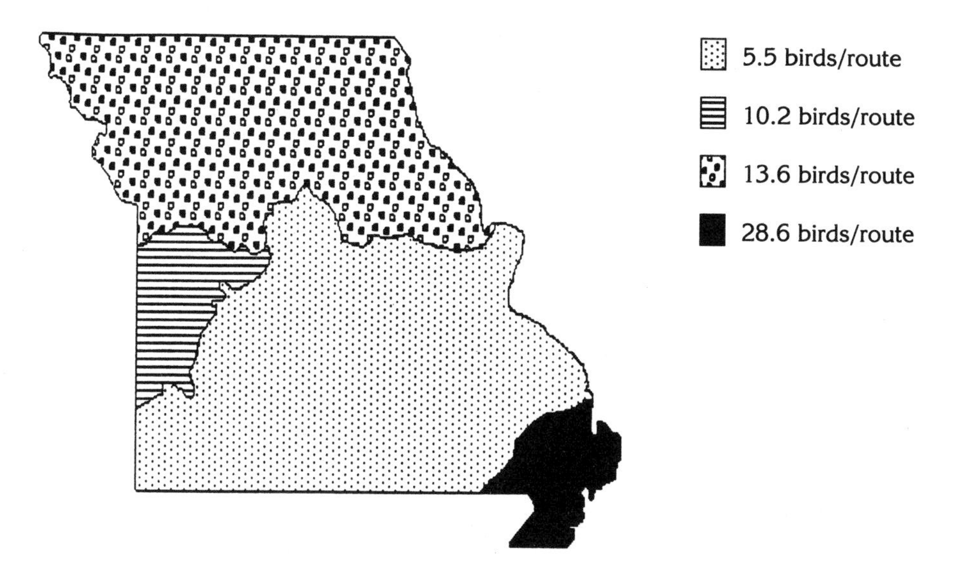

Summer: In the nonforested regions of the state it is the most common warbler (Map 56).

Fall Migration: It remains common through Sept (peak during the second and third week), with few birds remaining after the first week of Oct. During mild falls a few birds linger, especially in cattail marshes, until the end of the period. High counts, tower kills: 284, 24 Sept 1960, Columbia (**BB** 28[1]:9); 131, 19 Sept 1966, Columbia (Elder and Hansen 1967); 49, 6 Oct 1962, Cape Girardeau (Heye 1963).

Winter: Although there are at least fifteen records, most are during mild winters. Virtually all observations are concentrated during the last two weeks of Dec and the first week of Jan. Thus, this species is not a true winter resident. With two exceptions, all records are of single birds: 2, 19 Dec 1978, Squaw Creek CBC; and 2, 2 Jan 1989, Mingo CBC.

Hooded Warbler (*Wilsonia citrina*)

Status: Rare summer resident in Mississippi Lowlands and the southern and eastern portions of the Ozarks; casual transient in northern Missouri.

Documentation: Specimen: male, 11 May 1896, Dunklin Co. (MCZ 257065).

Habitat: In moist, understory of relatively large, mature tracts of woodland and forest.

Records:

Spring Migration: As with most of the migrant passerines, it arrives first in the Mississippi Lowlands where it is seen as early as the second week of Apr. Solitary birds are usually seen at St. Louis by the end of the third week of Apr. Peak in the southeast is during early May. It is rarest in the Glaciated and Osage plains. There are at least three records (all of males in May) in the extreme northwestern corner. Earliest dates: 2 Apr 1897, Dunklin Co. (Widmann 1907); 1, 15 Apr 1979, St. Louis (J. Dunham et al.-**NN** 51:45); 1, 15 Apr 1987, Kennett, Dunklin Co. (H. Schanda). High counts: 12, 9 May 1946, Big Oak Tree SP (WC-**AM** 48:114); 8, 19 May 1982, Eleven Point R., Oregon Co. (JW-**BB** 49[3]:22).

Summer: It appears that this species has also undergone a decline since the turn of the century. Widmann (1907) mentions that it was common in the Mississippi Lowlands, "fairly common in the Ozarks, Ozark border and in the bluff as well as bottom lands of the Mississippi and Missouri Rivers." He found the bird during the breeding season as far north as Platte Co.

Today it is quite local in the Mississippi Lowlands because this area was largely denuded in the early 1900s. It is not known to breed along the Missouri R. In the Mississippi R. drainage, it has not been found nesting north of northern Jefferson Co., although potential habitat exists at least as far north as Pike Co. It is most common along the Ozark streams and rivers in the southern two tiers of counties. High count: ca. 15, 26 June 1988, Eleven Point R., below Greer Crossing, Oregon Co. (RB). There is an extralimital record of a singing male at St. Joseph on 3–7 July 1976 (JHA, fide FL-**BB** 44[2]:21).

Fall Migration: Most of the summer residents have left by the end of Aug. Obvious migrants (birds seen away from breeding sites) are seen primarily at the end of Aug through the first half of Sept. There is only a single Oct observation. Extralimital record: 1, 21 Aug 1977, Trimble (SP, M. Myers-**BB** 45[1]:28). Latest dates: 1, 20 Oct 1984, Columbia (IA-**BB** 52[1]:33); 28 Sept 1895, Dunklin Co. (Widmann 1907); 3, 26 Sept 1968, Noel, McDonald Co. (V. Cronquist-**AFN** 23:65).

Wilson's Warbler (*Wilsonia pusilla*)

Status: Common transient.

Documentation: Specimen: ♂?, 9 Oct 1967, Cape Girardeau (SEMO 264).

Habitat: Thickets, brushy areas, woodland and forest edge.

Records:

Spring Migration: In the south a few appear by the end of Apr, but in the north it is often not seen until the end of the first week of May. Peak is usually between 10–18 May, with a few late birds encountered at the end of the month. Korotev recorded an average of 0.9 birds/hr at Forest Park between 8–25 May 1979–90 at Forest Park (n=243 hrs). Earliest dates: male, 17 Apr 1990, Tower Grove Park (JZ-**BB** 57[3]:140); 1, 20 Apr 1979, St. Louis (M. Scudder-**NN** 51:45). High counts: 10, 17 May 1976, Maryville area (MBR); 9, 15 May 1984, Forest Park (RK).

Summer: Late migrants are casually seen during the first week of Jun. Latest date: 1, 18 June 1979, Jackson Co. (KH-**BB** 46[4]:9).

Fall Migration: Birds begin to reappear during the final ten days of Aug, becoming relatively common at the end of the month through the first half of Sept. From mid-Sept through early Oct, only 1–2 birds/day are encountered. Thereafter, it is only casually observed. Earliest date: 1, 22 Aug 1981, St. Joseph (FL). High counts: 15–20, 3 Sept 1977, Maryville/Squaw Creek (MBR, TB); 10, 31 Aug 1967, St. Joseph (FL). Latest dates: 1, 17 Nov 1970, near Bourbon, Crawford Co. (D. Reger, S. Phillips-**NN** 43:7); 1, 20 Oct 1985, St. Louis (RA-**NN** 57:85).

Canada Warbler (*Wilsonia canadensis*)

Status: Uncommon transient.

Documentation: Specimen: male, 9 May 1889, Marion Co. (ANSP 42386).

Habitat: Thickets, brushy areas, woodland and forest edge.

Records:

Spring Migration: One of the latest migrants; it is rarely seen before the end of Apr in the extreme south and usually not until a week to ten days later in the north. Peak is between 18–25 of May. It is regularly seen through the end of the period. An average of 0.5 birds/hr was recorded at Forest Park between 8–25 May 1979–90 (RK; n=243 hrs). Earliest dates, both by Widmann (1907): 10 Apr 1893, New Madrid Co.; 28 Apr 1888, St. Louis. High counts: there are at least four dates when five birds/day were observed from both sides of the state.

Summer: Late migrants are not unusual during the first week of June. Latest date: singing male, 13 June 1984, Big Oak Tree SP (JH).

Fall Migration: Like the Wilson's Warbler, this species begins to reappear during the last ten days of Aug and peaks in early Sept. Birds are regularly seen until the final days of Sept. There are only two Oct records. Earliest date: 1, 21 Aug 1970, Maryville (MBR). High counts: 8, tower kill, 20–21 Sept 1963, Columbia (George 1963); 7, 5 Sept 1969, Big Oak Tree SP (JH). Latest dates: 1, 19–21 Oct 1944, Clay Co. (NF-**BB** 11[9]:53); 1, 11 Oct 1977, Cass Co. (JG-**BB** 45[1]:28).

Yellow-breasted Chat (*Icteria virens*)

Status: Common summer resident in Ozarks, Ozark Border, and Mississippi Lowlands; uncommon elsewhere.

Map 57.
Relative breeding abundance of the Yellow-breasted Chat among the natural divisions. Based on BBS data expressed as birds/route.

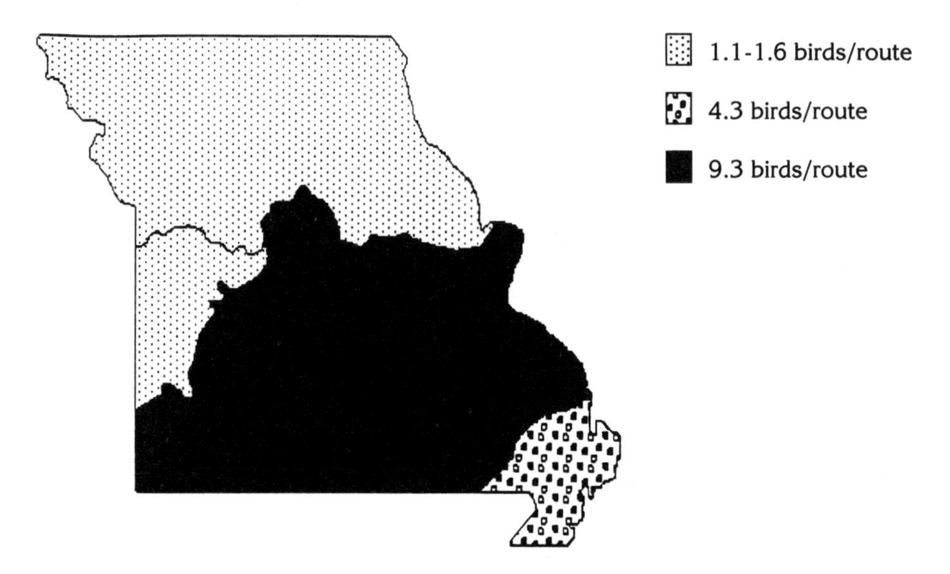

Documentation: Specimen: male, 12 May 1879, Charleston, Mississippi Co. (MCZ 45498).

Habitat: Dense brushy areas, thickets, and dense second-growth woodland.

Records:

Spring Migration: The first chats arrive during the final week of Apr in the south and about a week later in the north. Peak is at the end of the first week of May in the south and during mid-May in the north. Earliest date: 18 Apr 1903, St. Louis (Widmann 1907). High count: 12, 15 May 1975, Honey Creek WA, Andrew Co. (DAE, MBR).

Summer: Chats are most abundant in the Ozarks and Ozark Border and least common in the Glaciated and Osage plains (Map 57). Shirling (1920) reported a total of 18 singing males at Swope Park in 1916; however, none were recorded at this site in 1973 (Branan and Burdick 1981). BBS data for the past 23 years indicate the Missouri population is stable (Graph 19).

Fall Migration: Most breeders and migrants have left the state by late Aug. There are very few Sept reports. Latest dates: 25 Sept (year ?), St. Louis (Widmann 1907); 1, 20 Sept 1964, Amazonia, Andrew Co. (FL).

Summer Tanager (*Piranga rubra*)

Status: Common summer resident in Ozarks and Ozark Border, uncom-

Graph 19.
Missouri BBS data from 1967 through 1989 for the Yellow-breasted Chat. See p. 23 for data analysis.

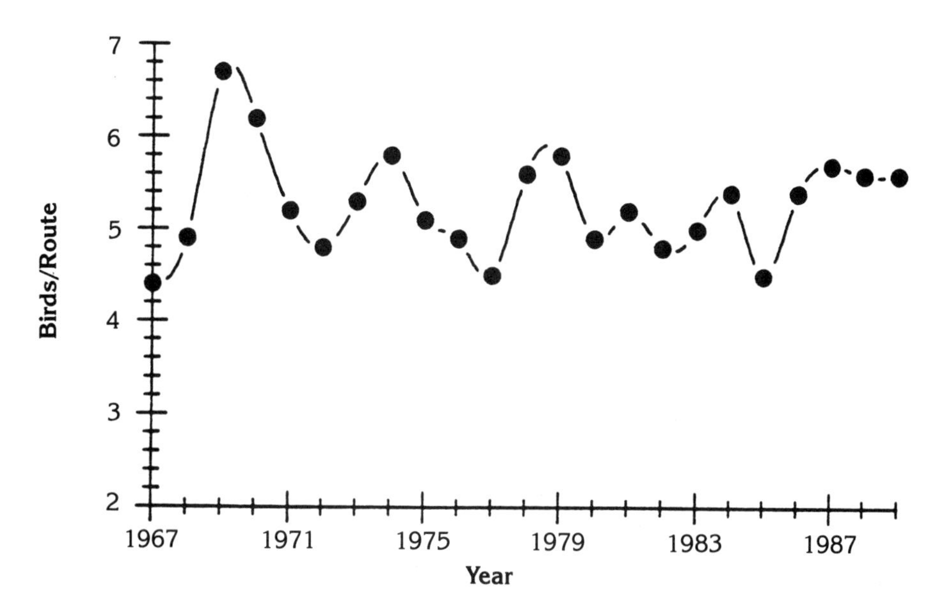

mon elsewhere; accidental winter visitant.

Documentation: Specimen: male, 22 May 1967, Gravois Mills (NWMSU, LCW 560).

Habitat: Most common in mature woodland and forest in floodplain.

Records:

Spring Migration: The first arrivals appear at the southern border during the third week of Apr, but this tanager is usually not seen in the north until the final days of Apr or the beginning of May. Peak is in mid-May. Earliest dates in south: 1, 14 Apr 1981, Farmington (BRE); in north: 1, 17 Apr 1971, Squaw Creek (MBR).

Summer: This tanager is most common in the Ozarks and Ozark Border, and not surprisingly it is least abundant in the more open regions of the state (Map 58). It is not uncommon in appropriate habitat even in the northernmost counties of the state. This species is certainly more common than the Scarlet Tanager in all regions of the state except perhaps the extreme northern section of the Glaciated Plains, although BBS data indicate that it outnumbers the Scarlet even there.

Fall Migration: It is most common during the first three weeks of Sept, with fewer birds (1–2/day) seen through the first week of Oct. It is accidental after mid-Oct. Latest dates: 1, 4 Nov 1979, near Roby, Texas Co. (D. Jones-**BB** 47[1]:26); male, 30 Oct 1987, Cuivre R. SP, Lincoln Co. (EL-**BB** 55[1]:13).

Map 58.
Relative breeding abundance of the Summer Tanager among the natural divisions. Based on BBS data expressed as birds/route.

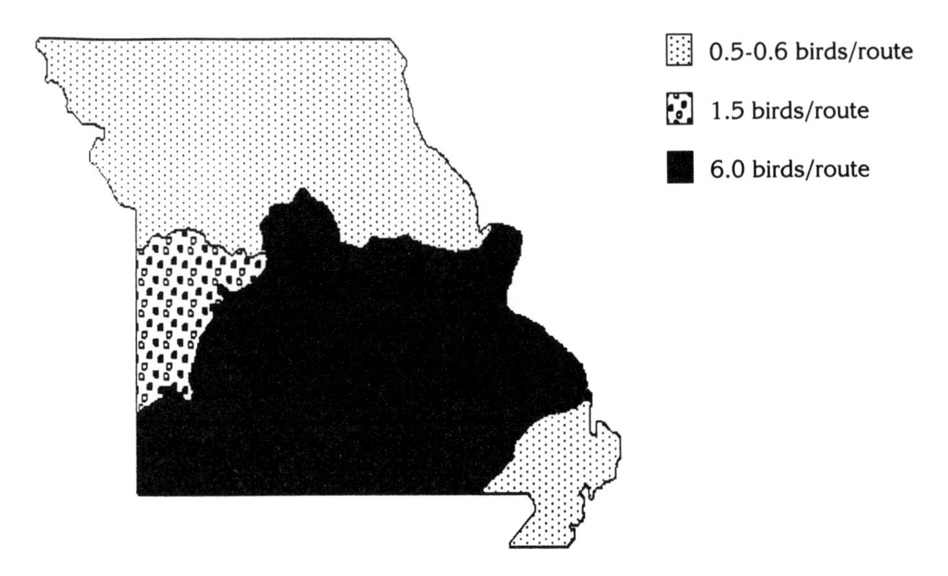

Winter: 1, photos, 3–10 Jan 1987, Springfield (C. Johnson-**BB** 54 [2]:14–15; VIREO x05/4/003).

Scarlet Tanager (*Piranga olivacea*)

Status: Common transient; uncommon summer resident statewide.

Documentation: Specimen: male, 19 May 1968, near Maryville (NWMSU, PER 50).

Habitat: Mature woodland and forest.

Records:

Spring Migration: Like the Summer Tanager, the first migrants appear in the south during the third week of Apr. Birds arrive in the north at the end of Apr. Peak is in mid-May. An average of 0.5 birds/hr was recorded at Forest Park between 1–20 May 1979–90 (RK; n=333 hrs). Earliest date: 1, 17 Apr 1977, St. Louis (CS et al.-**NN** 49:39).

Summer: BBS data indicate that this species is most common in the Ozarks and Ozark Border (average 0.4 birds/route) and least common in the Mississippi Lowlands (average < 0.1 birds/route).

Shirling (1920) recorded 22 males at Swope Park in 1916, whereas Branan and Burdick (1981) failed to find a single individual there in 1973. Over 50% of the upland forest had been lost at Swope Park in the intervening period between these two censuses.

Fall Migration: Most common during mid-Sept, with smaller numbers (1–2/day) through the first week of Oct. It is accidental thereafter. High count: 5, 11 Sept 1964, St. Louis (SHA-**BB** 31[4]:18). Latest dates: 14 Oct 1906, St. Louis (Widmann 1907); female, 12 Oct 1981, St. Joseph (LG-**BB** 49[1]:15).

Western Tanager (*Piranga ludoviciana*)

Status: Casual transient.
Documentation: Sight records only: see below.
Records:
Spring: There are four spring observations: male, 4 May 1977, near Charity, Dallas Co. (J. Smith-**BB** 44[4]:28); male, 7 May 1967, St. Louis (E. Hath-**BB** 34[2]:9); male, 9 May 1952, Kansas City (J. Myers-Rising et al. 1978); female, 21 May 1966, near Gray Summit, Franklin Co. (JEC-**BB** 33[2]:10).
Fall Migration: A single sighting: male, 21 Aug 1966, St. Louis (F. Erb-**BB** 34[1]:10).

Northern Cardinal (*Cardinalis cardinalis*)

Status: Common permanent resident.
Documentation: Specimen: male, 28 Dec 1961, Lincoln Co. (KU 40317).
Habitat: Brushy areas, thickets, second-growth woodland, and residential areas.
Records:
Summer: According to BBS data, the cardinal is most abundant in the Osage Plains and the Ozarks and is least common in the Mississippi Lowlands (Map 59). If data were available to separate the Ozark Border region from the Ozarks, it would probably reveal that this species reaches its greatest abundance in the former area, since there is more extensive optimal habitat in that region than in any other in the state.

A significant increase in the abundance of this species has coincided with the conversion of forest and prairie to parkland and residential areas at Swope Park between 1916 (n=70 territorial males; A. Shirling 1920) and 1973 (n=110 territorial males; Branan and Burdick 1981).

Winter: Early winter abundance for this species is greatest in the Ozark Border (6.6 birds/pa hr) and the Ozarks (6.0 birds/pa hr). Lower but similar average values (birds/pa hr) are recorded in the Mississippi Lowlands (4.8), Osage Plains (4.7), and Glaciated Plains (4.2). High counts: 1,117, 17 Dec 1983, Kansas City Southeast CBC; 965, 27 Dec 1987, Columbia CBC.

Map 59.
Relative breeding abundance of the Northern Cardinal among the natural divisions. Based on BBS data expressed as birds/route.

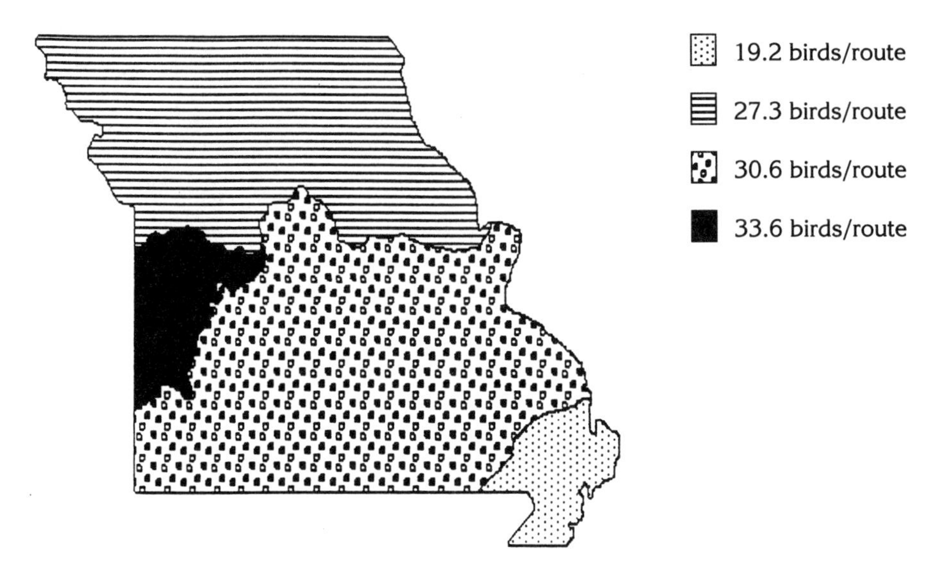

Rose-breasted Grosbeak (*Pheucticus ludovicianus*)

Status: Common transient; common summer resident in Glaciated Plains, rare in northern Osage Plains and Ozark Border and casual elsewhere; accidental winter visitant.

Documentation: Specimen: male, 15 June 1883, Jackson Co. (KU 39433).

Habitat: Openings and edge of woodland and forest.

Records:

Spring Migration: Common statewide; the initial birds appear at the beginning of the third week of Apr in the south but not until the final days of the month in the north. Peak is during the first week of May in the south and the second week in the north. An average of 0.9 birds/hr was recorded at Forest Park between 27 Apr–17 May 1979–90 (RK; n=359 hrs). Earliest date: female, 22 Apr 1964, St. Joseph (FL). High count: 100+, 6–7 May 1978, Roaring R. SP (JG-**BB** 45[3]:19).

Summer: Common in the Glaciated Plains (average 4.4 birds/route) except on the southwestern edge (south of the Missouri R.) where it becomes uncommon. It is rare in the Osage Plains (average 0.2) and becomes scarcer farther south in this region. In the Ozark Border, south of the Missouri R., it is also rare. In the heart of the Ozarks it is only casually encountered (some of these observations involve unpaired birds). The southernmost, definite breeding record is from near Freistatt, Lawrence Co., where adults were observed feeding a young on 6 July 1885 (Nehrling; Widmann 1907). More recently, an adult male and immature were observed in June 1986 at Springfield (C. Tyndall-**BB** 53[4]:17).

Fall Migration: Migrants begin to appear in nonbreeding areas at the end of Aug. Peak across the state occurs in mid-Sept. Numbers are greatly reduced by the end of Sept, with maximum of 1–2 birds/day seen through the first half of Oct. It is only casually seen thereafter. High counts: 50+, 16 Sept 1971, 10 miles north of Maryville (MBR); 30, tower kill, 20–21 Sept 1963, Columbia (George 1963). Latest dates: 1, 3 Nov 1973, St. Joseph (FL-**BB** 41[1]:4); 1, 28 Oct 1972, Columbia (BG-**BB** 40[1]:9).

Winter: At least two records of single birds: male, at feeder, 18 Dec 1973–Mar 1974 (BG-**BB** 41[2]:3); 1, 19 Dec 1982, Weldon Spring CBC.

Black-headed Grosbeak (*Pheucticus melanocephalus*)

Status: A rare and infrequent transient.

Documentation: Photograph: immature male at feeder, 28 Nov–3 Dec 1984, Roaring R. SP (J. Baugardt et al.-**BB** 52[1]:33; photo M. Rogers, VIREO r18/1/001; Pl. 8e).

Habitat: Woodland and forest edge, especially riparian areas.

Records:

Spring Migration: At least seven observations (five from the west, two the from east) of single birds; all but one recorded in late Apr or early May. Earliest date: 1, 24 Feb 1986 (winter resident?), Kansas City (KH-**BB** 53[2]:10). Latest date: male, 6 May 1973, St. Joseph (FL, SR-**BB** 40[3]:6).

Fall Migration: At least ten reliable records, which are evenly divided between east and west, range from early Sept through the end of the period; most are concentrated in Sept. Earliest date: immature male, 6 Sept 1962, St. Charles Co. (KA, RA-**BB** 29[4]:28). Latest dates: female, specimen (not preserved, fide JW), 2 Dec 1983, Meta, Osage Co. (J. Rowan-**BB** 51[2]:23); and the above photographic record.

Comments: Observers should be aware of the possibility of hybrids between this species and the Rose-breasted Grosbeak.

Blue Grosbeak (*Guiraca caerulea*)

Status: Common summer resident in Ozarks, Ozark Border, and Osage Plains; rare elsewhere.

Documentation: Specimen: male, 7 June 1968, north of Tarkio, Atchison Co. (NWMSU, DAE 1877).

Habitat: Brushy areas, hedgerows, and second growth.

Records:

Spring Migration: The first migrants appear in southern Missouri at the

Graph 20.
Missouri BBS data from 1967 through 1989 for the Blue Grosbeak.
See p. 23 for data analysis.

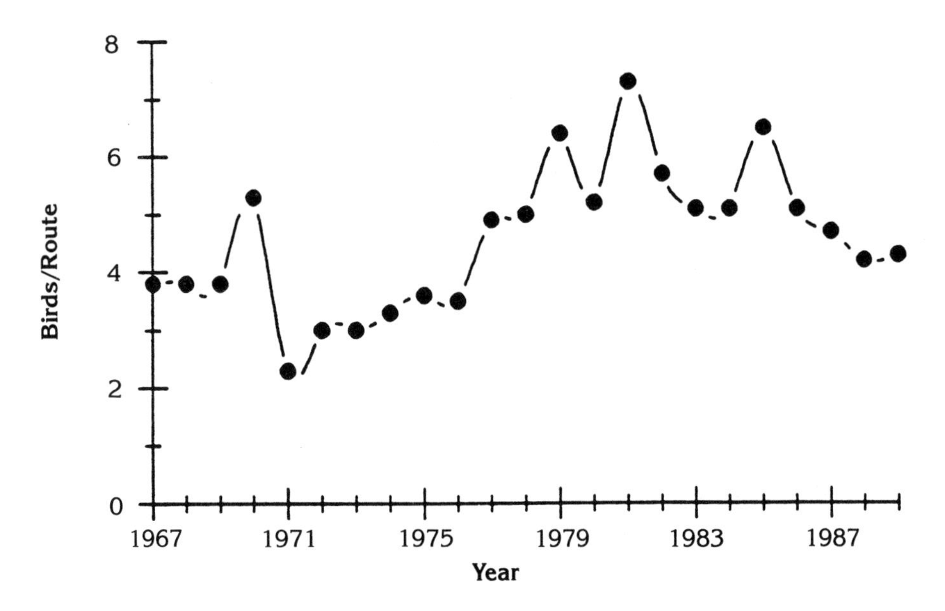

end of Apr and about a week later in the north. Peak is in mid-May. Earliest dates: no.?, 24 Apr 1904, Shannon Co. (W. Savage; Widmann 1907); 1, 25 Apr 1953, Mt. Grove, Wright Co. (D. McKinley-**BB** 20[6]:3). High count: 20+, 8 May 1989, St. Clair and Cedar counties (MBR, TE-**BB** 56[3]:94).

Summer: BBS data show that this species reached its greatest abundance in Missouri during the late 1970s through the mid-1980s (Graph 20). As Map 60 depicts, this grosbeak is most abundant in the Ozarks and Osage Plains. Within the Ozarks it is perhaps most abundant in cedar glades of the White River section (see Map 3). It is not known to breed in the Mississippi Lowlands (none have been recorded on the sole BBS route in that region; n=16 years).

Fall Migration: After mid-Sept very few birds are encountered, and virtually none remain after the first week of Oct. High count: 8, 4 Sept 1960, Trimble (FL). Latest dates: male, 4 Dec 1952, Baden, St. Louis Co. (RA-**BB** 20[3]:1); male, 21 Oct 1987, Jackson Co. (KH, K. Brobisky-**BB** 55[1]:13).

Lazuli Bunting (*Passerina amoena*)

Status: Casual spring transient in west, accidental in east; hypothetical summer resident and fall transient.

Documentation: Sight records only: see below.

Map 60.
Relative breeding abundance of the Blue Grosbeak among the natural divisions. Based on BBS data expressed as birds/route.

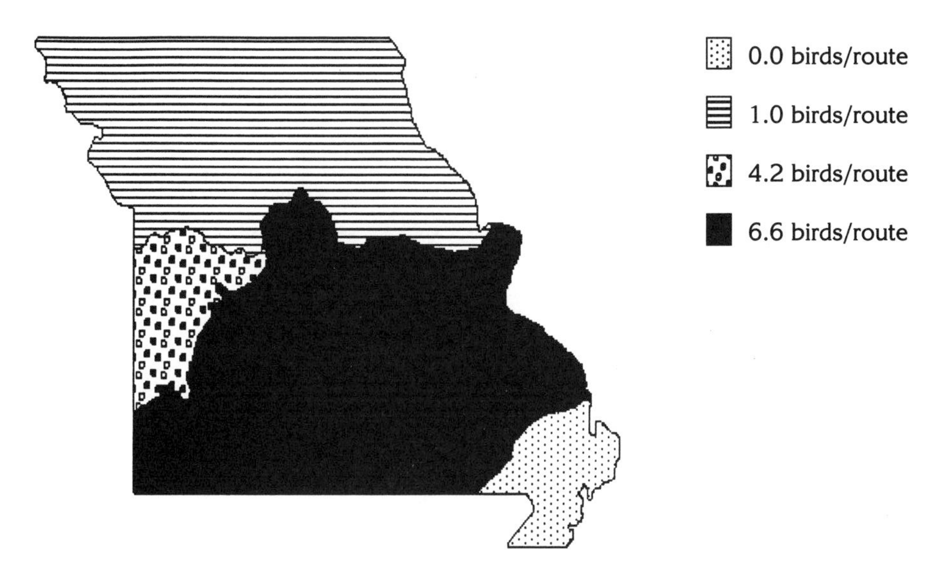

Habitat: Open brushy areas, hedgerows, thickets.

Records:

Spring Migration: There are a total of seven records (five from the west, two from the east). All sightings are between the final week of Apr and mid-May. The two eastern sightings are as follows: male, 7 May 1950, near Webster Groves, St. Charles Co. (A. Bolinger, JEC, JC et al.-**BB** 18[1]:1); male, 14 May 1983, St. Louis (RK, EL-**BB** 50[3]:14). Earliest date: 28 Apr 1959, Mt. Washington Cemetary, Jackson Co. (FB-**AFN** 13:376). Latest date: 13 May 1934, Swope Park (WC; Rising et al. 1978).

Summer: There is a report of a male Lazuli with a female (possibly a female Indigo?) that were observed on a least two occasions in Jackson Co. in 1939 (1 June and mid-June); however, neither nest nor young were found (Walters 1939).

Fall Migration: The report of a male and female taken from a flock composed of "all females and young males" on 13 Sept 1894 by S. Wilson must be viewed with skepticism, as Wilson reported other questionable observations (Widmann 1907; copy of Wilson's catalog). These birds may have been immature Indigo Buntings (the specimens are apparently no longer extant).

Comments: Hybridization between this species and the Indigo Bunting occurs in the Great Plains. At least one hybrid (a male) has been reported in the state: 14 June 1980, Marais Temps Clair (DJ).

Indigo Bunting (*Passerina cyanea*)

Status: Common summer resident; accidental winter visitant.
Documentation: Specimen: male, 4 July 1964, Platte Co. (KU 46817).
Habitat: Brushy areas, second growth, forest and woodland edge.
Records:
Spring Migration: A very common migrant. The first birds appear by mid-Apr in the south and during the final week of Apr in the north. Peak is in early May in the south and mid-May in the north. An average of 0.9 birds/hr was recorded at Forest Park between 25 Apr–25 May 1979–90 (RK; n=468 hrs). Earliest date: 2, 1 Apr 1980, Mississippi Co. (JH).
Summer: It is most abundant in the Ozarks and Mississippi Lowlands and least common in the Glaciated and Osage plains (Map 61). Until 1989 this species steadily increased over the course of the 23 years that the BBS routes have been run in the state (Graph 21).
Fall Migration: Migrants are commonly seen through the first ten days of Oct; primarily individuals or small groups (< 5 birds/flock) are seen through the third week of Oct. Thereafter, only single birds are casually seen. Latest dates: male, 3 Dec 1984, Poplar Bluff, Butler Co. (V. Moss-**BB** 52[2]:17; specimen apparently not preserved); male, 22 Nov 1973, near Warrensburg, Johnson Co. (T. Sappington-**BB** 41[1]:4).
Winter: Two records: 1, banded, 19 Jan 1958, Busch WA (JC-**BB** 25[3]:1); male, 29 Jan 1987, Tuscumbia, Miller Co. (JW-**BB** 54[2]:15; VIREO x05/4/002).

Painted Bunting (*Passerina ciris*)

Status: Rare summer resident in southwest; casual transient and accidental summer resident elsewhere.
Documentation: Specimen: male, 23 June 1928, Vernon Co. (CMC 129).
Habitat: Most common in fallow fields with scattered trees bordered by woodland, and open cedar glades.
Records:
Spring Migration: Usually not seen until the first of May, with the majority of the migrants not arriving until midmonth. Earliest dates: male, 19 Apr 1986, Hornesville, Dunklin Co. (H. Schanda-**BB** 53[3]:13); 1, 28 Apr 1984, Springfield (N. Vandenbrink-**BB** 51[3]:14). High count: 13 (10 males, 3 females), 17 May 1977, Protem, Taney Co. (W. West-**BB** 44[4]:28).

Extralimital records north of the Missouri R.: pair, 5 May 1985, Bluffwoods SF (D. Reynolds; fide FL); male, 6 May 1988, Watkins Mill SP, Clay Co. (J. Eldridge-**BB** 55[3]:90); male, 24 May 1930, Parkville, Platte Co. (J. Jackson; Bennitt 1932); 1, 30 May 1939, Iatan Marsh, Platte Co. (WC, A. Shirling-**NN** 6:59). A photograph of the purported female banded in Atchison Co. (Norris and Elder 1982a) was examined and proved to be a

Map 61.
Relative breeding abundance of the Indigo Bunting among the natural divisions. Based on BBS data expressed as birds/route.

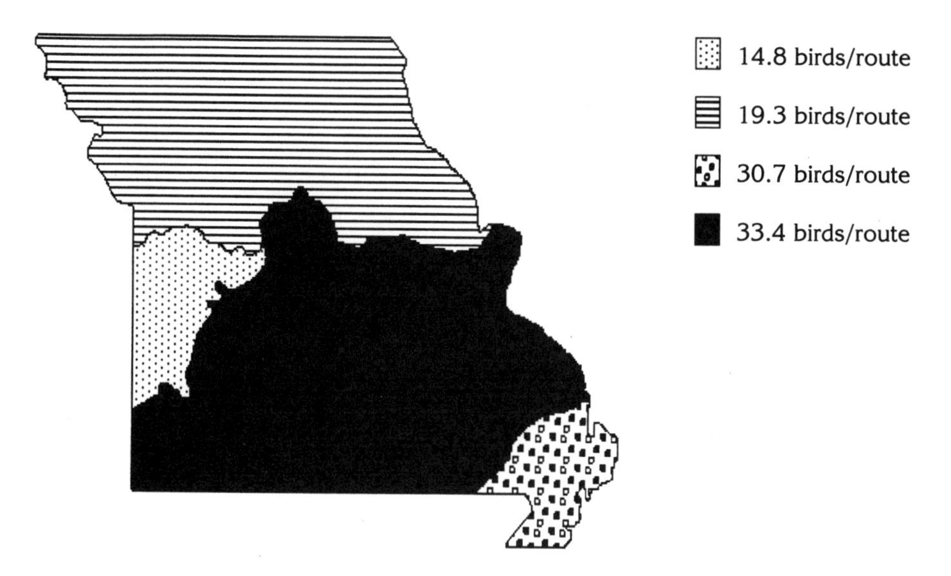

Graph 21.
Missouri BBS data from 1967 through 1989 for the Indigo Bunting. See p. 23 for data analysis.

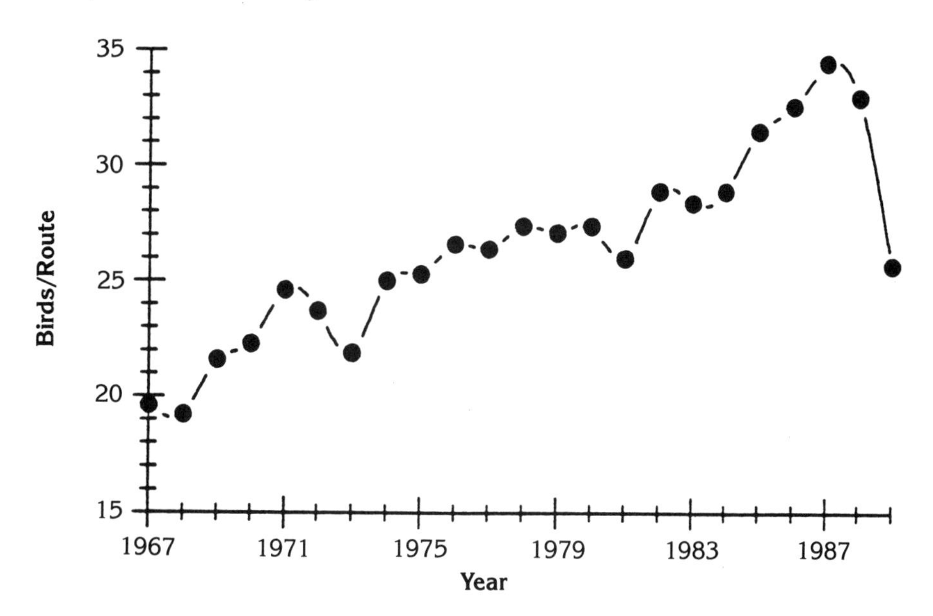

female Indigo Bunting (DAE, MBR).

Summer: The stronghold of this species' breeding range is the White River Glades region (Map 3). It is known to regularly breed as far north as the Springfield area. Although there are no definite nesting records east of Oregon Co. (J. Jokerst; Bennitt 1932), it may sparingly breed as far east as the Mississippi Lowlands, since it is apparently common in the Mississippi R. valley in Arkansas (James and Neal 1986).

Norris and Elder (1982a) gave characteristics of male Painted Bunting territories studied over a two-year period near Protem, Taney Co. They determined that there were at least eight and eleven territorial males in this area in 1979 and 1980, respectively.

Extralimital records of nesting birds: pair, summer 1956, Gladstone, Clay Co. (Rising et al. 1978); pair, banded, 1973, St. Clair Co (Norris and Elder 1982a). The following represent the more northern observations of single birds away from known breeding areas: 2 June 1972, along Missouri R., Kansas City (Rising et al. 1978); male, 22 June 1986, Meramec SP, Franklin Co. (WL-**BB** 53[4]:17); male, 23 June 1984, Ray Co. (SP-**BB** 51[4]:18); male, 5 July 1988, near Macks Creek, Camden Co. (TE); male, 11 July 1930, Osage Co. (J. Peeler; Bennitt 1932).

Fall Migration: All but two records are during Aug at or near breeding localities. Latest date: 1, 19 Sept 1952, Big Oak Tree SP (JEC-**AFN** 7:20). Extralimital record: 1, photo, 5 Aug 1982, Hannibal, Marion Co. (A. Allmon-**BB** 50[1]:27).

Comments: The Missouri population has erroneously been attributed to the nominate subspecies (A.O.U. 1957, Paynter 1970). Four adult males (in Missouri collections) are clearly referable to interior *pallidior.*

Dickcissel (*Spiza americana*)

Status: Common summer resident statewide; casual winter resident.

Documentation: Specimen: female, 23 Nov 1966, west Maryville (NWMSU, DAE 1182).

Habitat: Fallow fields with scattered bushes, prairie, and cultivated fields.

Records:

Spring Migration: A very common migrant that usually arrives in the south by the beginning of the third week of Apr and a few days later in the north. Peak is in mid-May. Earliest dates: 1, 10 Mar 1987, near Forsyth, Taney Co. (PM-**BB** 55[1]:14); 1, 20 Mar 1975, Cape Girardeau (T. Daugherty-**BB** 42[3]:14).

Summer: This ubiquitous bird of open country is most abundant in the Osage Plains and least common in the Ozarks (Map 62). Although it is less uniformly distributed and common in the Ozarks and Mississippi Lowlands than in the other natural divisions of the state, it nevertheless has

Map 62.
Relative breeding abundance of the Dickcissel among the natural divisions. Based on BBS data expressed as birds/route.

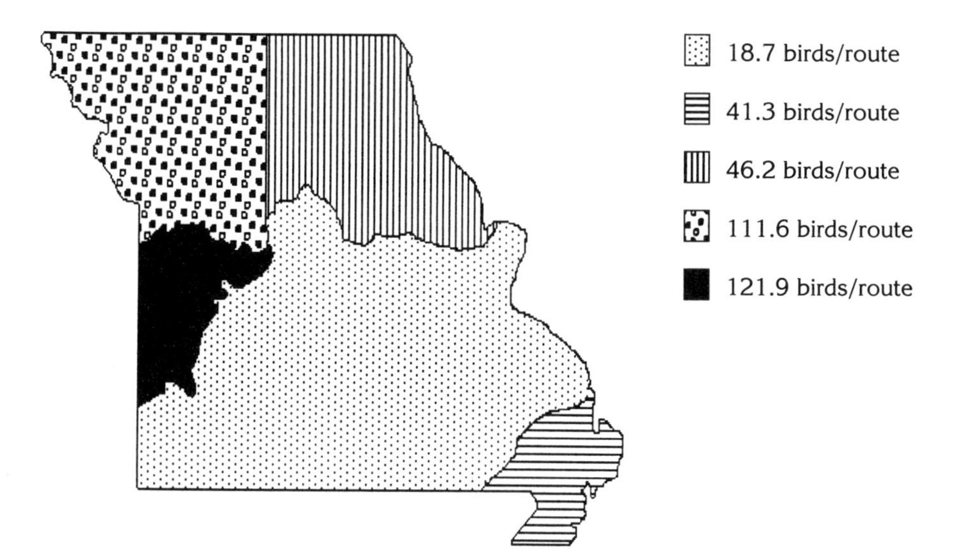

increased considerably in these two regions since presettlement. Prior to the extensive clearing of the state's most heavily forested areas, the Dickcissel was "rare in the Ozarks where only in open, long settled localities; sparingly on the cultivated ridges of the southeast [Mississippi Lowlands]" (Widmann 1907).

Even though this species is now more common in two of the state's natural divisions than it was in presettlement Missouri, it undoubtedly has significantly decreased since European man converted most of its primary habitat to cultivation.

Fall Migration: By the latter half of Aug, small flocks consisting of 10 to 20 birds are seen. Although these groups are less frequently encountered during the last two weeks of Sept and the first week of Oct, they nevertheless are still present, e.g., 10–12, 7 Oct 1978, 10 miles north of Maryville (MBR). After mid-Oct usually only solitary birds are casually observed. High counts: 42+, 13 Sept 1984, Taberville Prairie (MBR, FL); 32, tower kill, 20–21 Sept 1963, Columbia (George 1963).

Winter: There are at least seven records scattered throughout the state, with all but one involving single birds: 2, at feeder, 18 Dec 1971, Springfield (R. Matthews-**BB** 39[1]:7). Northernmost record: immature male, found dead at bird feeder, 26 (not 28) Jan 1980, Maryville (DAE-**BB** 47[3]:18; NWMSU, MBR 1429).

Graph 22.
Missouri BBS data from 1967 through 1989 for the Dickcissel.
See p. 23 for data analysis.

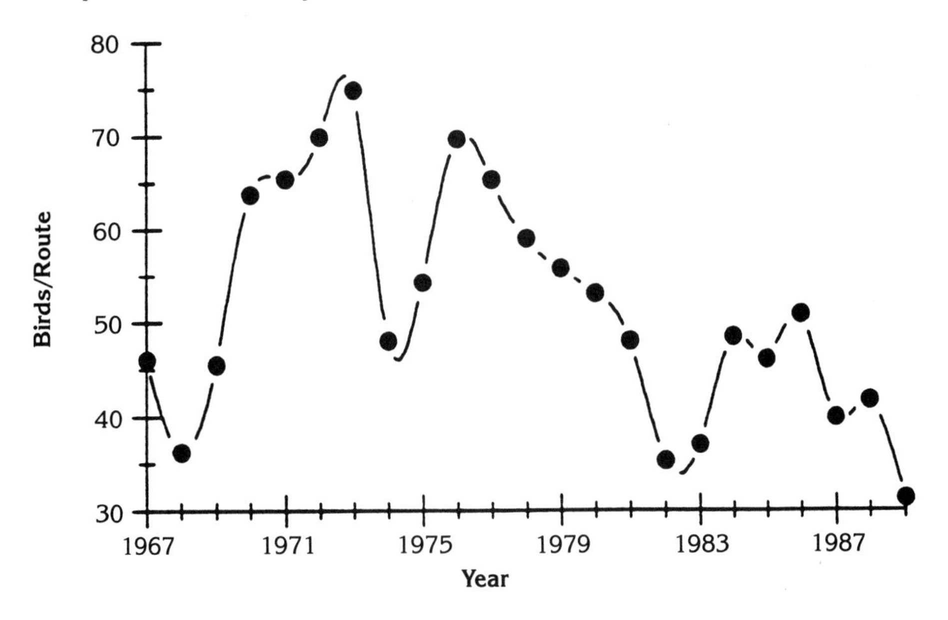

Green-tailed Towhee (*Pipilo chlorurus*)

Status: Accidental transient and winter resident.
Documentation: Specimen: 1, 12 Mar 1920, Lexington, Lafayette Co. (C. Salyer-**BL** 22:231; CMC 121).
Habitat: Brushy areas and thickets.
Records:
Spring Migration: Only three records, including the one above: 1, 1 Mar 1975, Bennett Spring SP, Dallas Co. (JG-**BB** 42[3]:14); 1, 23 Mar 1947, near Liberty, Clay Co. (M. Diemer-**BB** 14[6]:3). A bird reported at Busch WA in Apr 1957 (**BB** 24[8–9]:4) was later retracted (**NN** 28:83); this retraction has been overlooked by others.
Fall Migration: A single observation: 1, 23 Nov–26 Dec 1952, Busch WA (JC et al.-**BB** 20[3]:1).
Winter: Two observations: 1, at feeder, 18 Dec–25 Jan 1976–77, Springfield (D. Hagewood et al.-**BB** 44[3]:26; photographed); 1, 3 Feb 1976, Busch WA (DJ et al.-**NN** 48:32).

Rufous-sided Towhee (*Pipilo erythrophthalmus*)

Status: Common summer resident in all areas, except the Mississippi

Map 63.
Relative breeding abundance of the Rufous-sided Towhee among the natural divisions. Based on BBS data expressed as birds/route.

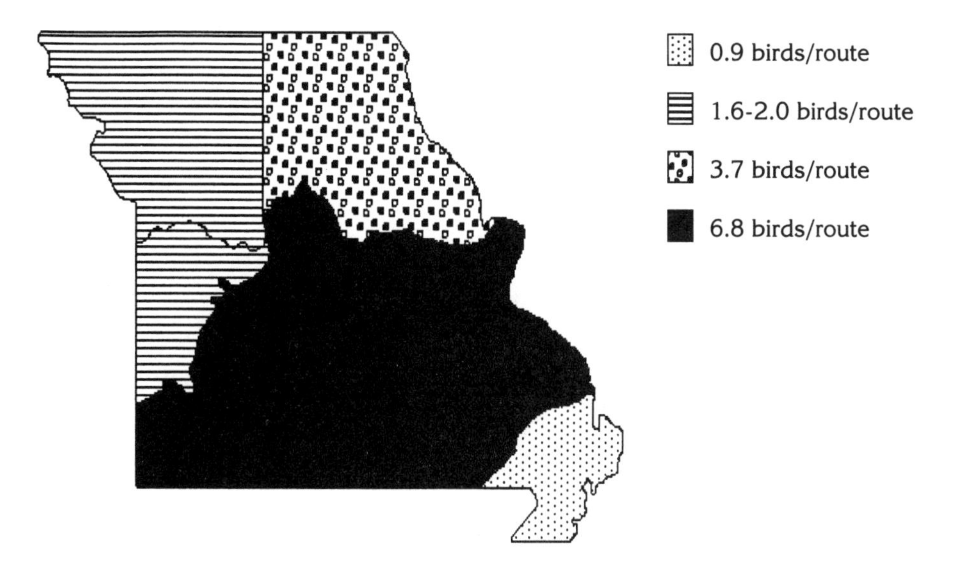

Lowlands where it is uncommon; uncommon to locally common winter resident in southeastern section of Ozarks, rare elsewhere.

Documentation: Specimen: male, 12 Mar 1965, Platte Co. (NWMSU, DAE 855).

Habitat: Brushy areas, thickets, woodland and forest edge.

Records:

Spring Migration: Migrants begin supplementing the winter population in the south by early Mar; birds first appear in the north by mid-Mar. Peak is in early Apr in the south and a week later in the north.

Summer: BBS data indicate that it is considerably more common in the Ozarks than in any other area of the state (Map 63).

Fall Migration: Although numbers are reduced from those present in mid-Sept, it remains fairly common across the northern half of the state until mid-Oct and in the south until the beginning of Nov. Thereafter, primarily winter residents remain.

Winter: Generally a rare and irregular winter resident in the state (absent during severe winters) except the southeastern section of the Ozarks where it is generally uncommon. Map 64 presents CBC data. High counts, both on the Mingo CBC: 121, 28 Dec 1965; 108, 30 Dec 1964.

Comments: The distinctive race, *P. e. arcticus* of the western "Spotted" Rufous-sided Towhee complex, is an uncommon migrant in the extreme western section of the state and rare elsewhere. The vast majority of the

Map 64.
Early winter relative abundance of the Rufous-sided Towhee among the natural divisions. Based on CBC data expressed as birds/10 pa hrs.

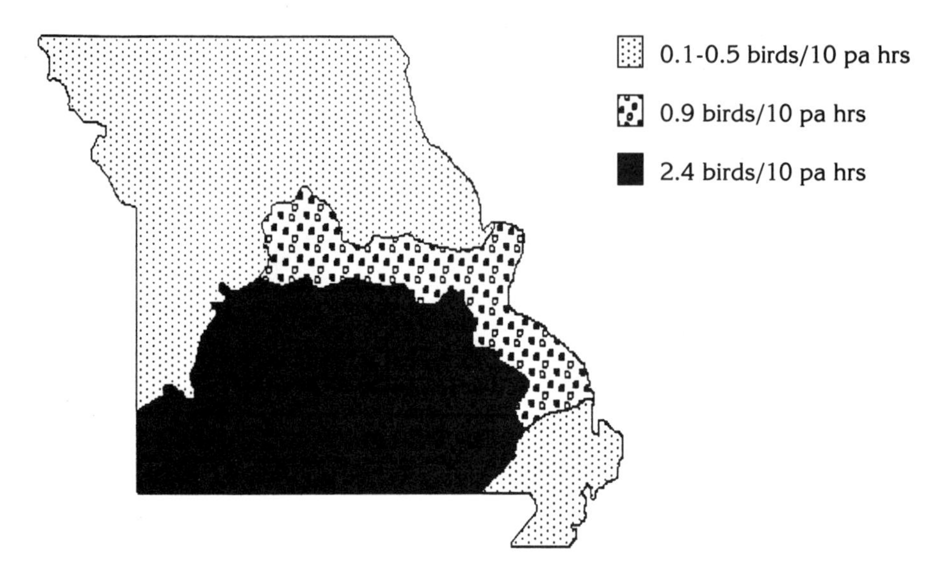

wintering towhees from at least the Kansas City area northward to the Iowa border are of this race, and not of the unspotted, summer resident race (*P. e. erythrophthalmus*). However, with the exception of an occasional *arcticus,* the nominate race is the bird that winters throughout the remainder of the state.

The first fall migrants of the Spotted Towhee usually appear by the beginning of Oct. Most have left by mid-Apr. Earliest fall date: male and female, 24 Sept 1990, Brickyard Hill WA (MBR). Latest spring date: 1, 10 May 1964, Cooley L. (FB-**BB** 31[2]:15). High count: 25, 26 Dec 1938, Parkville, Platte Co. (WC, HH et al.-**BB** 6[2]:14).

Bachman's Sparrow (*Aimophila aestivalis*)

Status: Very rare and local summer resident in the Ozarks, and at least formerly in the Ozark Border.

Documentation: Specimen: male, 4 June 1968, Gravois Mills (NWMSU, DAE 1919).

Habitat: Overgrown grassy fields with scattered scrubs and trees; open pinewoods.

Records:

Spring Migration: The first migrants arrive at the very end of Mar and early Apr. Most are on the breeding grounds by the end of Apr. Earliest dates: male, 19 Mar 1907, Shannon Co. (Woodruff 1908; AMNH 229745); 1, singing, 29 Mar 1974, McCormack L., Oregon Co. (JC-**BB** 41[2]:3); 3, 29 Mar 1976, Caulfield, Howell Co. (R. Tweit-**AB** 30:727). When this species was more common and bred further north (see below), it was seen in areas where it is now considered a vagrant. For example, there were a minimum of five observations within the city limits of St. Louis up through the mid-1960s. There are only two extralimital records north of the Missouri R. at this season: 1, 20 Apr 1941, near Ashland, Boone Co. (m.ob.-**BB** 8[4]:35); 2, 1-5 May 1889, along Mississippi R., Marion Co. (O. Poling; Widmann 1907).

Summer: Until relatively recently the Bachman's occurred in the Ozarks from the southwestern corner (Lawrence and Barry counties) east to Butler, Wayne, and St. Genevieve counties and north into the Ozark Border (Morgan and St. Louis counties). It apparently has undergone a retraction from the northern end of its range, and it appears to have decreased in the more southern areas during the past fifteen to twenty years. Up until the mid-1960s, it bred as far north as western St. Louis Co. (Anderson and Bauer 1968). Adults were observed feeding young on 4 June 1968, ca. 10 miles south of Gravois Mills in oldfield habitat (DAE). It was also recorded on BBS route 20 in Morgan Co. through 1981, but it has not been found there since. Likewise, this species appeared to decline during the latter part of the 1970s in the Salem, Dent Co. area (D. Plank, pers. comm.). Although the above gives the impression of a dramatic decline, it may actually represent no more than a local extirpation due to the succession of primary Bachman's Sparrow habitat. A concerted effort is needed to determine if the Bachman's has truly disappeared from some of the above regions of the state.

In 1976, 21 singing males were recorded in the Mark Twain NF of Taney and Ozark counties at the southern border (Hardin et al. 1982). These birds bred in glades with a relatively dense herbaceous cover, but since that time the species has been difficult to find in this unique habitat. Hardin et al. (1982) suggest that overgrazing and drought have a detrimental effect on the herbaceous growth in the glades, thus depressing sparrow densities. This coupled with the above apparent declines, which may be related to succession, underscores the rather narrow habitat preferences of this species.

Prior to settlement, it was probably most common in the Shortleaf Pine area of the state, as Woodruff (1908; AMNH 29740-229747) found it "common throughout the mixed pine and oak woods" in Shannon Co. before this area was entirely logged. Certainly the unnatural prevention and containment of forest fires in the pine region of the state has been detrimental to this species, as very few mature stands of pine remain where there is not an extensive deciduous undergrowth. Extralimital record: 1, 18 June 1966, Cuivre SP, Lincoln Co. (RA-**NN** 38:65).

Fall Migration: All but two observations for this period are of birds on the breeding grounds in Aug. Latest dates: immature male, 7 Sept 1964, near Salem, Dent Co. (DAE, D. Plank; NWMSU, DAE 691); 4, 2 adults and 2 immatures, 1 Sept 1947, Bourbon, Crawford Co. (JEC-**NN** 19:35).

[**Cassin's Sparrow** (*Aimophila cassinii*)]

Status: Hypothetical; accidental spring transient.
Documentation: Sight record only: see below.
Comments: There is a single observation: 1, 9 May 1987, south of Camden, Ray Co. (RF-**BB** 54[3]:15). Although the bird was meticulously described, it was seen by only a single observer and was not photographed.
There are now several spring records in the upper midwest of birds "overshooting" the breeding grounds. It should be looked for in late Apr and May in the tallgrass prairie region of the state.

American Tree Sparrow (*Spizella arborea*)

Status: Common winter resident in Glaciated and Osage plains, uncommon in Ozarks and Mississippi Lowlands.
Documentation: Specimen: female, 3 Apr 1965, Cooley L. (NWMSU, DAE 859).
Habitat: Weedy fields, hedgerows, brushy areas.
Records:
Spring Migration: The majority leave the state during the first half of Mar. However, it can still be common, even in the southwest, until the beginning of the second week of Apr. But in most years, it is rare statewide by the beginning of the third week of Apr, and only single birds can be casually observed during the final days of Apr. Latest dates: 1, 1 May 1939, Webster Groves, St. Louis (JEC-**BB** 6[6]:51); 1, 30 Apr 1974, St. Louis (RA-**BB** 41[3]:5).
Fall Migration: Normally the first arrivals appear at the very end of Oct. It does not become common until mid-Nov, with numbers increasing into Dec. Earliest dates: 10 Oct 1894, St. Joseph (Widmann 1907); 1, 14 Oct 1981, Bluffwoods SF (FL).
Winter: This is the common sparrow in open country in the Glaciated and Osage plains. It is decidedly less common in the Ozarks and Mississippi Lowlands (Map 65). High counts: 5,018, 2 Jan 1955, Weldon Spring CBC; 3,587, 17 Dec 1983, Kansas City Southeast CBC.

Chipping Sparrow (*Spizella passerina*)

Status: Common summer resident in Ozarks, more local, but still common elsewhere; casual winter resident in south.

Map 65.
Early winter relative abundance of the American Tree Sparrow among the natural divisions. Based on CBC data expressed as birds/10 pa hrs.

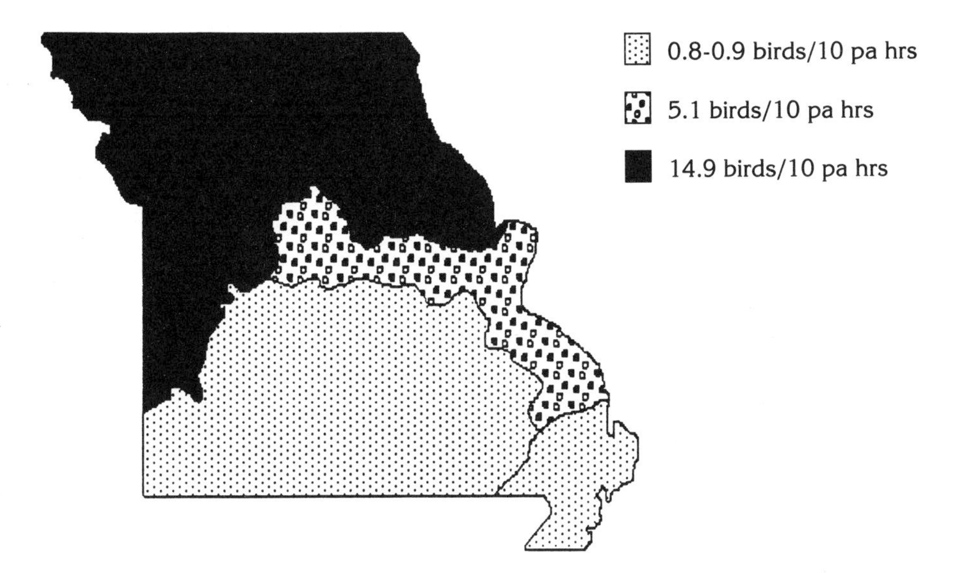

Documentation: Specimen: male, 12 Mar 1907, Black Valley, Shannon Co. (Woodruff 1908; AMNH 229727).

Habitat: Most common in open stands of conifers, but also found in entirely deciduous areas; also common in residential areas.

Records:

Spring Migration: Birds begin to reappear in extreme southern Missouri by early Mar but not until the first week of Apr in the north. Peak is in early Apr in the south and late Apr in the north. Relatively large-sized, migrating flocks are seen through the second week of May, e.g., 30, 14 May 1990, west of Maryville (MBR-**BB** 57[3]:141). An average of 0.6 birds/hr was recorded at Forest Park between 15 Apr–8 May 1979–90 (RK; n=332 hrs). High counts: 125, 20 Apr 1988, Tower Grove Park, (SRU-**BB** 55[3]:90); 40, 29 Apr 1979, St. Joseph (FL). Earliest arrival at St. Louis: 10 Mar 1887 (Widmann 1907); FL's earliest arrival at St. Joseph (n=35 years), 27 Mar 1963.

Summer: It is most common in the Ozarks, especially in the Shortleaf Pine region. The overall average for the Ozarks is 5.4 birds/route; however, BBS routes 8 and 10 in the pine region have averaged 18.0 and 15.5 birds/ route, respectively. It is significantly less common in the Osage and Glaciated plains (Map 66). The sole route in the Mississippi Lowlands has never recorded the species, but it does breed there locally. Apparently it was once common in the cypress swamps of the latter region at the turn of the century (Widmann 1907).

Map 66.
Relative breeding abundance of the Chipping Sparrow among the natural divisions. Based on BBS data expressed as birds/route.

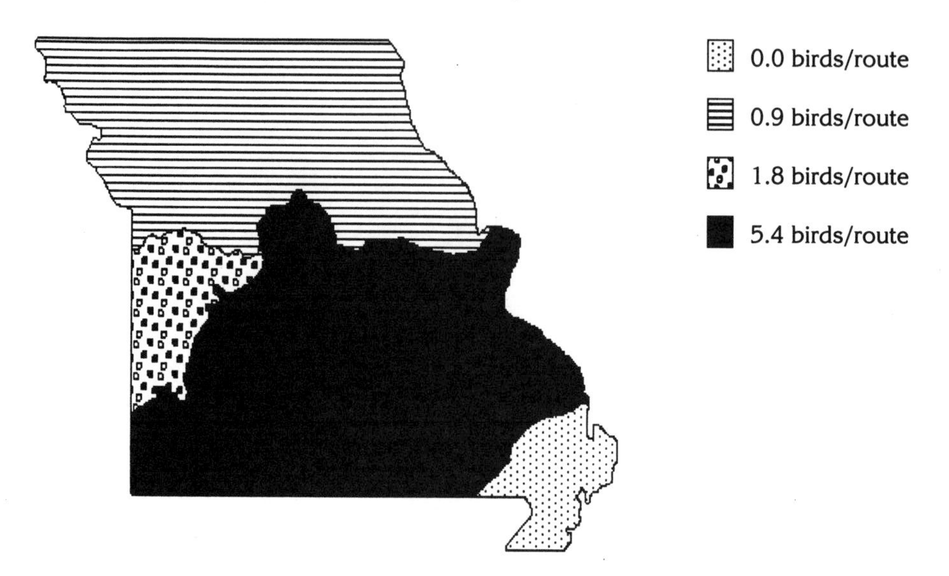

Fall Migration: This sparrow remains common and widespread across the state through Sept. By Oct, it is more local, but relatively large flocks are not uncommon (10–25 birds/flock; comprised largely of immatures). After the third week of Oct, numbers drop off dramatically. High counts: 50+, 21 Oct 1983, St. Joseph (FL-**BB** 51[1]:46); 50, 30 Oct 1989, Dallas Co. (L. Kennard-**BB** 57[1]:50). Latest date by FL at St. Joseph (n=35 years): 1, 24 Nov 1978.

Winter: There are no more than eight reliable records for the state. All but one of these are near the southern border. Only northern record: 1, 25 Jan–7 Feb 1967, St. Joseph (FL-**BB** 34[1]:24). High counts: 2, 30 Dec 1964, Mingo CBC; 2, 18 Dec 1976, Springfield CBC.

Clay-colored Sparrow (*Spizella pallida*)

Status: Uncommon transient in west, rare in east; accidental winter visitant.

Documentation: Specimen: male, 21 Apr 1961, Columbia (NWMSU, DAE 337).

Habitat: Weedy fields, hedgerows, and open parkland.

Records:

Spring Migration: Much more common and regular in the western half of the state than in the east. It is not recorded every spring in the east, and nearly all eastern observations are of single birds. In the west, the Clay-colored first appears during the final week of Apr and peaks during the second week of May. There are no reliable records after the third week of May. Earliest date: 1, 11 Apr 1976, St. Joseph (FL). High counts: 20, 12 May 1974, Pigeon Hill WA, Buchanan Co. (FL-**BB** 41[3]:5); 12, 6 May 1979, Squaw Creek (TB-**BB** 46[3]:12); 12, 14 May 1982, Honey Creek WA, Andrew Co. (FL-**BB** 49[3]:23). Latest date: 2, 19 May 1965, Squaw Creek (DAE-**BB** 32[1–2]:14).

Fall Migration: More frequently encountered at this season than in spring in the eastern half of the state. Individuals begin arriving during the final days of Sept. Peak is in mid-Oct. Virtually none are seen after the first week of Nov. Earliest date: 1, 10 Sept 1977, near Maryville (DAE); 1, tower kill, 10–11 Sept 1964, Kansas City (DAE). Latest date: 1, 6 Nov 1974, Busch WA (JC-**BB** 42[2]:7).

Winter: Two records: 1, 22–26 Dec 1985, Squaw Creek CBC (TB, MBR, DAE et al.); 1, 29 Dec 1963, Weldon Spring CBC.

Field Sparrow (*Spizella pusilla*)

Status: Common summer resident; uncommon winter resident in Mississippi Lowlands, Ozarks, and Ozark Border; rare elsewhere.

Documentation: Specimen: male, 12 Jan 1976, Maryville (DAE; NWMSU, JWG 229).

Habitat: Overgrown fields with scattered trees and hedgerows; woodland and forest edge.

Records:

Spring Migration: Migrants begin supplementing the winter population in the south at the beginning of Mar, but birds do not appear in the north until the final week of Mar. Peak in the south is during late Mar and in early to mid-Apr in the north. High counts in north, both by FL in the St. Joseph/Squaw Creek area: 20, 14 Apr 1968; 20, 26 Apr 1970.

Summer: The Field Sparrow is the most common sparrow in the state at this season. It is most abundant in the Ozarks, Ozark Border, and the Osage Plains (Map 67). The highest densities of this species may be in the cedar glades of the White River section of the Ozarks (see Map 3).

Fall Migration: It is still common in the north through most of Oct, although numbers are less than in Sept. By mid-Nov few remain there. Numbers remain relatively high in the Ozarks and Ozark Border and Mississippi Lowlands through early Dec.

Winter: Most wintering birds are found in the Mississippi Lowlands and in the southern and eastern portions of the Ozarks and Ozark Border (Map 68). It is rarest in the Osage and Glaciated plains, especially the northern part of the latter region. High counts, both on the Taney Co. CBC: 378, 2 Jan 1988; 240, 27 Dec 1986.

Map 67.
Relative breeding abundance of the Field Sparrow among the natural divisions. Based on BBS data expressed as birds/route.

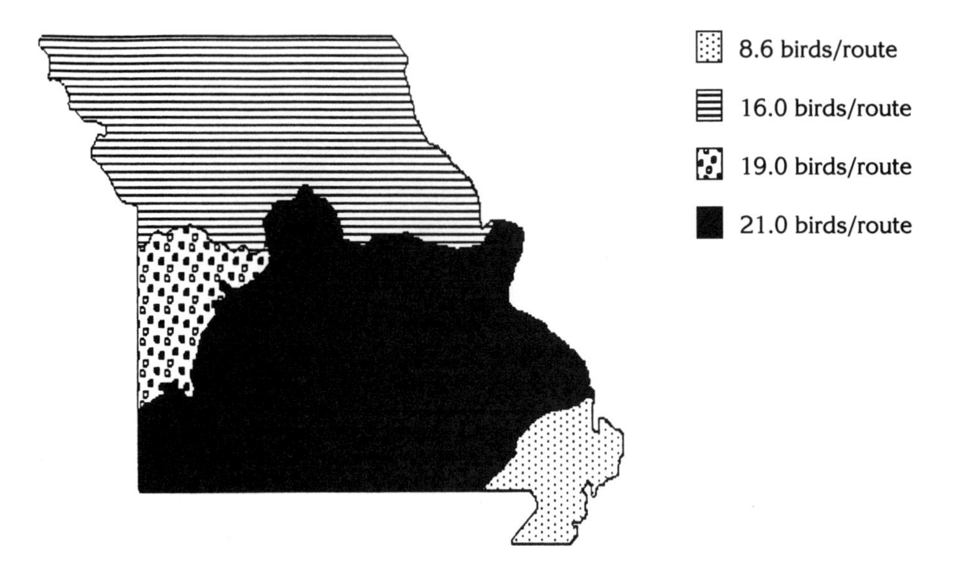

Map 68.
Early winter relative abundance of the Field Sparrow among the natural divisions. Based on CBC data expressed as birds/10 pa hrs.

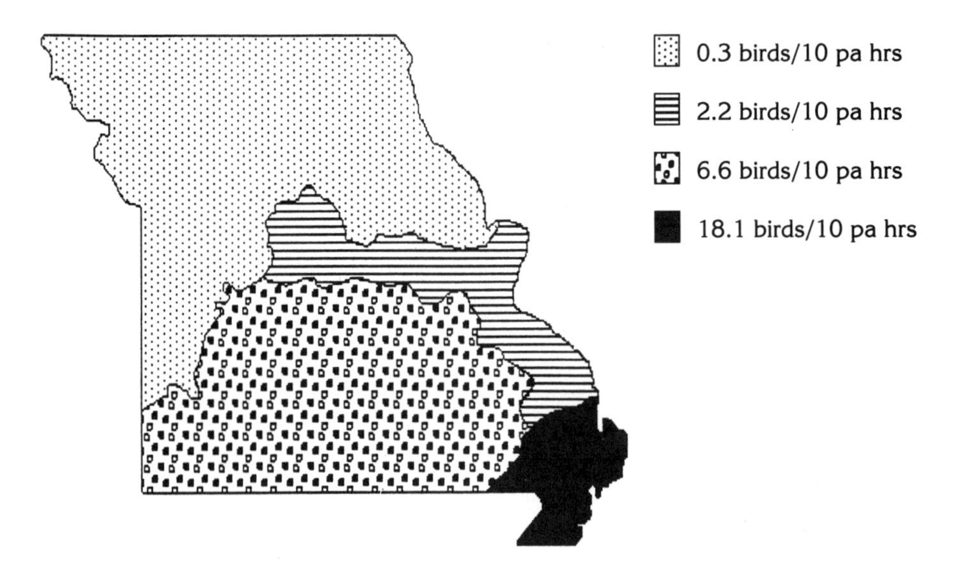

Vesper Sparrow (*Pooecetes gramineus*)

Status: Uncommon transient; rare to locally uncommon summer resident in Glaciated Plains, casual elsewhere; casual winter resident.

Documentation: Specimen: male, 7 June 1968, north of Rock Port, Atchison Co. (Easterla 1969a; NWMSU, DAE 1757).

Habitat: During the breeding season found in short-grass fields and pasture with scattered trees; during migration found in bare and overgrown fields and along hedgerows.

Records:

Spring Migration: The first arrivals appear by the second week of Mar at the southern border and by the final week of Mar in the north. Peak is during the first week of Apr. Only breeding birds remain by the beginning of May. Early in the season high count: 35+, 24 Mar 1968, Mississippi Co. (JH). High counts at peak: 50, 4 Apr 1965, near Salem, Dent Co. (D. Plank-**BB** 32[1-2]:14); 50, 6 Apr 1980, St. Joseph/Squaw Creek (FL).

Summer: A rare to locally uncommon summer resident across the Glaciated Plains, and at least formerly in the Ozark Border and at the extreme northern edge of the Ozarks and the Osage Plains (bred at Warrensburg in 1874, Scott 1879). Up until the late 1950s birds were found breeding as far south as Crawford Co. (Bourbon, Steeville; JEC-**NN** 31:49, **BB** 6[8]:67). In addition, breeding was documented in 1944 in St. Louis, Franklin, and Washington counties (WC, E. Knapp-**BB** 11[10]:58–61). Since the above records there are the following sightings, which open the possibility that this species still breeds—at least occasionally—south of the Missouri R.: 1, singing, 8 June 1976, L. Springfield, Greene Co. (CB et al.-**BB** 44[2]:22); 1-2 birds on BBS routes 3 (1976, 1985) and 21 (1978); and 1, 17 July 1988, near Charleston, Mississippi Co. (JH).

"Breeding Bird Atlas" surveys in 1990 indicate that this species is more widespread and abundant than previously appreciated. Jack Hilsabeck (pers. comm.) recorded this species in 16 of 18 atlas blocks from Andrew Co. east to Sullivan Co., and M. McNeely (pers. comm.) found a dozen territorial males in a single atlas block in Harrison Co.

Fall Migration: There are several late Aug and early Sept observations, and all of them are from areas where breeding may have occurred. Migrants begin to appear at the end of Sept and peak during the last half of Oct. It remains relatively common through early Nov, with a sharp drop in numbers thereafter. It is rare but regular until the end of Nov. The Vesper is only casually seen in Dec.

Winter: A minimum of eleven observations, all of single birds, are scattered throughout the state. All but three sightings are during the final two weeks of Dec. Midwinter records: 1, 24 Jan 1978, Busch WA (B. Boesch-**BB** 45[2]:15); 1, 4–10 Feb 1984, Greene Co. (C. Johnson-**BB** 51[2]:23).

Map 69.
Relative breeding abundance of the Lark Sparrow among the natural divisions. Based on BBS data expressed as birds/route.

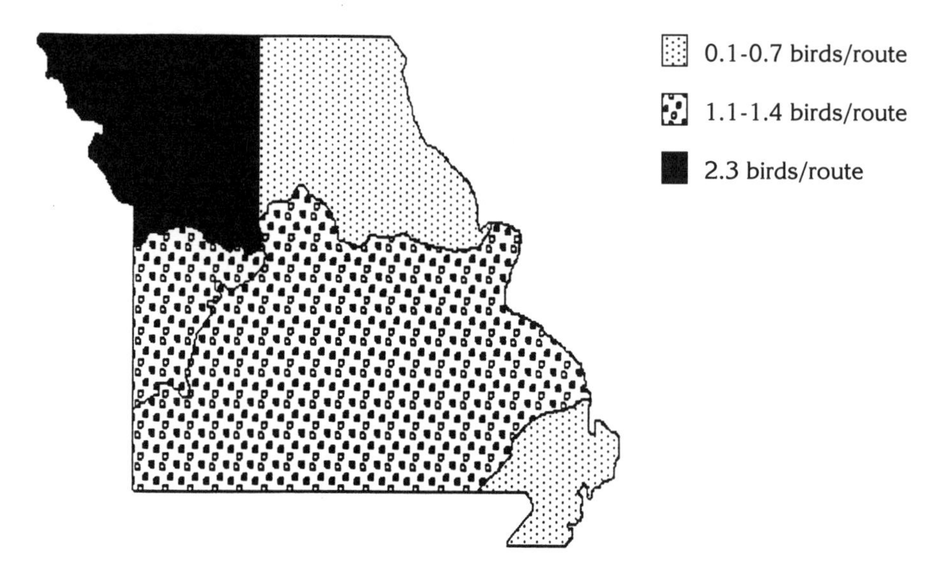

Lark Sparrow (*Chondestes grammacus*)

Status: Uncommon summer resident; accidental winter visitant.

Documentation: Specimen: male, 25 May 1986, 11 miles north of Maryville (ANSP 178068).

Habitat: Cultivated fields, pastures, short-grass fields, and woodland edge.

Records:

Spring Migration: The first appear by mid-Apr, but the bulk arrives in early May. Earliest dates: 1, 26 Mar 1988, Laumeier Park, St. Louis (EL-**BB** 55[3]:90); male, 4 Apr 1926, Fayette, Howard Co. (CMC 321). High counts, both by FL in the St. Joseph/Squaw Creek areas: 15, 11 May 1984; 15, 23 May 1965.

Summer: Relatively recent data (BBS) support Widmann's assertion (1907) that this species is a "generally distributed summer resident"; it is most common in the western half of the Glaciated Plains and least so in the eastern section of that division and the Mississippi Lowlands (Map 69).

Fall Migration: The Lark Sparrow is rarely seen after mid-Sept, with less than ten records after the first of Oct. Latest dates: 1, 2 Nov 1938, Lebanon, Laclede Co. (P. Draper-**BB** 5[12]:117); 1, 30 Oct 1937, Joplin area, Jasper Co. (G. Banner-**BB** 4[12]:136).

Winter: One reliable observation: 1, 27 Dec 1986, northwest of Lowry City, St. Clair Co. (SH-**BB** 54[2]:15).

Comments: Like the Bell's Vireo and the Harris' Sparrow, this species was first made known to science through its initial discovery in Missouri. The type was secured in 1819, ca. 4 miles from the mouth of the Missouri R. at Bellefontaine in St. Louis Co. (Say 1823).

Lark Bunting (*Calamospiza melanocorys*)

Status: Casual transient in northwest, accidental elsewhere; accidental summer resident in northwest.

Documentation: Specimen: male, 25 May 1971, 8.5 miles west of Spickard, Grundy Co. (KU 65112).

Habitat: Short-grass fields, pastures, and hedgerows.

Records:

Spring Migration: Only six records, with all but two from the northwestern corner. Dates range from mid-May through the end of the month. Earliest date: 2 males, 11 May 1973, Buchanan Co. (J. Fairlie). Only record, besides the above specimen, outside northwestern Missouri: 1, 12 May 1950, near Clinton, Henry Co. (S. Hughes-**BB** 18[4]:4).

Summer: The Lark Bunting is a sporadic breeder at the eastern edge of its range, with periodic incursions well outside its normal distribution. Such a phenomenon occurred in 1969 when two colonies were found in northwestern Missouri (summarized below from Easterla 1970a). One colony, consisting of at least twelve birds (10 males, 2 females), nested near L. Contrary. Birds were seen from at least 2 June through 1 July (J. Fairlie et al.). Another colony was located on 5 June, north of Tarkio, Atchison Co. This colony also consisted of about twelve birds (9 males, 3 females; male, NWMSU, CLJ 147; nest photographed, DAE). It seems surprising that no additional nesting colonies have been discovered since 1969.

There are only two additional, reliable sightings for this period: male, 5 June 1969, 2 miles east of the above Tarkio nesting colony (DAE); female, 14 June 1964, NW of St. Joseph (FL, SR et al.-**BB** 31[3]:20).

Fall Migration: Five of the six records are during the first half of Sept. Earliest date: male, 13 Aug 1920, Lexington, Lafayette Co. (J. Salyer; CMC 113). Latest dates: female plumaged bird, 16 Sept 1967, east of St. Joseph (FL-**BB** 35[1]:13); immature male, 16 Sept 1976, Maryville SL (DAE, TB-**BB** 44[3]:22; NWMSU, DAE 3098). High count: 2 (1 male), 10 Sept 1984, 3 miles west of Forest City, Holt Co. (MBR-**BB** 52[1]:33; NWMSU, MBR 1427). Only eastern Missouri record: 1, 5 Sept 1983, St. Charles Co. (CS et al.-**BB** 51[1]:46).

Savannah Sparrow (*Passerculus sandwichensis*)

Status: Common transient; accidental summer visitant; hypothetical summer resident; rare to locally uncommon winter resident in south, casual in north.

Documentation: Specimen: o?, 24 Dec 1963, Clinton Co. (JRI; **AFN** 18:244; KU 46840).

Habitat: Fallow fields, prairie, pastures, cultivated fields, and wet meadows.

Records:

Spring Migration: Migrants begin arriving in early Mar in the south and by mid-Mar in the north. Peak is in mid-Apr, but it remains common through the first week of May. Numbers rapidly drop off by the beginning of the second week of May; however, a few individuals are regularly seen until the final days of the month. High counts: 150, 6 May 1962, Holt Co. (FL); 100+, 22 Apr 1972, Squaw Creek (MBR).

Summer: Scott (1879) mentions breeding records from near Warrensburg, Johnson Co., in 1874, and Widmann (1907) also gives records for Pierce City, Lawrence Co. (no date given; but presumably these are of Nehrling who was there in the mid–1880s). It is difficult to assess these records, but both observers apparently were reliable. Perhaps, this species did indeed breed in the state at that time; however, these observations may have involved late migrants (often singing individuals are encountered until the beginning of June). Surprisingly there have been no definite breeding records for the northern section of the state, since the Savannah is an uncommon breeder throughout much of Iowa (Dinsmore et al. 1984) and a rare nester as far south as central Illinois (Bohlen and Zimmerman 1989).

There are the following recent summer records: 1, June 1971, BBS route 34; male, nonbreeding condition, 24 June 1976, Maryville SL (MBR, DAE-**BB** 44[2]:22; NWMSU, DAE 2981); 1, 31 July 1978, Springfield (CB-**BB** 45[4]:14).

Fall Migration: Normally the first arrivals are seen after mid-Sept, with a gradual increase during late Sept and early Oct. By the second week of Oct it is common, with peak during the last ten days of Oct. It is uncommon until the final days of Nov; thereafter it is rare (especially in the north). Earliest dates: 1, 29 Aug 1971, Squaw Creek (MBR-**BB** 39[1]:7); 1, 30–31 Aug 1974, Maryville SL (MBR, DAE-**BB** 42[2]:7). High counts: 400+, 28 Oct 1984, Montrose–Schell-Osage–Taberville Prairie areas (CH et al.-**BB** 52[1]:33); 85, 30 Oct 1937, Joplin area, Jasper Co. (G. Banner-**BB** 4[12]:136).

Winter: Rare but regular in the Mississippi Lowlands (average 4.2 birds/10 pa hrs); scarcer and less regular in other regions (average 0.1–0.3 birds/10 pa hrs), especially in the northern half of the Glaciated Plains where it is only casually found. Nevertheless, it can be locally uncommon in prime habitat; for example, it is regularly found wintering as far north as the St. Charles and St. Louis counties in relatively large numbers (over one hundred birds/day found during mild winters [SRU, RG, RA, pers. comm.]). High counts: 116, 2 Jan 1983, Orchard Farm CBC; 30, 31 Dec 1971, Big Oak Tree SP.

[**Baird's Sparrow** (*Ammodramus bairdii*)]

Status: Hypothetical; casual transient.
Documentation: Sight records only: see below.
Habitat: Short-grass prairie, fallow fields.
Records:

Spring Migration: There are very few records that are not suspect. Observations by Widmann (see below) and S. Wilson (1896) are very questionable. Moreover, revelations concerning the "documentation" of an observation in the spring of 1977 (**BB** 44[4]:28) make the sighting unacceptable.

The following represent possible observations: 1, 17 Apr 1954, St. Charles Co. (RA; Easterla 1967a); 1 singing, 1 May 1956, St. Joseph (FL); 1, 7 May 1959, Trimble (EC et al.-**AFN** 13:376).

Fall Migration: Again there are few sightings that we feel merit listing. Field marks reported for a bird seen in St. Charles Co. in the fall of 1894 are inconsistent with those of a Baird's (Widmann 1895b); this not only casts doubt on this record but also on the 1895 spring observation (Widmann 1907). Furthermore, two reported specimens have been reexamined and have proved to be pale, juvenile LeConte's (Easterla 1967a; **BB** 44[3]:23).

A bird seen on 30 Oct 1971, Reed WA, was thoroughly described (CH, KH-**BB** 39[1]:7). The following represent possible sightings: 1, 6 Oct 1960, Trimble (FB); 1, 19 Oct 1974, near Golden City, Barton Co. (m.ob.-**BB** 42[2]:7); 1, 29 Oct 1961, Busch WA (JC-**BB** 28[4]:21).

Comments: Specimen or photographic confirmation is highly desirable given the frequency that other sparrows are confused with Baird's; thus, we treat this species as hypothetical until such evidence is gathered.

Grasshopper Sparrow (*Ammodramus savannarum*)

Status: Common summer resident; accidental winter resident.
Documentation: Specimen: male, 15 Apr 1956, Gravois Mills (MU 2148).
Habitat: Prairie, fallow fields.
Records:

Spring Migration: The first arrivals are detected during the second week of Apr in the south and about a week later in the north. It is common statewide by the final days of Apr. Earliest dates: "several," one collected (spec-

Map 70.
Relative breeding abundance of the Grasshopper Sparrow among the natural divisions. Based on BBS data expressed as birds/route.

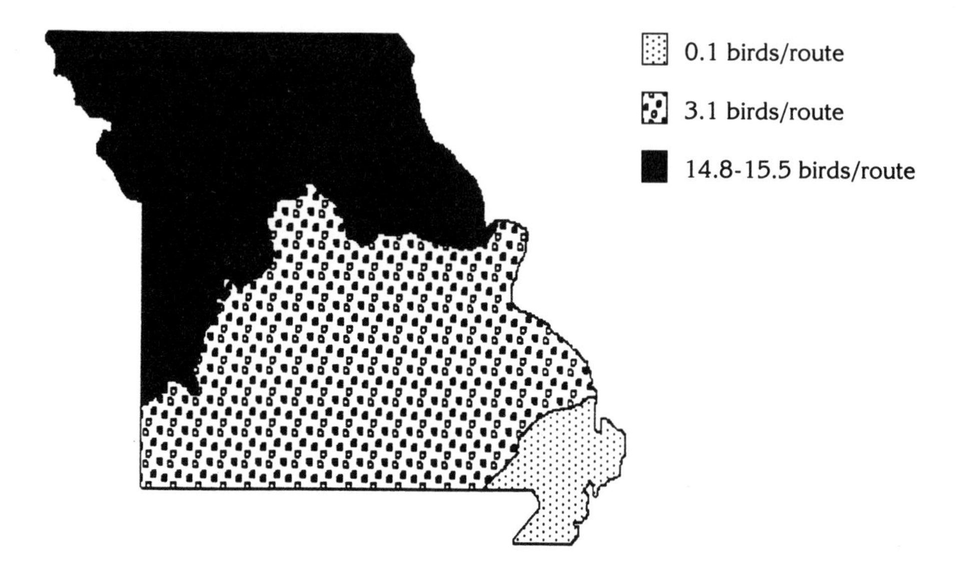

imen apparently no longer extant), 19 Mar 1907, Shannon Co. (Woodruff 1908); 1, 5 Apr 1956, near Columbia (DAE). High counts: "wave of hundreds," 21 Apr 1965, Platte Co. (DAE-**BB** 32[1–2]:13).

Summer: Most abundant in the Osage and Glaciated plains, and significantly less common in the Mississippi Lowlands (Map 70).

Fall Migration: The Grasshopper Sparrow remains common through the third week of Sept; but thereafter, numbers drop off considerably, with small numbers (< 5 birds/day) seen through the first week of Oct. It is only casually observed thereafter. High count: 36, tower kill, 22 Sept 1965, Columbia (Elder and Hansen 1967). Latest dates: immature female, 3 Nov 1972, Maryville SL (Easterla 1975; NWMSU, DAE 2736); 1, first week of Nov 1960, Tucker Prairie, Callaway Co. (Easterla 1962a).

Winter: Surprisingly, only one record, involving two birds: 14–24 Jan 1961, near Cole Camp, Benton Co. (Easterla 1962d; male, NWMSU, DAE 347).

Henslow's Sparrow (*Ammodramus henslowii*)

Status: Locally uncommon summer resident in the Osage Plains; scarcer and extremely local in the Ozark Border; very rare and local in the Glaciated Plains.

Documentation: Specimen: male, 1 June 1966, Taberville Prairie (Eas-

Graph 23.
Missouri BBS data from 1967 through 1989 for the Grasshopper Sparrow. See p. 23 for data analysis.

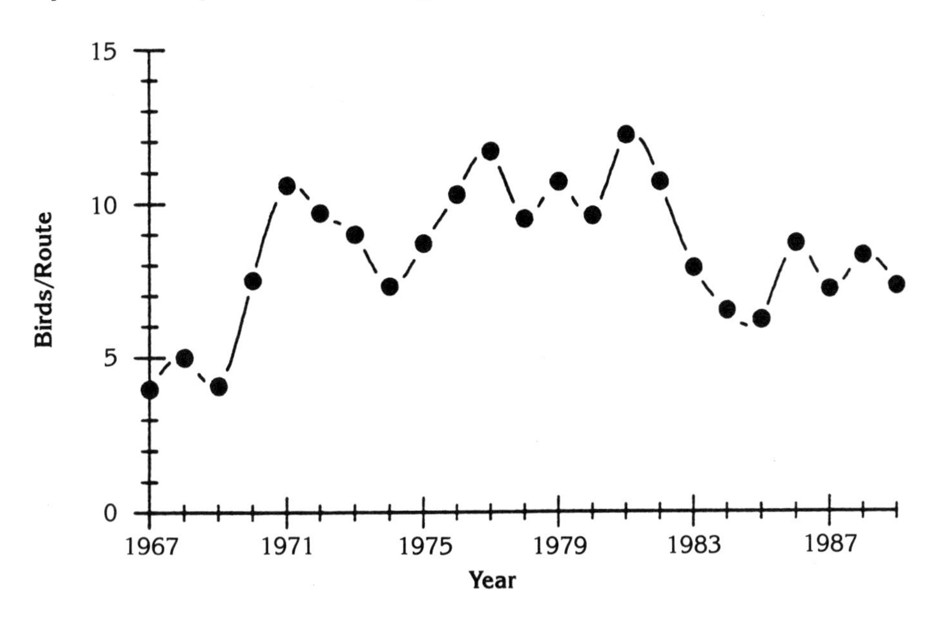

terla 1967b; USNM 529734).

Habitat: Prairie, fallow weedy and viney fields.

Records:

Spring Migration: Migrants are rarely encountered away from breeding localities. The first usually appear during the first week of Apr, and it is uncommon at nesting sites by the end of Apr. Earliest date: male, 19 Mar 1907, Shannon Co. (Woodruff 1908; AMNH 229727); 28 Mar 1916, Kansas City (Harris 1919b).

Summer: Formerly much more common and widespread throughout much of the Osage and Glaciated plains and the Ozark Border (Widmann 1907). Today it is most common on the virgin prairies of the southern part of the Osage Plains (see Map 3). Easterla (1967b), between 1961–66, recorded this species as the most abundant avian species on Taberville Prairie (1680 acres). Two hundred pairs were estimated breeding there annually during that period—mainly in prairie containing abundant dewberry (unburned prairie). There are now very few known breeding localities within the Glaciated Plains, and most of those have no more than a few pairs. "Breeding Bird Atlas" surveys in 1989 uncovered three sites in the latter region: 20–30 birds, late May, Harrison Co. (BJ, pers. comm.); adult feeding young on 8 June, ca. 4 miles NW of Helena, Andrew Co. (LG-**BB** 56[4]:149) and in Grundy Co. (M. McNeely). At least one bird was found during the Atlas work near Milan, Sullivan Co., in 1990 (JHI, pers. comm.).

Fall Migration: Birds are encountered in relatively high numbers at breeding sites through the first half of Sept. A few are still seen into early Oct. Most migrants that are observed away from nesting localities are seen during Oct, e.g., 1, 18 Oct 1963, Busch WA (KA et al.-**NN** 35:62); 1, 18–23 Oct 1980, Busch WA (RK, PS-**BB** 48[1]:10); 1, 21 Oct 1955, St. Louis (JEC-**BB** 22[11]:2).

Winter: No definite record; a bird reported in Dec–Jan 1960–61 (**BB** 28[1]:16) was later retracted.

LeConte's Sparrow (*Ammodramus leconteii*)

Status: Common transient; locally uncommon winter resident in south, rare in north.

Documentation: Specimen: male, 26 Nov 1966, Versailles, Morgan Co. (NWMSU, DAE 1184).

Habitat: Prairie, fallow fields, wet meadows, and marshes.

Records:

Spring Migration: By early Mar migrants begin to appear at nonwintering areas. Peak is in late Mar and early Apr, with birds commonly seen through Apr. Very rarely a relatively large concentration may be seen in early May, e.g., 9, 7 May 1977, Bigelow Marsh (MBR, DAE-**BB** 44[4]:28). Latest date: 1, 16 May 1979, Squaw Creek (TB-**BB** 46[3]:12).

Fall Migration: The first birds usually appear during the last week of Sept, with numbers increasing and peaking in late Oct. It remains locally common through the end of the period. Earliest dates: pale, juvenile female, 9 Sept 1976, Maryville SL (DAE-**BB** 44[3]:23—originally reported as a Baird's Sparrow; NWMSU, DAE 2980); 1, 15 Sept 1968, Squaw Creek (FL). High counts: 50, 15 Nov 1952, Busch WA (JEC-**NN** 24:54); 35, 30 Oct 1971, Reed WA (KH, CH-**BB** 39[1]:7); 25–30, 17 Oct 1954, Busch WA (JEC-**AFN** 9:31).

Winter: Very local, but present in relatively high numbers in prime habitat. Much rarer in the northern half of the state. High counts: 44, 2 Jan 1955, Busch WA (JEC-**AFN** 9:262); 42, 30 Dec 1964, Mingo CBC; 40, 14 Jan 1966, Lowry City, St. Clair Co. (SH-**BB** 33[1]:15). Northernmost observations: 3, 20 Dec 1975, Maryville CBC (DAE; NWMSU, JWG 72); 5, 27 Dec 1980, Squaw Creek CBC.

Sharp-tailed Sparrow (*Ammodramus caudacutus*)

Status: Rare transient.

Documentation: Specimen: o?, 13 Oct 1956, Gravois Mills (MU 2151).

Habitat: Marshes, wet meadows, and prairies.

Records:

Spring Migration: With the exception of fewer than five Apr records, all sightings are in May. Peak is during mid-May. Earliest dates: 1, 7 Apr 1957, Trimble (JRI; Rising et al. 1978); 1, 18 Apr 1953, Busch WA (JC, JEC-**BB** 20[5]:1). High counts: 10+, 18–20 May 1968, Squaw Creek (DAE, FL, JHA-**BB** 36[1]:27); 9, 17 May 1970, Bigelow Marsh (DAE). Latest date: 1, 24 May 1954, Trimble (BK; Rising et al. 1978).

Fall Migration: More commonly encountered at this season than in the spring. Birds begin to appear by mid-Sept, with peak during the last week of Sept. It is regularly seen through the third week of Oct, but thereafter it is accidental. Earliest date: 2, 11 Sept 1977, Thomas Hill Res. (JR-**BB** 45[1]:29). High counts: 6, 22 Sept 1977, 10 miles north of Maryville (MBR-**BB** 45[1]:29); 5, 22 Sept 1965, Busch WA (SHA-**BB** 33[1]:10). Latest dates: 1, 2 Nov 1963, L. Contrary (FL); 3, 23 Oct 1954, near Palmyra, Marion Co. (HH-**BB** 21[11]:3).

Comments: All Missouri observations are referable to the inland race *nelsoni.*

Fox Sparrow (*Passerella iliaca*)

Status: Uncommon transient; uncommon winter resident in all regions except the Glaciated and Osage plains where it is rare.

Documentation: Specimen: o?, 31 Dec 1916, Jackson Co. (KU 39563).

Habitat: Thickets, brush piles, second growth at woodland and forest edge.

Records:

Spring Migration: Migrants begin supplementing the winter population by early Mar, with numbers gradually increasing to a peak in late Mar in the south and in early Apr in the north. It is regularly seen in small numbers until the beginning of the final week of Apr. High counts: 100, 14 Mar 1975, Bennett Spring SP, Dallas Co. (FL, MBR, DAE); 98, 2 Apr 1979, Squaw Creek (TB-**BB** 46[3]:12). Latest dates: 1, 26 Apr 1981, Bluffwoods SF (FL); 1, 26 Apr 1983, St. Louis (m.ob.-**NN** 55:51).

Fall Migration: Usually first seen at the beginning of Oct. Peak is during the last week of Oct and early Nov. By the end of Nov, numbers have dropped off considerably. Earliest dates: 1, 24 Sept 1990, Brickyard Hill WA (MBR); 1, 25 Sept 1988, Big L. SP (MBR-**BB** 56[1]:16). High counts: 200, 25 Oct 1941, Missouri R. bottoms, near Parkville, Platte Co. (WC-**BB** 8[11]:81); 30+, 24 Oct 1974, east of Maryville (MBR).

Winter: The Fox Sparrow is locally uncommon in the southern section of the state; it is rare in the Osage and Glaciated plains regions (Map 71). High counts: 116, 21 Dec 1980, Columbia CBC; 69, 27 Dec 1980, Weldon Spring CBC.

Comments: The nominate race, *P. i. iliaca,* is the only one recorded for the state.

Map 71.
Early winter relative abundance of the Fox Sparrow among the natural divisions. Based on CBC data expressed as birds/10 pa hrs.

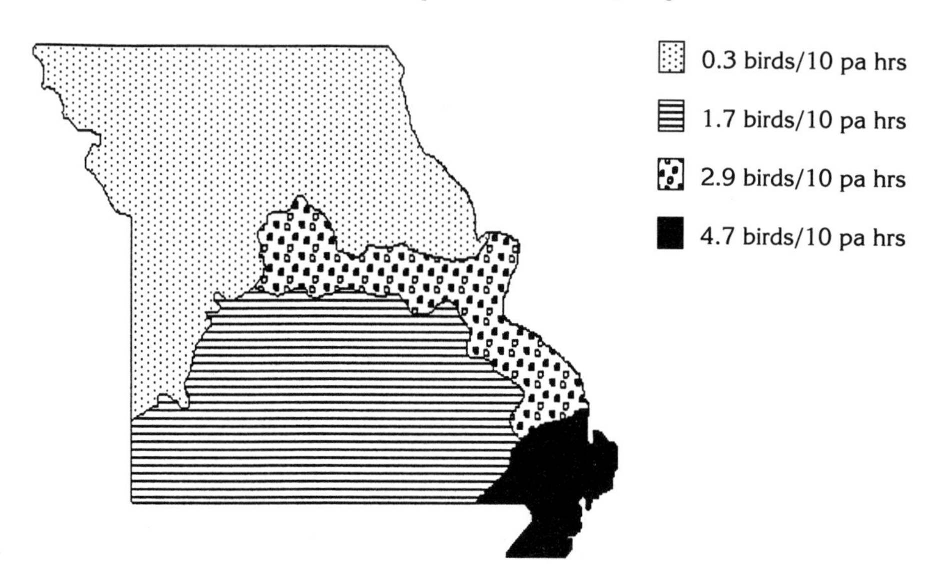

Song Sparrow (*Melospiza melodia*)

Status: Common summer resident from the Missouri R. drainage northward and along the Mississippi R. drainage; casual elsewhere; common winter resident statewide.

Documentation: Specimen: female, 29 Dec 1948, Jackson Co. (KU 9595).

Habitat: During migration and winter found in a wide range of relatively open habitats; in summer, along streams, rivers, thickets, and second-growth woods at edge of marshes and swamps.

Records:

Spring Migration: Migrants begin supplementing winter populations as early as the end of Feb. Peak is during mid-Mar in the south and at the end of the month in the north. Few are seen away from nesting areas after mid-Apr. High count: 50, 14 Mar 1975, Bennett Spring SP, Dallas Co. (FL, MBR, DAE).

Summer: This sparrow is principally found from the Missouri R. northward and along the Mississippi R. floodplain south to the Arkansas border. BBS data indicate that it is most common (at least based on the sole route there) in the Mississippi Lowlands (average 7.0 birds/route), with an average of 1.8 birds/route across the Glaciated Plains. It is much more local and

Map 72.
Early winter relative abundance of the Song Sparrow among the natural divisions. Based on CBC data expressed as birds/10 pa hrs.

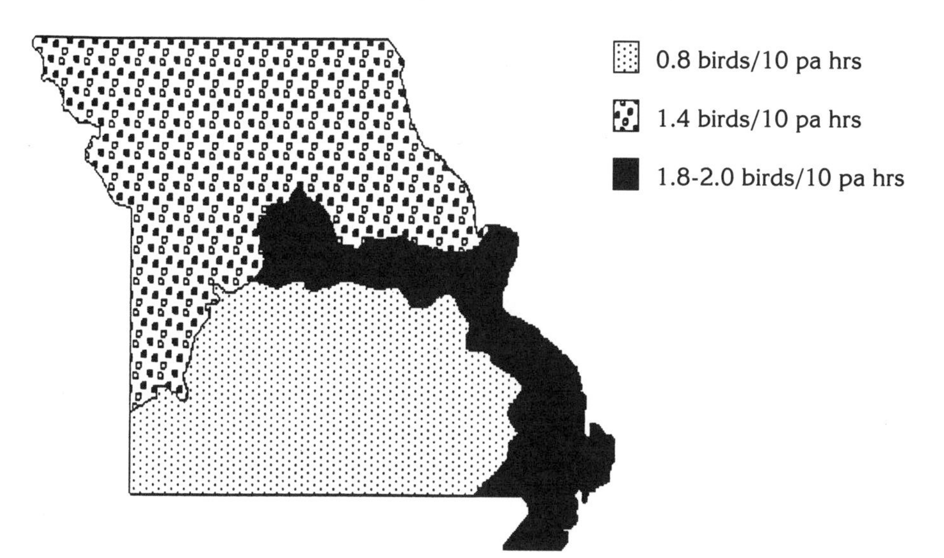

rarer away from the above areas. For example, it has been recorded only twice (4 birds each year) on nine Cass Co. Summer counts (1978–86; fide JG). Since 1986, 3–4 singing birds/summer have been found at Schell-Osage (MMH et al.-**BB** 53[4]:17).

Fall Migration: Migrants begin to reappear by mid-Sept, with numbers increasing to a peak in late Oct and early Nov. Thereafter it is common statewide. High counts: 500, 25 Oct 1941, Missouri R. bottoms, near Parkville, Platte Co. (WC-**BB** 8[11]:81); 35, 2 Nov 1983, near St. Joseph (FL).

Winter: CBC data indicate that it is most common in the Mississippi Lowlands and Ozark Border and least abundant in the Ozarks (Map 72). High counts: 534, 2 Jan 1955, Weldon Spring CBC; 438, 20 Dec 1980, Kansas City CBC.

Lincoln's Sparrow (*Melospiza lincolnii*)

Status: Uncommon transient; rare and irregular winter resident.

Documentation: Specimen: male, 28 Dec 1931, Jackson Co. (KU 18511).

Habitat: Thickets, brushpiles, hedgerows, and edge of woodland and forest.

Records:

Spring Migration: The first birds are seen in the south at the end of Mar, with numbers building to a peak at the end of Apr and early May. It is regularly seen until the beginning of the third week of May. High counts: several counts of 15 birds/day at the end of Apr and early May. Latest dates: 1, 28 May 1897, St. Louis (Widmann 1907); 1, 27 May 1989, Big L. SP (MBR, DAE-**BB** 56[3]:94).

Fall Migration: A few appear by mid-Sept, with relatively large numbers (20+ birds/day) in the north by the final week of Sept. Peak is during the second week of Oct. By the beginning of Nov, most have left the state. It is rarely seen thereafter. Earliest dates: 1, 10 Sept 1977, Maryville (DAE); 1, 13 Sept 1980, Columbia (BG-**BB** 48[1]:10). High counts: 27, 13 Oct 1974, east of Maryville (MBR); 25+, 13 Oct 1990, Squaw Creek (TB, CH).

Winter: During most winters it is scarce, with a few individuals seen scattered across the state (primarily in the south). In mild years, particularly during the initial two weeks of the period, it is more common, with low numbers (1–3/day) being encountered even in the north. High counts: 17, 19 Dec 1982, Trimble CBC; 16, 18 Dec 1982, Kansas City Southeast CBC.

Swamp Sparrow (*Melospiza georgiana*)

Status: Common transient; formerly a summer resident; uncommon winter resident.

Documentation: Specimen: male, 7 Oct 1927, Saline Co. (KU 39578).

Habitat: Swamps, marshes with scattered brushes, wet meadows, and fields with brushpiles.

Records:

Spring Migration: Migrants begin supplementing the winter population by early Mar in the south and by mid-Mar in the north. Peak is during early Apr in the south and mid-Apr in the north. By early May, usually only 1–2 birds/day are seen, with virtually all having left by mid-May. Latest dates: 25 May 1874, Warrensburg (Scott 1879); 1, 21 May 1990, Bigelow Marsh (DAE-**BB** 57[3]:141).

Summer: This species has not been conclusively found breeding in the state since the early 1900s. Widmann (1907) gives the three following records: nest with young, 2 June 1905, St. Charles Co. (P. Smith); "nesting near Montgomery City" (no date given; E. Parker); near Wayland, Clark Co. (E. Currier).

Two birds were present at Smithville L. on 3 June 1984, but definite breeding was not established (NJ-**BB** 51[4]:18). Seemingly suitable habitat exists in the northern part of the state, and there is a recent (1979) breeding record as far south as Pike Co., Illinois (opposite Ralls and Pike counties in Missouri; Bohlen and Zimmerman 1989).

Fall Migration: A few begin to trickle back into the state by mid-Sept, with relatively large numbers (20+ birds/day) occasionally seen during the final week of Sept. However, peak is usually not until the end of Oct. Relatively large numbers typically remain into early Dec. Earliest dates: 1, 17 Sept 1980, L. Contrary (FL); 1, tower kill, 19 Sept 1966, Columbia (Elder and Hansen 1967). High count: 75, 24 Oct 1979, Squaw Creek (FL).

Winter: Generally distributed in small numbers, but in prime habitat it can be locally abundant, e.g., at Squaw Creek, Swan L., and Mingo. However, during severe winters numbers are greatly reduced even at the above areas. CBC data reveal the following early winter pattern: an average of 3.5 birds/10 pa hrs in the Ozarks and Ozark Border, 5.2 birds/10 pa hrs in the Glaciated Plains, and 11.9 birds/10 pa hrs in the Mississippi Lowlands. High counts: 470, 19 Dec 1978, Squaw Creek CBC; 272, 21 Dec 1978, Orchard Farm CBC.

White-throated Sparrow (*Zonotrichia albicollis*)

Status: Common transient; accidental summer visitant; common winter resident in Mississippi Lowlands and Ozarks, uncommon elsewhere.

Documentation: Specimen: female, 2 Nov 1960, near Tucker Prairie, Callaway Co. (NWMSU, DAE 333).

Habitat: Woodland and forest edge, thickets, and brushy areas.

Records:

Spring Migration: The wintering population begins to be enhanced by migrants in the south in early Mar, but numbers do not significantly increase there until early Apr. Peak is in mid-Apr. Birds first appear at non-wintering sites in the north at the end of Mar. Peak in the north is during the final days of Apr and the first week of May. Few birds remain after mid-May. High counts: 500+, 23 Apr 1990, Tower Grove Park (JZ-**BB** 57[3]:141); 100+, 28 Apr 1963, St. Joseph/Squaw Creek area (FL). Latest dates: 1, 26 May 1964, Springfield (NF-**BB** 31[2]:15); no.?, 24 May 1883, St. Louis (Widmann 1907).

Summer: There are at least four observations during this season as follows: 1, 1 June 1975, Eminence, Shannon Co. (DAE, FL-**AB** 29:863); 2, 6 June 1962, Beverly L., Platte Co. (JRI-**BB** 29[3]:23); 1, 10 June 1944, "present several days," near Liberty, Clay Co. (NF-**BB** 11[8]:48); 2, 10 July 1982, Springfield (GD et al.-**BB** 49[4]:21).

Fall Migration: The first birds arrive at the beginning of the final week of Sept, but it does not become common until the end of the first week of Oct. Peak is during the final two weeks of Oct. Numbers are reduced considerably in the north by the second week of Nov. Earliest dates: 12, 20 Sept 1936, Mud L., Buchanan Co. (WC-**BB** 3[11]:89); 1, 21 Sept 1982, Bluffwoods SF (FL-**BB** 50[1]:27). High counts: 80+, 9 Oct 1987, Brickyard Hill WA (MBR-**BB** 55[1]:13); 80+, 13 Oct 1974, east of Maryville (MBR).

Winter: The White-throat is decidedly more common in the Mississippi Lowlands than in any other region (Map 73). Numbers vary considerably, especially in more northern areas, with the severity of the winter. Although

Map 73.
Early winter relative abundance of the White-throated Sparrow among the natural divisions. Based on CBC data expressed as birds/10 pa hrs.

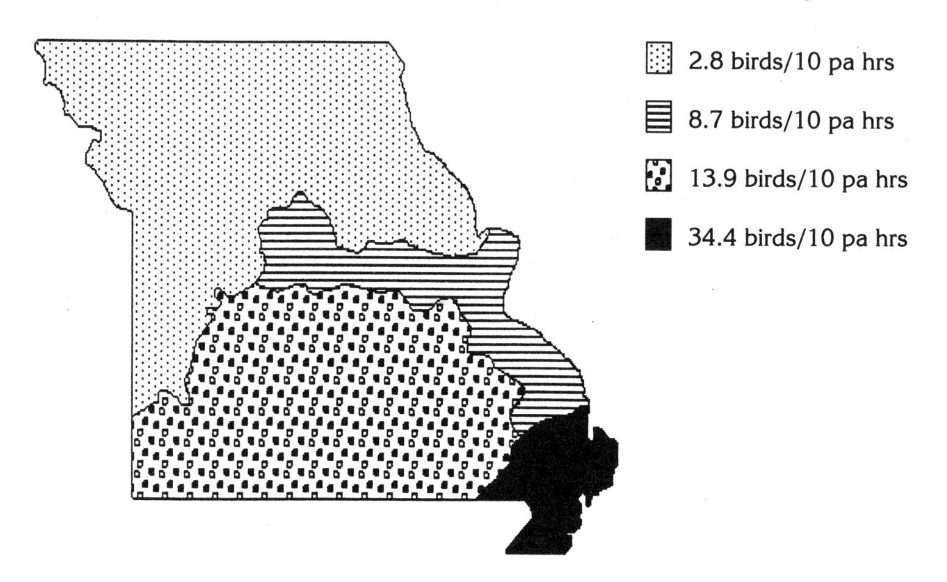

the CBC data indicate that it is more common than the Song Sparrow in the Glaciated and Osage plains, the Song is more widespread and common throughout these two regions. High counts, all on the Mingo CBC: 1,756, 28 Dec 1962; 897, 30 Dec 1964; 811, 28 Dec 1965.

White-crowned Sparrow (*Zonotrichia leucophrys*)

Status: Common transient; locally common winter resident in all regions but the Glaciated Plains where it is uncommon.

Documentation: Specimen: female, 7 Oct 1927, Saline Co. (KU 39551).

Habitat: Brushy areas, thickets, hedgerows, woodland and forest edge.

Records:

Spring Migration: Migrants begin appearing by mid-Mar, but it does not become prevalent until mid-Apr. Peak is during late Apr and early May. By mid-May few remain. Latest date: no.?, 22 May 1907, St. Louis (Widmann 1907).

Fall Migration: A few begin to appear at the beginning of Oct, with numbers increasing rapidly by the end of the second week of Oct. Peak is in late Oct and early Nov. Numbers remain relatively high until mid-Nov in the north and until the end of the month in the south. Earliest dates: immature, 25 Sept 1988, Big Lake SP (MBR-**BB** 56[1]:16); at least 3 sightings for 30 Sept.

Winter: It is most common in the Mississippi Lowlands (average 1.8 birds/10 pa hrs) and the Ozarks (1.0 birds/10 pa hrs) and least common in the Glaciated Plains (0.5 birds/10 pa hrs). High counts: 267, 3 Jan 1985, Mingo CBC; 258, 26 Dec 1971, Kansas City Southeast CBC.

Comments: The northwestern race, *Z. l. gambelii,* is an uncommon transient and winter resident (at least in the western half of the state). Occasionally, it even outnumbers the nominate race. It has been detected as early as 10 Oct 1920, Kansas City (D. Teachenor-**BL** 22:360), with 6 May 1956, Squaw Creek (FL), being the latest spring sighting.

Harris' Sparrow (*Zonotrichia querula*)

Status: Common transient in west, rare in east; uncommon winter resident in west, rare in east.

Documentation: Specimen: male, 11 Mar 1961, Columbia (NWMSU, DAE 329).

Habitat: Brushy areas, thickets, and hedgerows.

Records:

Spring Migration: Much more common in the western half of the state, with usually only individuals or small groups (< 5 birds/season) encountered in the east. Like the preceding species the winter population is augmented with migrants by mid-Mar, e.g., 35, 19 Mar 1967, St. Joseph/Squaw Creek (FL). However, the large waves of birds do not appear until the end of Apr and early May. Although it is seen in much fewer numbers after the first week of May, it nevertheless is regularly seen through the third week of May. High counts: "thousands," 22 Apr 1922, Kansas City (HH-**BL** 24:360); 100, 6 May 1962, St. Joseph/Squaw Creek (FL). Latest dates: 1, 25 May 1968, northern Atchison Co. (JHA-**BB** 36[1]:28); 1, 23 May 1970, Bigelow Marsh (DAE-**BB** 38[3]:9).

Fall Migration: More frequently seen in the eastern half at this season than in spring. Usually first seen at the end of the first week of Oct in the north. It does not become numerous until mid-Oct. Peak is at the end of the month or in early Nov. Few migrants remain by late Nov. Earliest dates: adult, 26 Sept 1990, Brickyard Hill WA (MBR); 1, 3 Oct 1970, Squaw Creek (FL). High count: 60+, 2 Nov 1974, east of Maryville (MBR).

Winter: It is most prevalent in the extreme western section of the Glaciated and Osage plains (Map 74). High counts: 266, 18 Dec 1977, Trimble CBC; 228, 14 Dec 1974, Kansas City Southeast CBC; 228, 29 Dec 1974, Kansas City North CBC.

Comments: The Harris' Sparrow, like the Bell's Vireo and the Lark Sparrow, was first made known to science by its discovery in Missouri. Thomas Nuttall secured a specimen on 28 Apr 1834 a few miles west of Independence (Harris 1919a).

Map 74.
Early winter relative abundance of the Harris' Sparrow among the natural divisions. Based on CBC data expressed as birds/10 pa hrs.

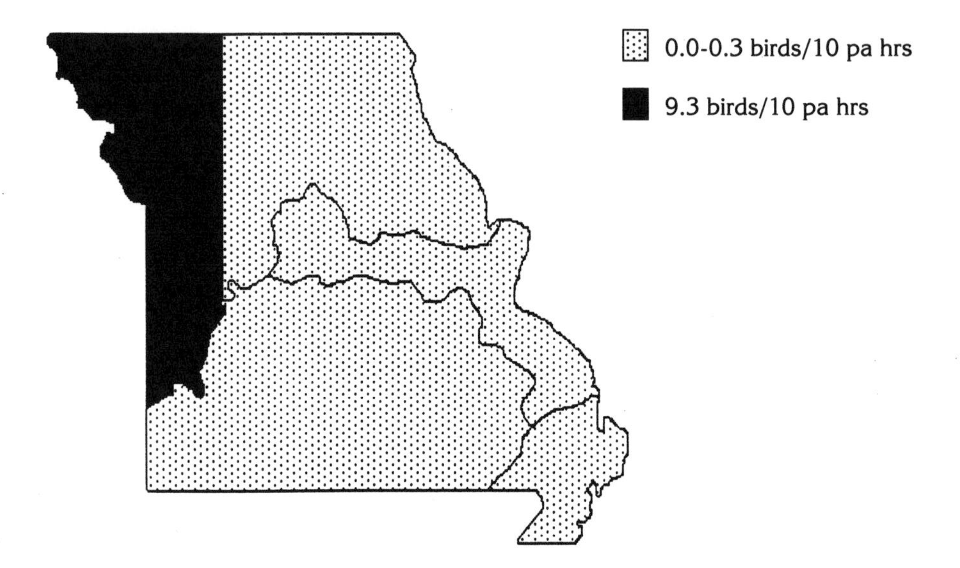

Dark-eyed Junco (*Junco hyemalis*)

Status: Common winter resident; accidental summer visitant.
Documentation: Specimen: male, 10 Dec 1926, Jackson Co. (KU 39510).
Habitat: Weedy fields, brushy areas, hedgerows, edge of woodland and forest.
Records:
Spring Migration: The greatest numbers are seen during the last two weeks of Mar. By the beginning of the second week of Apr, there are considerably fewer birds present. After the first few days of May, virtually none are encountered. Latest dates: 1, 31 May 1990, Squaw Creek (TB-**BB** 57[3]:141); 1, 29 May 1882, St. Louis (Widmann 1907).
Summer: Two records of presumed late migrants: 1, 10 June 1987, near Bennett Spring SP, Dallas Co. (B. Handy-**BB** 54[4]: 29); 1, photographed, 20 June 1978, 10 miles east of Salem, Dent Co. (D. Wallan-**BB** 45[4]:15).
Fall Migration: The first arrivals usually appear during the final week of Sept in the north. Numbers increase dramatically by mid-Oct and peak at the very end of Oct and early Nov. Thereafter it continues to be common. Earliest dates: 1, 2 Sept 1972, St. Louis (A. Bromet-**BB** 40[1]:9); 1, 17 Sept 1956, St. Louis (O. Peterson-**BB** 23[10]:4).
Winter: This species is one of the most common birds in the state in

Map 75.
Early winter relative abundance of the Dark-eyed Junco among the natural divisions. Based on CBC data expressed as birds/pa hr.

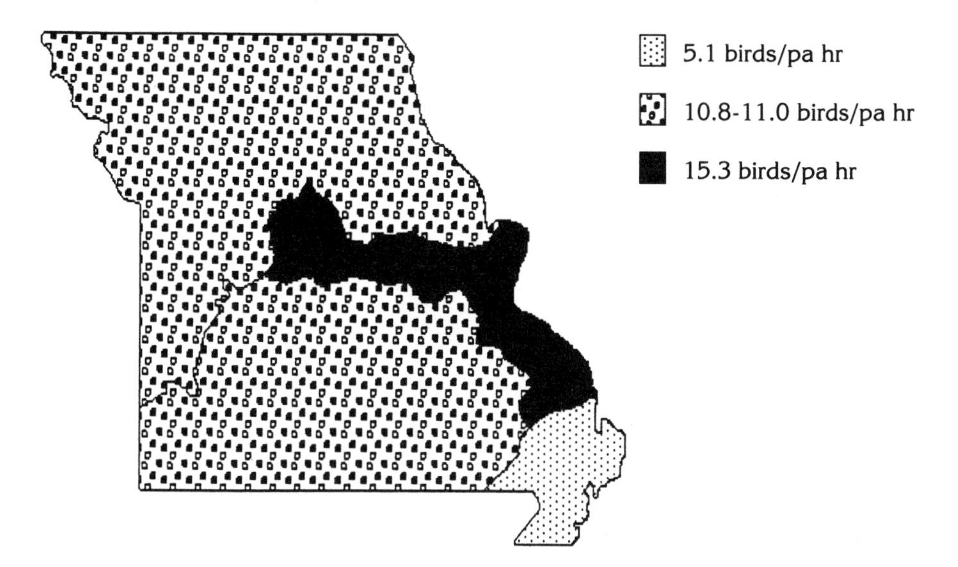

winter. CBC data reveal that it is most common in the Ozark Border and least abundant in the Mississippi Lowlands (Map 75). High counts: 3,882, 16 Dec 1973, Weldon Spring CBC; 2,894, 17 Dec 1983, Kansas City Southeast CBC.

Comments: Of the western races only the "Oregon" (*J. h. montanus*) has been positively recorded in the state. *Montanus* is a rare but regular transient and winter resident. The earliest fall occurrence of this race is of a male on 26 Oct 1968, Maryville (MBR-**BB** 36[3]:4). The latest spring sighting is of 1, 5 May 1952, Kansas City (**BB** 19[6]:1). *Shufeldti* has been reported (Bennitt 1932) but needs confirmation.

The Black Hills race, the so-called "White-winged Junco" (*J. h. aikeni*), has been reported on several occasions; however, this subspecies is easily confused with aberrant examples of the nominate race, which occasionally have white-tipped wing coverts (giving the appearance of wing-bars) and an unusual amount of white in the tail. A good example of one of these atypical birds was one that was banded and photographed on 20 Mar 1981 at Independence (M. Solomon; VIREO x05/1/019; Fig. 30). Superficially this bird looked like a typical "White-winged Junco," because it had white wing-bars and a proportionally long tail (a characteristic of *aikeni*). Fortunately, Solomon carefully examined the bird, noting wing length, mass, and the extent of white in the outer rectrices. The bird was a typical Dark-eyed Junco in all of the above characteristics. Thus, all suspected birds

Fig. 30. This aberrant plumaged Dark-eyed Junco was netted and photographed on 20 March 1981 at Independence, Jackson Co., by Marguerite Solomon. Although the bird superficially looks like a "White-winged" Junco from the Black Hills, the bird proved to be an abnormally pigmented "Slate-colored" Junco.

of this race should be examined in the hand and documented with photographs.

McCown's Longspur (*Calcarius mccownii*)

Status: Accidental transient and winter resident.

Documentation: Specimen: female, 16 Mar 1975, Maryville SL (MBR, DAE-**BB** 42[3]:14; NWMSU, DAE 2943).

Habitat: Cultivated fields, pastures, and short-grass fields.

Records:

Spring Migration: The above is the only record for this season.

Fall Migration: A male and two unsexed birds were flushed at Taberville Prairie on 15 Nov 1986 (WL-**BB** 54[1]:34).

Winter: Two sightings: 2, 28 Dec 1951, near Squaw Creek (JB-**BB** 19[3]:3—we believe the date, 29 Dec, in Rising et al. 1978 is incorrect); 1, 11–16 Jan 1979, Busch WA (PS, CP, BR et al.-**BB** 46[2]:14; Rudden 1980a).

Lapland Longspur (*Calcarius lapponicus*)

Status: Common transient; uncommon winter resident in all regions, except the Ozarks and Ozark Border where it is rare.

Documentation: Specimen: male, 28 Mar 1965, Platte Co. (NWMSU, DAE 865).

Habitat: Cultivated fields, pastures, and short-grass fields.

Records:

Spring Migration: Most common during Mar when enormous-sized flocks are occasionally encountered. It is most common in the Glaciated and Osage plains. After the first few days of Apr, only small groups (< 10 birds/flock) and individuals are observed. It is only accidentally seen during the final week of Apr and early May. High counts: 50,000, 4 Mar 1962, Platte Co. (JRI, D. Cole-**BB** 29[2]:20); 10,000, 28 Mar 1965, Platte Co. (DAE-**BB** 32[1–2]:14). Latest dates: 1, 6 May 1979, Taberville Prairie (CH, KH-**BB** 46[3]:12); 1, 5 May 1978, Taberville Prairie (KH, N. McLaughlin-**BB** 45[3]:20).

Fall Migration: The first birds are usually detected at the beginning of the final week of Oct. It can be occasionally found in large numbers at the beginning of Nov, e.g., 500, 5 Nov 1980, St. Joseph (FL-**BB** 48[1]:10); but normally the largest flocks are not observed until the end of Nov. Earliest dates: 1, 5 Oct 1988, 3 miles west of Forest City, Holt Co. (MBR-**BB** 56[1]:16); 1, 9 Oct 1977, 10 miles north of Maryville (MBR-**BB** 45[1]:29). High count: 8,000+ (5,000+, W of Rich Hill, Bates Co.; 3,000+, NW corner of Jasper Co.), 28 Nov 1984 (FL, MBR-**BB** 52[1]:24).

Winter: Less common in winter, but occasionally relatively large-sized flocks are encountered at this time of year. As expected, it is most abundant in the relatively open regions of the state. High counts: 1,900, 21 Dec 1968, Squaw Creek CBC; 1,400, 18 Dec 1983, Orchard Farm CBC.

Smith's Longspur (*Calcarius pictus*)

Status: Uncommon transient in west; uncommon spring, rare fall transient in east; casual winter resident in west.

Documentation: Specimen: female, 15 Apr 1920, Lexington (CMC 457).

Habitat: Prairie, short-grass and cultivated fields.

Records:

Spring Migration: A few begin to pass through the state at the beginning of Mar and gradually increase in numbers to peak at the end of the third or the beginning of the fourth week of Mar. Small-sized flocks (5–20 birds) are regularly seen through mid-Apr. It is accidental by the beginning of the final week of Apr. Earliest date: 1, 1 Mar 1958, St. Charles Co. (JEC-**BB** 25[4]:3). High counts: 5,000–6,000, 21 Mar 1978, east of Columbia (fide BG-**BB** 45[3]:20); 300, 19 Mar 1967, near Ford City, Gentry Co. (DAE et al.-**AFN** 21:426); 300, 22 Mar 1964, near Trimble WA (FL-**BB** 31[2]:15). Latest dates: Harris (1919b) mentions that 3 were collected on 1 May 1905 in the Kansas City area; female, 28 Apr 1984, near Camden, Ray Co. (KH, RF, MMH-**BB** 51[3]:15).

Fall Migration: The first are usually seen at the beginning of Nov and peak during the middle of the month. It is only accidentally observed after the first of Dec. Earliest dates: 5, 28 Oct 1978, Maryville SL (MBR-**BB** 46[1]:28); 30, 31 Oct 1977, Taberville Prairie (MBR, JHI). High counts in west: 150+ (100+ Taberville Prairie, 50+ Osage Prairie, Vernon Co.), 11 Nov 1978 (MBR, TB, FL); 40, 23 Nov 1969, near St. Joseph (FL, DAE). High counts in east: "few," 27 Nov 1976, St. Charles Co. (JE-**BB** 44[3]:23).

Winter: Only five reliable observations as follows: 2, 15 Dec 1984, Kansas City Southeast CBC (CH); 1, 21 Dec 1965, Montrose CBC (DAE, SH); 5 (including two adult males), 29 Dec–25 Jan 1989–90, Prairie State Park, Barton Co. (MBR, BRE, SD-**BB** 57[2]:105); 1, 22 Jan 1978, near Amazonia, Andrew Co. (FL-**BB** 45[2]:15); 150, 15 Feb 1953, Windsor, Henry Co. (K. Miller, DJ, JC et al.-**BB** 20[4]:2).

Chestnut-collared Longspur (*Calcarius ornatus*)

Status: Casual spring transient in west, accidental in east; accidental fall transient in west; accidental winter visitant in east.

Documentation: Specimen: male, 26 Mar 1966, Maryville (DAE-**BB** 33[2]:11; NWMSU, DAE 1082).

Habitat: Pastures, short-grass and cultivated fields.

Records:

Spring Migration: At least eight definite records (including the one above), five of which are from the western side of the state. At least two of the western records followed very strong WSW winds. Earliest date: 1, 22–23 Feb 1957, St. Charles Co. (S. Springer et al.-**BB** 23[4]:3). High count: 2 males, 27 Mar 1975, Maryville SL (MBR-**BB** 42[3]:14; NWMSU, DAE 2944). Latest date: male, 5 Apr 1960, Lowry City, St. Clair Co. (SH-**BB** 27[4]:23).

It is difficult to assess whether Scott (1879) actually found this species to be "rather common" in Apr 1874 on the prairies near Warrensburg, Johnson Co. Intriguingly, there are two male specimens, one in nuptial and the other in basic plumage, in the Hurter Collection (now deposited in STCS), which were taken on 9 Mar 1875 at St. Louis (fide RA). Reports of relatively large numbers in Iowa during the late 1800s (Dinsmore et al. 1984) also lend some credence to this species having been more common in the midwest prior to the turn of the century.

Fall Migration: Only one definite record: female, 20–22 Oct 1989, St. Joseph (MBR, DAE et al.-**BB** 57[1]:51). Rising et al. (1978) list two records, 3 Oct 1954 and 7 Nov 1953, without any additional details.

Winter: One record: 1, 14 Feb 1959, St. Charles Airport (A. Bolinger, JEC et al.-**BB** 27[1]:16).

Snow Bunting (*Plectrophenax nivalis*)

Status: Rare winter resident in Glaciated Plains, accidental elsewhere.
Documentation: Specimen: male, 1 Nov 1975, Maryville SL (NWMSU, DAE 2973).
Habitat: Along rocky and grassy edges of sewage lagoons, ponds and lakes, and cultivated fields and pastures.
Records:
Spring Migration: Accidental during the final week of Feb, with only two Mar observations: no.?, 24 Mar 1885, Audrain Co. (Mrs. Musick; Widmann 1907); 6, 3 Mar 1978, near Clearmont, Nodaway Co. (TB-**BB** 45[3]:20).
Fall Migration: The vast majority of the observations are from early Nov through the end of the period at the Maryville SL where it is regular. At this locality, the Snow Bunting has been recorded every year but one between 1972 and 1989 (DAE). The high counts there are usually during Nov and early Dec. Earliest dates: 1, 20 Oct 1977 (DAE, MBR, TB-**BB** 45[1]:29); immature male, 20 Oct 1989 (DAE, MBR-**BB** 57[1]:51). High counts: 16, 8 Dec 1984 (DAE); 11, 12 Nov 1985 (DAE).
Winter: A rare and irregular winter resident throughout the Glaciated Plains, accidental elsewhere. Southernmost records: 2–6, 25–26 Jan 1982, Washington Co. (SD et al.-**BB** 49[2]:17); 1, 29 Dec 1958, Springfield CBC. High counts: 70+, Feb 1982, St. Charles Co. (RK et al.); 60, 1 Feb 1982, near Columbia (TB).

Bobolink (*Dolichonyx oryzivorus*)

Status: Uncommon to locally common transient; rare to locally uncommon summer resident in the Glaciated Plains.

Fig. 31. This Snow Bunting was photographed in November 1987 at the Maryville Sewage Lagoons by Scott Laughlin. This rare species regularly appears at this locality in late fall.

Documentation: Specimen: male, 2 May 1964, near Lexington, Ray Co. (NWMSU, DAE 554).

Habitat: Tallgrass fields, pastures, clover fields, wet meadows, and prairies.

Records:

Spring Migration: Usually not seen until the final week of Apr, with numbers increasing to peak in mid-May. Earliest dates, both at Marais Temps Clair by G. and T. Barker: 1, 4 Apr 1982 (**BB** 49[2]:22); 1, 20 Apr 1980. High counts: 300, ? May 1980, Licking, Texas Co. (D. Hatch-**BB** 47[3]:14). Latest migrant date: 1, 30 May 1966, Charleston, Mississippi Co. (JH).

Summer: The Bobolink is a rare to locally uncommon breeder across the Glaciated Plains north of the Missouri R. It is most prevalent in the northern half of this region (BBS data; 2.5 birds/route in north vs. 1.0 birds/route in south). This species is only casually found south of the Missouri R., e.g., near Warrensburg in 1983 (fide JW-**BB** 50[4]:35), Cass Co. (pair, summer 1980 and 1983; fide JG); and in single "Breeding Bird Atlas" blocks in Bates and Vernon counties (*Missouri Breeding Bird Atlas Newsletter,* 1990:4).

The hiatus of breeding records between Widmann (1907) and the late 1960s (see below; Watkins 1968; Easterla 1969a) was undoubtedly related to the paucity of observers in the northern half of Missouri during this period. An unpublished record in the late 1960s is of a nest with five eggs found on 6 June 1967, near Hurdland, Knox Co. (G. Campbell; North American Nest Record Program, Cornell Univ.).

Fall Migration: Relatively large-sized flocks (in the low hundreds) are encountered during the last week of Aug through the second week of Sept. Only small-sized groups or individuals are seen after mid-Sept; however, a few are regularly seen until early Oct. High counts: 500, 2 Sept 1968, Squaw Creek (FL-**BB** 36[4]:12); 500, 10 Sept 1967, near Squaw Creek (FL, BB-**BB** 35[1]:12). Latest dates: o?, 9 Oct 1967, Cape Girardeau (SEMO 262); 2, 8 Oct 1978, 10 miles north of Maryville (MBR, TB-**BB** 46[1]:28).

Red-winged Blackbird (*Agelaius phoeniceus*)

Status: Common summer resident; locally common winter resident.

Documentation: Specimen: male, 13 Apr 1917, Buchanan Co. (KU 39343).

Habitat: Breeds in marshes, swamps, wet meadows, and along ditches. During migration seen in a wide array of habitats.

Records:

Spring Migration: This is one of the most abundant birds at all seasons in the state. In late Feb, large flocks are already seen returning north. Enormous sized flocks are observed during the last two weeks of Mar into early Apr. Although less frequent and in smaller numbers, groups are seen pass-

Map 76.
Relative breeding abundance of the Red-winged Blackbird among the natural divisions. Based on BBS data expressed as birds/route.

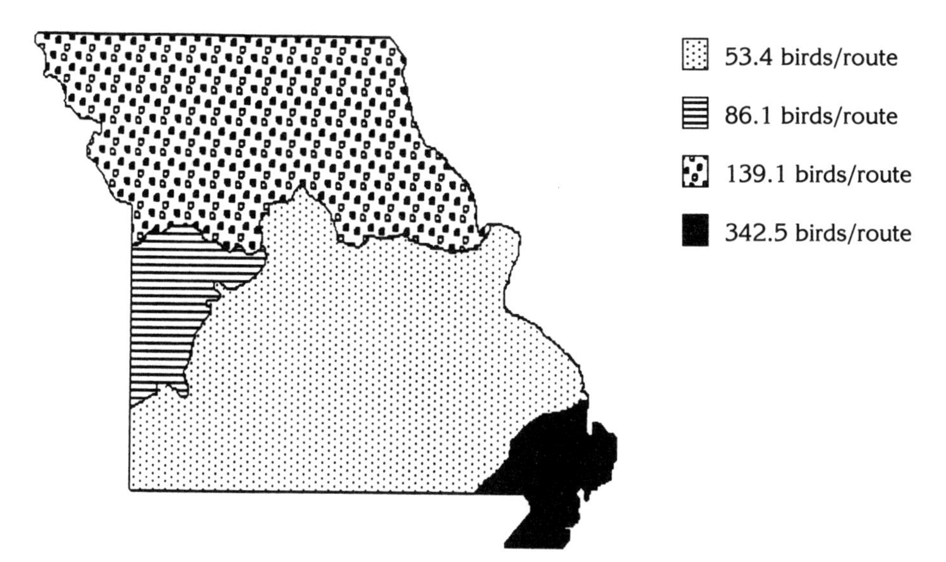

ing through the state into May. High count: 5 million, 13 Mar 1966, Squaw Creek (FL-**BB** 33[2]:10).

Summer: BBS data indicate that this is the most abundant bird in the state at this season. It is most common in the Mississippi Lowlands and least so in the Ozarks (Map 76).

Fall Migration: Southbound flocks are seen during the later part of Sept, with numbers increasing as the season progresses. The largest concentrations are usually seen at the end of the period at prime roosting sites, such as Squaw Creek, when several million birds may be present.

Winter: At this season the species is considerably more local; nevertheless, it can be quite abundant at some areas (especially early in the season). High count: 8,019,000, 19 Dec 1978, Squaw Creek CBC. A very similar, independent estimate was made during the same week by a blackbird biologist (fide B. Heck). Some earlier estimates for this species (and other icterids) on this count are inaccurate as a result of poor estimating techniques; however, observers who participated in some of the earlier counts agree that this species was as common, if not more so, than the 1978 total.

Eastern Meadowlark (*Sturnella magna*)

Status: Common permanent resident.

Map 77.
Relative breeding abundance of the Eastern Meadowlark among the natural divisions. Based on BBS data expressed as birds/route.

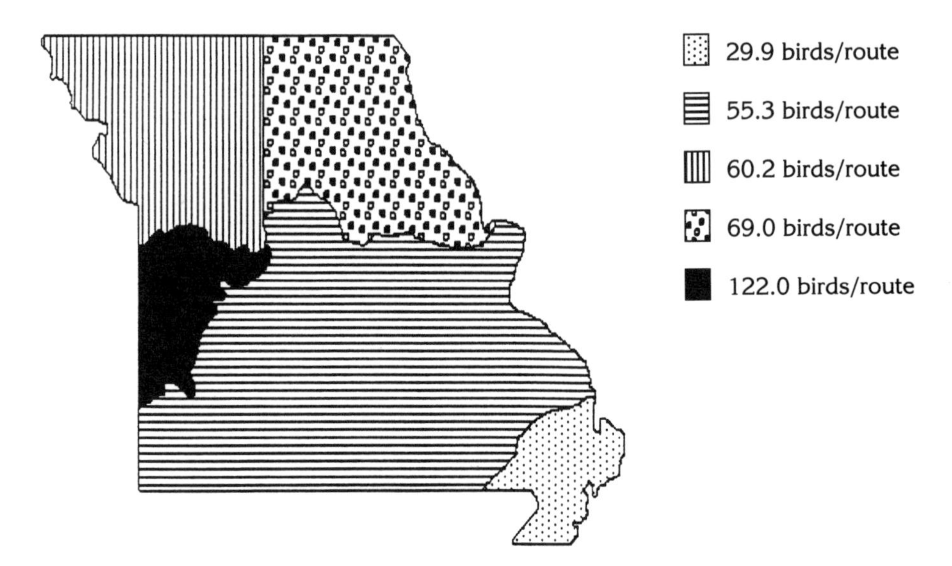

Documentation: Specimen: male, 15 June 1934, Arbyrd, Dunklin Co. (MU 308).

Habitat: Fallow fields, pastures, prairies, and cultivated fields.

Records:

Spring Migration: There is a definite movement of birds during Mar, which continues until at least mid-Apr. Detailed observations by FL, during the 1960s and 1970s, clearly show that this species does not begin to reappear in numbers until early Mar in northwestern Missouri. Up until mid-Mar, in some years not until the end of Mar, the Western Meadowlark is more common there; however, thereafter the Eastern is more numerous, e.g., the same route of St. Joseph/Squaw Creek was used on the following two dates: 12 Eastern, 3 Western, 13 Mar 1967; 75 Eastern, 3 Western, 2 Apr 1967 (FL).

Summer: Most abundant in the Osage Plains and least common in the Mississippi Lowlands (Map 77). Undoubtedly, it has increased dramatically in the Ozarks and the Mississippi Lowlands since these regions were cleared or "opened up" at the turn of the century. Widmann (1907) mentions that it was "scarce" in both these areas. It outnumbers the Western Meadowlark even in the northwestern corner.

Fall Migration: Migration occurs principally in Oct and early Nov, with at least some movement until the end of the period. In northwestern Missouri this species outnumbers the Western until the end of Oct; thereafter the

Western is the predominant bird (FL).

Winter: Winter abundance is imperfectly understood because of birds being less vocal and observers not carefully scrutinizing them. Nevertheless, the Eastern outnumbers the Western in all regions except the western quarter of the Glaciated Plains. Data are insufficient for the relative status of these two species in the remainder of the latter region.

CBC data (using only data where birds were positively identified; thus these estimates are minimums, especially for the Glaciated Plains and Ozark Border) indicate that it is considerably more common in the Mississippi Lowlands (average 4.1 birds/pa hr) than in any other region (average 1.5 Ozarks, 1.2 Ozark Border, 0.5 Glaciated Plains). However, when data for all meadowlarks (both species, plus unidentified birds) are lumped, the Glaciated Plains averages more meadowlarks/pa hr (1.9) than do either the Ozarks (1.7) or Ozark Border (1.4). High counts, both on the Columbia CBC: 463, 28 Dec 1975; 356, 1 Jan 1972.

Comments: See Western Meadowlark account.

Western Meadowlark (*Sturnella neglecta*)

Status: Locally common permanent resident in northwest; very local and quite rare summer resident elsewhere; rare winter resident elsewhere.

Documentation: Specimen: male, 15 Apr 1916, Jackson Co. (KU 39333).

Habitat: Prefers drier, shorter grass areas than the Eastern Meadowlark, especially on slopes and hilltops.

Records:

Spring Migration: At the beginning of this period the Western Meadowlark is the predominate species in the northwestern corner, and it is rare, but regularly recorded, in all other regions of the state. By mid- to late Mar it becomes less common in the northwest, and most of the wintering individuals in other regions of the state are last seen and heard at this time, too. Late dates for birds at nonbreeding sites: 1, 3 May 1964, St. Charles Co. (SHA, J. Willet-**BB** 31[2]:14); 1, 8 Apr 1967, St. Charles Co. (JEC et al.-**NN** 39:42); 1, 2 Apr 1980, Taberville Prairie (MBR); 1, 28 Mar 1975, Mississippi Co. (JH).

Summer: Locally common in dry, short-grass areas, especially along well-drained slopes and ridges in northwestern Missouri. As one progresses eastward across the Glaciated Plains it becomes more local and scarce. It is extremely local, irregular, and quite a rare breeder in other regions.

At least since the very early 1900s, this species has bred in the northwest corner. Widmann (1907) gives accounts of it nesting there, and Lanyon's (1956) work in the early 1950s established that the distribution of this species had not changed since the turn of the century. He showed that the principal area of nesting was limited to the extreme northwestern corner, with isolated pockets farther east. In 1986, a transect across the northern two tiers of counties did not produce any Westerns east of Harrison Co., and it was quite local in that county. Nevertheless, it was found to be locally common from the western half of Worth Co. on westward (MBR). Thus, it appears that the distribution of this meadowlark has been stable for at least the past 90 years.

BBS data show that the Western is quite common at least as far south as southern Clinton Co., and it is uncommon in Ray and western Carroll counties. However, south of the Missouri R. only a single bird has been recorded on route 24 (Jackson and Cass counties; n=16 years). An average of 0.3 birds/route has been recorded on routes 33, 34, and 35 in north central and northeastern Missouri. It is nearly three times as common on route 37 in Atchison Co. (average 59.0 birds/route) than on the route with the next highest average.

Single (paired?) birds have recently (1987) been found in southwestern Missouri in the following areas: Bois D'Arc, Greene Co. (L. Kennard); near Shell City, Vernon Co. (E. Powell); and at two sites in Bates Co. (JJ-**BB** 54[4]:29). There is also an observation for the Mississippi Lowlands: 1 singing, 3 June 1977, Scott Co. (PH, BE-**BB** 44[4]:31).

Fall Migration: By mid-Oct this species becomes more common in the northwestern corner, and it begins to appear at nonbreeding localities. Early dates at nonbreeding locales: 1 singing, 3 Oct 1981, Busch WA (TP et al.-**NN** 53:65); 1 singing, 19 Oct 1967, Busch WA (KA-**BB** 35[1]:13); "reached St. Louis area by 24 Oct" in 1964 and 1966 (**BB** 31[4]:18; 33[1]:10).

Winter: The predominate meadowlark in the northwestern corner at this season. It is rare but regular in all other areas, with some even wintering in the Mississippi Lowlands. Many of the meadowlark totals on CBCs are inaccurate, as all too often observers report what is "expected" in an area. Unless birds are heard singing or carefully examined they should be reported as "meadowlark sp."

Comments: Apparently this species occasionally hybridizes with the Eastern Meadowlark in northwestern Missouri (NWMSU, LLH 116, 117).

Yellow-headed Blackbird (*Xanthocephalus xanthocephalus*)

Status: Locally common summer resident in the northwestern counties bordering the Missouri R.; rare and very local elsewhere in the Glaciated Plains; rare transient outside northwestern corner; rare winter resident in northwest, accidental elsewhere.

Documentation: Specimen: male, 30 Jan 1987, Bigelow (ANSP 178190).

Habitat: Cattail marshes and swamps; in winter found in cultivated fields and feed lots.

Records:

Spring Migration: The first arrivals typically are seen at the beginning of Apr. Numbers gradually increase, and by the end of Apr it is common. Peak is during early to mid-May. The Yellow-headed is decidedly more common in northwestern Missouri, especially at Squaw Creek, than in any other area of the state. It is rare and irregularly encountered in other regions. Earliest dates: there are several Mar observations, and some of these probably represent birds that overwintered. High counts: 300+, 29 Apr 1990, near Squaw Creek (DAE-**BB** 57[3]:141); 229, 7 May 1989, Squaw Creek area (DAE-**BB** 56[3]:93). High count in east: 15, 24 Apr 1975, Williamsburg, Callaway Co. (RW-**BB** 43[1]:21).

Summer: Principally breeds in the westernmost tier of counties bordering the Missouri R., from Platte Co. northward to the Iowa border. The largest population is invariably found at Squaw Creek where it can be abundant in some years; the population is densest during years when water levels are high.

It does breed very locally across the Glaciated Plains; however, breeding birds are quite scarce and irregular there. Widmann (1907) gives breeding records for Jasper, Johnson, Saline, Franklin, and Clark counties for the late 1800s and the very early 1900s. The 3 egg sets which he states were taken in Clark Co., by E. Currier, were actually collected on 4 June 1895, not on 28 May (all 3 sets are deposited in WFVZ).

Fall Migration: Most have left breeding areas by early Sept. Only individuals and very small-sized groups (usually a maximum of 2–3 birds/day) are encountered after mid-Sept. It is regularly seen in small numbers with other blackbirds in cultivated fields and feed lots throughout the season in the extreme western part of the Glaciated and Osage plains.

Winter: It is a rare but regular winter resident in the westernmost tier of counties in the Glaciated Plains. There are less than five observations away from this region. This is in part a reflection of the paucity of observers in the westernmost counties in the Osage Plains, as it surely winters there as well. Extralimital sightings: male, 4 Jan 1981, Swan L. CBC; 1, 18 Dec 1982, Mingo CBC. High counts, both on the Trimble CBC: 50, 18 Dec 1977; 7, 27 Dec 1986.

Rusty Blackbird (*Euphagus carolinensis*)

Status: Locally common transient; locally uncommon winter resident in south, rare in north; hypothetical summer visitant.

Documentation: Specimen: male, 13 Mar 1965, near Tucker Prairie, Callaway Co. (NWMSU, DAE 854).

Habitat: Swampy woodland, flooded forest, cultivated fields, and feed lots.

Records:

Spring Migration: Migrants begin to reappear at the end of Feb, and their numbers increase to a peak in late Mar and early Apr. It remains fairly common until the end of the second week of Apr, but thereafter it is rare, with no records after the third week of Apr. High counts: 2,000+, 21 Feb 1916, Kansas City (Harris 1919b). Latest dates: 23 Apr 1874, Johnson Co. (Widmann 1907); female, 23 Apr 1977, Bigelow Marsh (DAE-**BB** 44[4]:28).

Summer: One suspect record from Springfield.

Fall Migration: Usually reappears by the second week of Oct; by late Oct it is common. This species remains fairly common through mid-Nov; it becomes rare in the north by the end of Nov. Earliest date: 28 Sept 1896, Clark Co. (E. Currier; Widmann 1907). High counts, both at Squaw Creek by FL: 500, 13 Nov 1966; 500, 15 Nov 1981 (**BB** 49[1]:15).

Winter: Uncommon in the southern half of the state; perhaps most abundant and regular in the Mingo and Duck Creek areas. It is rare and irregular in the north, although it can be locally common during some years in the early part of the season (for example, at Squaw Creek). High counts, both at Squaw Creek: 20,250, 19 Dec 1978; 14,682, 19 Dec 1982. See comments under Red-winged Blackbird concerning early CBC totals at Squaw Creek.

Brewer's Blackbird (*Euphagus cyanocephalus*)

Status: Rare to locally uncommon transient and rare winter resident.

Documentation: Specimen: male, 10 Apr 1965, near Tucker Prairie, Callaway Co. (NWMSU, DAE 862).

Habitat: Pastures, short-grass fields and prairies, cultivated fields, and feed lots.

Records:

Spring Migration: More common in the western half of the state at all seasons. The first arrivals appear during the last week of Feb. Numbers increase through Mar and peak at the end of the month and early Apr. It is accidental after the third week of Apr. Earliest date: 5, 21 Feb 1970, near St. Joseph (FL). High counts: 2,000, mid-Apr 1974, Busch WA (JC-**BB** 41[3]:5); 300, 5 Mar 1961, Squaw Creek (FL). Latest dates: 8, 29 Apr 1979, St. Joseph (FL); at least 2 sightings for 18 Apr.

Summer: No reliable records; we consider all old (pre–1965) reports from the Maryville area erroneous.

Fall Migration: Normally not encountered until the beginning of the second week of Oct. Peak is not until mid-Nov. It is regularly seen until early Dec. Earliest dates: 1, 22 Sept 1965, Busch WA (KA, SHA et al.-**NN** 37:6); 4 (2 males, 2 females), 6 Oct 1987, at Langdon, Atchison Co. (MBR-**BB** 55[1]:13). High count: 500, 9 Nov 1969, near St. Joseph (FL-**BB** 38[3]:6).

Winter: Much rarer at this season than during the migratory period. Most *bona fide* records are from the Osage Plains and the western part of the

Glaciated Plains. Reports of thousands on CBCs are erroneous and pertain to other icterid species (Rusty Blackbirds, Common Grackles, and cowbirds). Observers should bear in mind that Common Grackles do not have keeled tails in the fall and winter. Nevertheless, the Brewer's does winter in the prairie region in relatively small numbers. It is often seen at feed lots and cultivated fields during the day; it often roosts on prairies. High counts: 350, 27 Dec 1975, Kansas City North CBC; 150, 20 Dec 1986, Jefferson City CBC.

Great-tailed Grackle (*Quiscalus mexicanus*)

Status: Rare and local summer resident in extreme west, accidental elsewhere; rare transient and winter resident in extreme west, accidental elsewhere.

Documentation: Specimen: female, 8 May 1976, Bigelow Marsh (DAE, JHI-**BB** 44[2]:18; NWMSU, DAE 2979).

Habitat: Primarily breeds in cattail marshes; during winter and migration found in marshes, cultivated fields, pastures, and feed lots.

Records:

Spring Migration: Migrants appear by the beginning of the period. Numbers increase through Mar. Peak is in early to mid-Apr. The first definite eastern record was of a single bird east of Farmington on 18 May 1987 (SD-**BB** 55[1]:14). High counts: 100, 21 Feb 1983, west of Adrian, Bates Co. (LG-**BB** 50[2]:27); 100, 4 Apr 1984, near St. Joseph (FL-**BB** 51[3]:15).

Summer: Breeding was first documented in 1979 at a cattail marsh adjacent to Big L. SP in the northwestern corner (Robbins and Easterla 1986). It was known to breed only at the above site and another nearby locality until 1986 when it was discovered nesting in Cass Co. (JJ-**BB** 54[3]:13). Several additional colonies, ranging in size from 3–30 birds, were located in 1987 at the following areas: near Excelsior, Ray Co. (L. Burgess), east of Warrensburg (G. Gremaud), and at three sites in Newtown Co. (JW-**BB** 54[4]:29). Young were also seen at L. Contrary in Aug 1987 (DAE-**BB** 55[1]:13). On 10 June 1990, a pair feeding 3 young was located at a pond with cattail northeast of Gilman, Harrison Co. (M. McNeely). Undoubtedly, the expansion will continue, with the species likely to colonize areas throughout the western section of the state. High count: 60, 13 June 1987, Cass Co. (JG-**BB** 54[4]:29).

Fall Migration: Birds are increasingly seen throughout the period in cultivated fields, feed lots, and marshes in the extreme west. The fall of 1987 was the first time birds were encountered away from the western border: "several" were still present at Big Oak Tree SP until 18 Nov (SD-**BB** 55[1]:13). The only other eastern record is of a female in St. Charles Co. on 10 Dec 1989 (JZ-**BB** 57[1]:51). High count: 100, 11 Nov 1984, western Bates Co. (K. Jackson, FL-**BB** 52[1]:34).

Winter: As early as 4 Jan 1958, an individual of the *Quiscalus mexicanus* complex was observed in St. Charles Co. (JEC; Anderson and Bauer 1968). Although its identity was not established, since the Great-tailed and Boat-tailed (*Q. major*) were considered conspecific at that time, it was very likely a Great-tailed because the Boat-tailed is not known to wander inland. The second record finally came in Dec 1972 when a bird of the *mexicanus* complex (the specific identity was undetermined) was found in a large blackbird roost in Springfield (NF et al.-**BB** 40[1]:9).

The single bird on the 1979 Squaw Creek CBC represented the first winter record where a positive identification was made. Numbers continue to increase on the Squaw Creek count; a high of 293 was recorded in Dec 1986. Birds are now overwintering at various sites in the western section. Most are seen at feed lots and in cultivated fields.

Comments: This species has shown a dramatic population increase and range expansion throughout the western United States, particularly in the Great Plains. The expansion is largely the result of an increase in surface water, which increases nesting opportunities, coupled with an increase in year-round food provided by agriculture. In years where there has been a drought in spring and summer in the Great Plains, grackles have bred in greatly reduced numbers or failed to breed at all, e.g., 1981 (Robbins and Easterla 1986), 1988, and 1989.

All specimens (n=4; above, ANSP) taken in the state are referable to *Q. m. prosopidicola*.

Common Grackle (*Quiscalus quiscula*)

Status: Common summer resident; locally common winter resident in south, uncommon in north.

Documentation: Specimen: male, 7 Dec 1949, Vernon Co. (KU 39409).

Habitat: Open areas with scattered trees, woodland and forest edge; in winter, in cultivated fields and feed lots.

Records:

Spring Migration: The winter population begins to be supplemented by migrants at the end of Feb. The state's population swells during the latter half of Mar and peaks at the end of the month or early Apr. By mid-Apr mostly summer residents remain.

Summer: Today it is most common in the Mississippi Lowlands and least common in the Ozarks (Map 78), although prior to 1900 it was much less common in both of these areas. The clearing of forest has significantly aided this species in these two regions. Widmann (1907) described the Common Grackle as "rare as a breeder" in the Ozarks. With the conversion of upland forest and prairie to parkland and residential areas, this species was documented to have increased dramatically at Swope Park between 1916 (n=9 territorial males; A. Shirling 1920) and 1973 (n=149 territorial males; Branan and Burdick 1981).

Map 78.
Relative breeding abundance of the Common Grackle among the natural divisions. Based on BBS data expressed as birds/route.

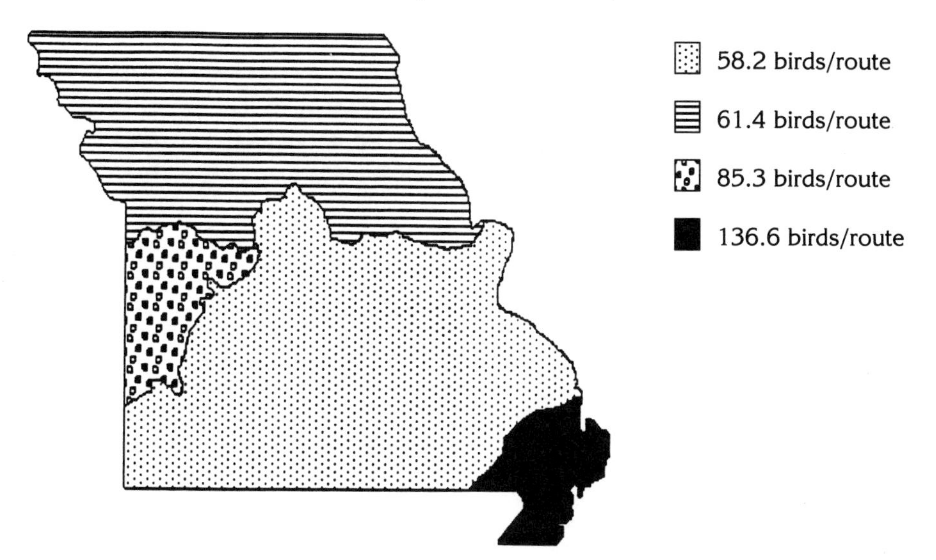

Fall Migration: Large concentrations, presumed to be mostly composed of summer residents, are seen from the onset of this period. At the end of Sept large flocks are seen moving southward. Numbers continue to build through Oct and peak at the very end of the month and early Nov. However, relatively large numbers (in the low thousands) continue through the end of the period. High count: 1 million, 26 Oct 1966, St. Joseph (FL-**BB** 34[1]:10).

Winter: In all but the severest late fall and "early" winters, the grackle population continues to be seen in relatively large numbers, although flocks are more local than in the fall. Concentrations are generally larger and more widespread in the south. By early Jan most of the huge flocks in the north have diminished considerably in number or have disappeared completely; nevertheless, during mild winters relatively large groups (in the hundreds or thousands) remain there throughout the winter. High counts, both on the Springfield CBC: 330,000, 19 Dec 1981; 300,000, 15 Dec 1979.

Bronzed Cowbird (*Molothrus aeneus*)

Status: Accidental winter visitant.

Documentation: Specimen: male, 5 Jan 1979, Squaw Creek (B. Heck et al.; Robbins and Easterla 1981; NWMSU, DAE 3094; VIREO x05/4/004; 8f).

Comments: The above record involved a bird that was visiting a bird feeder with other blackbirds at the refuge headquarters.

Brown-headed Cowbird (*Molothrus ater*)

Status: Common summer resident; uncommon and local winter resident.

Documentation: Specimen: male, 6 Sept 1933, Jackson Co. (KU 20210).

Habitat: Open areas, especially grassland, pastures, woodland and forest edge.

Records:

Spring Migration: Like the other blackbirds, migrants are first noted at the end of Feb. Numbers increase up until early Apr when it peaks in the south; peak follows about a week later in the north. High count: 1,500, 21 Mar 1984, St. Joseph (FL).

Summer: It is most common in the western half of the Glaciated Plains and least abundant in the Mississippi Lowlands (Map 79). The Brown-headed Cowbird has increased since presettlement in the Ozarks, Ozark Border, and the Mississippi Lowlands because these areas were "opened up" and cleared. The BBS data exhibit an increase for this species in the state over the past 23 years (Graph 24).

Fall Migration: It is commonly seen in flocks through early Nov statewide. It is more local and seen in smaller numbers thereafter.

Winter: As with the other blackbirds, winter numbers continue to increase mainly because food availability at this time of the year has increased (such as grains). Even in the north relatively large groups are seen in early winter, and during mild winters they may remain throughout the period. High counts, both at Springfield: 165,000, 19 Dec 1981; 150,000, 15 Dec 1979.

Orchard Oriole (*Icterus spurius*)

Status: Common summer resident.

Documentation: Specimen: male, 19 June 1932, Jackson Co. (KU 19368).

Habitat: Relatively young, second-growth woodland, brushy areas, and open areas with scattered, dense thickets.

Records:

Spring Migration: In the south it is seen by the third week of Apr, and in the north it typically arrives during the final days of the month. Peak is in early and mid-May in the south and north, respectively. Earliest dates: 5 Apr 1988, Farmington (BL, SD-**BB** 55[3]:90); two sightings by JH in Mississippi Co. on 17 Apr (1963 and 1965).

Map 79.
Relative breeding abundance of the Brown-headed Cowbird among the natural divisions. Based on BBS data expressed as birds/route.

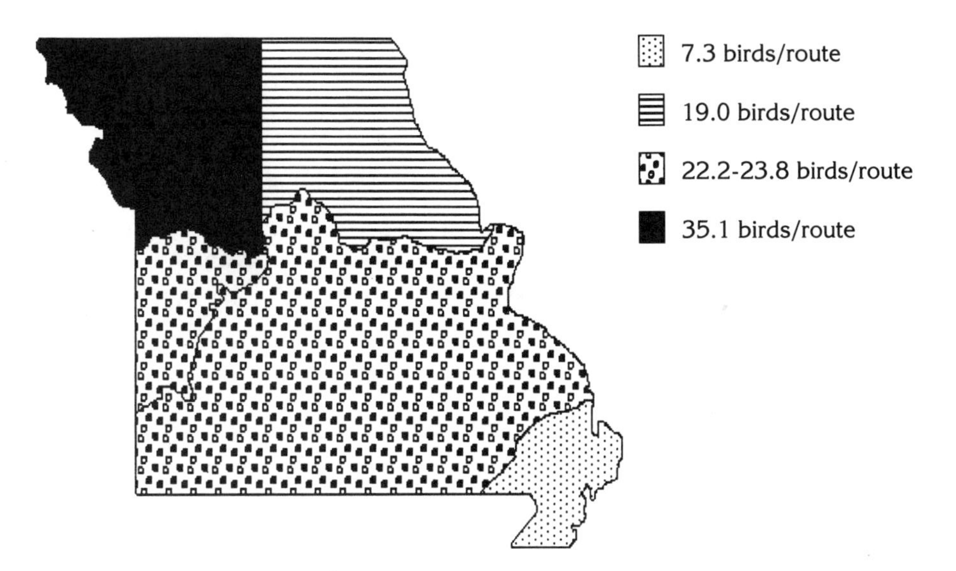

Graph 24.
Missouri BBS data from 1967 through 1989 for the Brown-headed Cowbird. See p. 23 for data analysis.

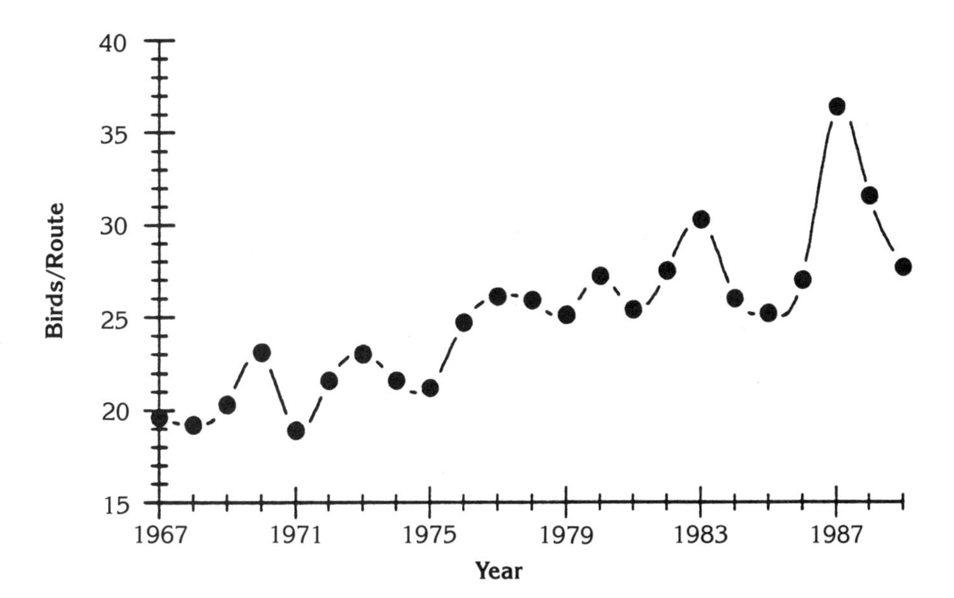

Graph 25.
Missouri BBS data from 1967 through 1989 for the Orchard Oriole.
See p. 23 for data analysis.

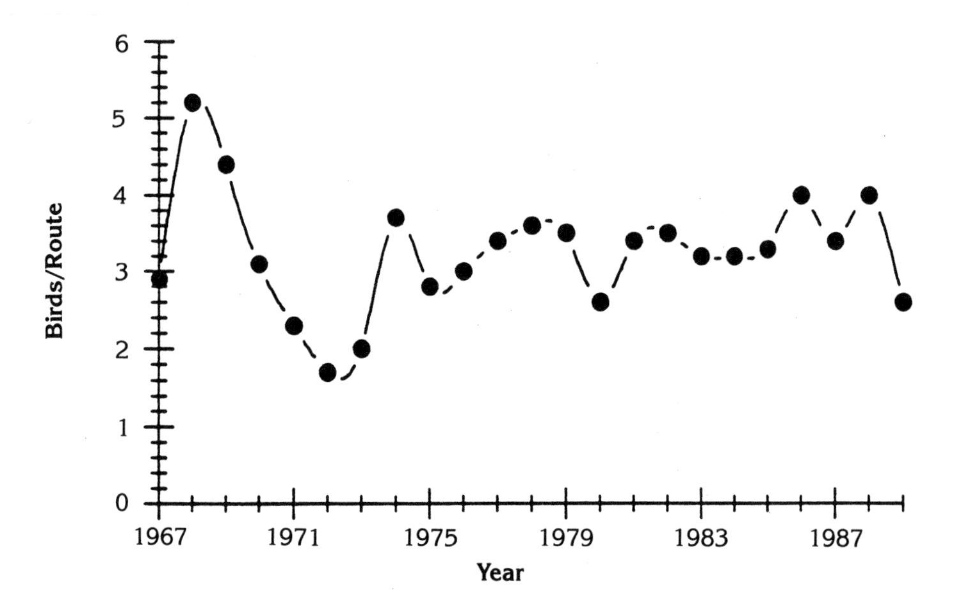

Summer: A common breeder statewide. Although the overall average for the Glaciated Plains is 2.1 birds/route, when the region is subdivided in west vs. east, it becomes apparent that the Orchard Oriole is decidedly more common in the west (3.3 birds/route vs. 0.8 birds/route). The Mississippi Lowlands and the Ozarks surprisingly show the highest averages, 3.5 and 3.6, respectively. An average of 2.9 birds/route is recorded in the Osage Plains. BBS data indicate that the Missouri population for this species is stable (Graph 25).

Fall Migration: This icterid leaves relatively early, and most have departed by late Aug. Very few are seen after mid-Sept, and it is accidental after the end of Sept. Latest date: 1, 6 Nov 1965, Little Dixie L. (J. Roller-**BB** 33[1]:10).

Northern Oriole (*Icterus galbula*)

Status: Common summer resident in all regions, except the Ozarks where it is uncommon; casual winter visitant.

Documentation: Specimen: male, 15 May 1882, Jackson Co. (KU 39404).

Habitat: Open woodland and forest, parks, and residential areas.

Records:

Spring Migration: The first birds are seen in mid-Apr at the southern

Map 80.
Relative breeding abundance of the Northern Oriole among the natural divisions. Based on BBS data expressed as birds/route.

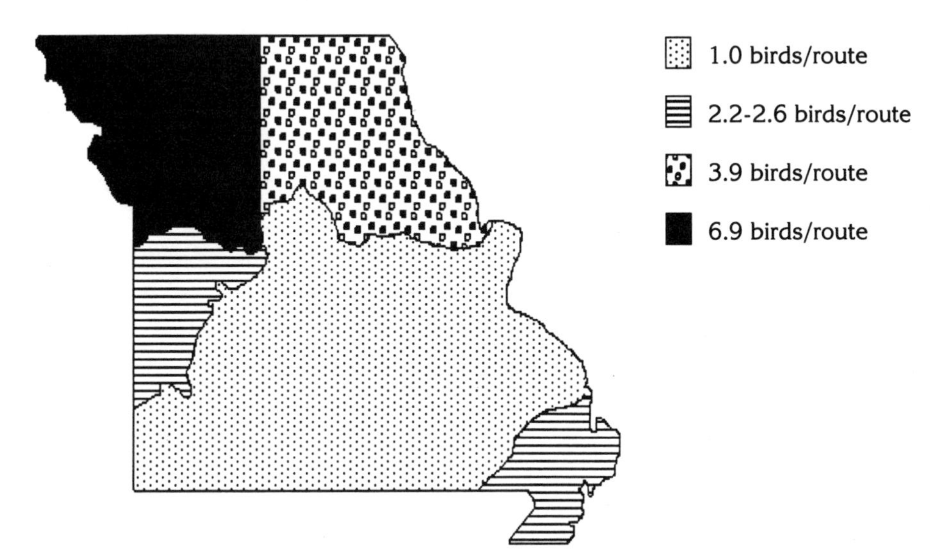

border and at the beginning of the final week of Apr in the north. Peak is in early May in the south and during mid-May in the north. An average of 0.5 birds/hr was recorded at Forest Park between 29 Apr–18 May 1979–90 (RK; n=343 hrs). Earliest dates: 1, 21 Apr 1987, Pemiscot Co. (FL, LG); 1, 21 Apr 1990, Taney Co. (PM, JH-**BB** 57[3]:141). High count: 40, 3 May 1964, St. Joseph/Squaw Creek (FL).

Summer: This oriole reaches its greatest abundance in the western half of the Glaciated Plains (Map 80). It is least common in the Ozarks where it is primarily found in the broader riparian valleys and in residential areas.

Fall Migration: Most leave the state by the end of Aug, with peak of migrants at the very end of Aug and early Sept. It is rarely encountered after the third week of Sept, and it is only casually found after mid-Oct. High count: 75, 2 Sept 1981, St. Louis (MP-**BB** 49[1]:15). Latest dates: there are at least four observations for the first two weeks of Dec.

Winter: There are no less than five sightings of single birds (usually at feeders) during the last two weeks of Dec and in very early Jan: 1, 9–23 Dec 1961, Kansas City (Thom-**AFN** 16:335); adult male, first appeared 14 Dec 1987 (present at least two weeks), Columbia (D. and N. Witten-**BB** 55 [2]:62); 1, first appeared at feeder on 25 Dec and stayed for several weeks in 1966–67 at St. Louis (D. Vogel-**NN** 39:17); adult male, 25 Dec 1969, Independence (CH); female, 29 Dec 1982, Rosendale, Andrew Co. (JHI).

Graph 26.
Missouri BBS data from 1967 through 1989 for the Northern Oriole. See p. 23 for data analysis.

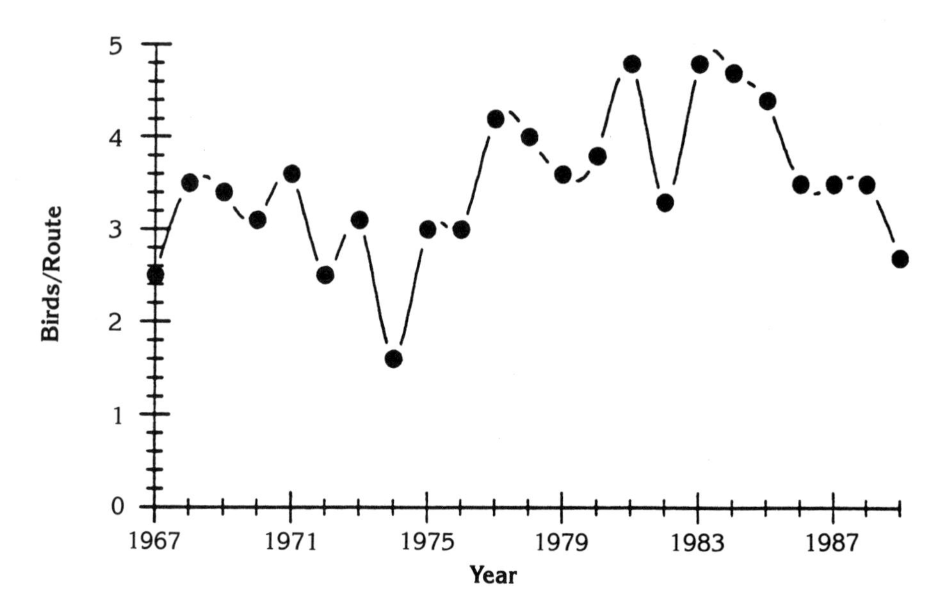

Comments: The bird reported as a possible female hybrid (*I. g. galbula* x *I. g. bullockii*) at Bigelow Marsh on 26 May 1970 (DAE-**BB** 38[3]:9; NWMSU, DAE 2381) proved to be a second-year male *galbula.* A possible female *bullockii* was present at a feeder from 20 Nov–1 Feb 1978–79 in St. Louis (DJ et al.; VIREO x08/16/001), and an adult male *bullockii* was reported on 30 May 1971, just north of LaGrange, Lewis Co. (R. Randall-**AB** 25:753).

Observers should document all sightings of hybrids and *bullockii.*

Family Fringillidae: Fringilline and Cardueline Finches and Allies

Pine Grosbeak (*Pinicola enucleator*)

Status: Casual winter resident in north, accidental in south.

Documentation: Specimen: female, one of three present, 2 Dec 1962, Kansas City (Rising 1965; KU 41460).

Habitat: Usually seen at bird feeders but found at fruiting bushes and trees and conifers.

Records:

Spring Migration: Only three observations, two of which are extraordinary because of their lateness and locality: female, 25 Feb 1978, St. Joseph (FL-**BB** 45[2]:14); 9, 24–25 Mar 1986, near Farmington, St. Genevieve Co.

Fig. 32. This immature male Pine Grosbeak was present in Maryville, Nodaway Co., from 8 January–24 February 1978. The Pine Grosbeak is of casual occurrence in Missouri. Photograph by Mark Robbins.

(G. Wylie, BRE, K. Adams-**BB** 53[3]:14); male, 15 May 1983, Busick Woods SP, Christian Co. (GD- **BB** 50[3]:14).

Fall Migration: There are only six records, with the first record being a female at LaGrange, Lewis Co., on 3 December 1903 (Johnson 1908). Earliest date: female, 13–14 Nov 1965, Maryville (Easterla 1966a; NWMSU, DAE 1033). High count: 3 (adult male and 2 females), 29 Nov–2 Dec 1962, Kansas City (Rising 1965; photographed, DAE).

Winter: Two records: male, 20 Dec 1980, Maryville CBC (photographed DAE, TE); immature male, 8 Jan–24 Feb 1978, Maryville (MBR et al.-**BB** 45[2]:14; Fig. 32).

Comments: All four specimens taken in the state are referable to the nominate race.

Purple Finch (*Carpodacus purpureus*)

Status: Common transient; uncommon to locally common winter resident in Ozarks and Ozark Border, rare elsewhere.

Documentation: Specimen: male, 24 Apr 1920, Fayette, Howard Co. (CMC 345).

Habitat: Woodland and forest.

Records:

Spring Migration: The winter population begins to be supplemented with migrants by the end of Feb, but numbers do not increase dramatically until mid-Mar. Peak is at the end of Mar in the south and in early Apr in the north. Small-sized flocks (< 20 birds) are regularly encountered into early May, and an occasional bird is seen into the second week of May. Latest dates: no.?, 23 May 1983, Springfield (CB-**BB** 50[3]:14); 19 May 1907, St. Louis (Widmann 1907).

Map 81.
Early winter relative abundance of the Purple Finch among the natural divisions. Based on CBC data expressed as birds/10 pa hrs.

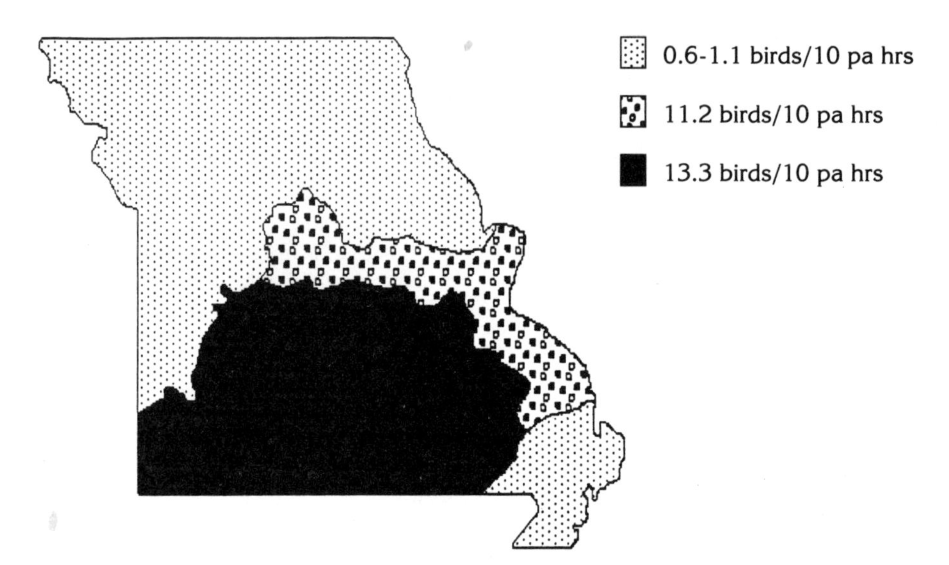

Fall Migration: The first arrivals appear during the last week of Sept; however, it usually is not detected until the second week of Oct and does not become numerous until the end of the third week of Oct. This species remains fairly common through mid-Nov, but by the end of the month most migrants have left. Earliest dates: immature, 16 Sept 1990, Swan L. (RF et al.); 2, 24 Sept 1979, Cleveland, Cass Co. (C. Swink, M. Stephens-**BB** 47[1]:26).

Winter: Most common in the Ozarks and Ozark Border and least abundant in the Mississippi Lowlands, Glaciated and Osage plains (Map 81). High counts: 1,500–2,000, 7 Jan 1981, Lewis Co. (R. DeCoster-**AB** 35:06); 309, 31 Dec 1976, Sullivan CBC.

House Finch (*Carpodacus mexicanus*)

Status: Rare permanent resident.

Documentation: Photograph: immature male, 1 Dec 1977–23 Apr 1978, St. Louis Co. (B. Whitner; VIREO m0025-01).

Habitat: Suburban areas, open areas with scattered trees and bushes.

Comments: House Finches from California (*C. m. frontalis*) were released onto Long Island, New York, about 1940 (Elliot and Arbib 1953). Initially the population grew relatively slowly, with ca. 280 birds estimated in the above area in 1951 (Elliot and Arbib 1953). Bock and Lepthien

(1976), using CBC data, demonstrated that this species underwent nearly a tenfold increase between 1962 and 1971 in the east. It continues to proliferate and expand its range throughout the eastern half of the country.

It was first detected in Missouri in 1974 when a single bird was observed at a feeder in St. Louis on 23 Nov (B. Erickson-**NN** 47:3; photographed). The next record did not occur until an immature male was found, again at St. Louis, from about 1 Dec–17 June 1977–78 (Jones 1978). There was another gap in observations until a couple of sightings in 1983. The first nesting record was established at St. Louis in 1983: an adult was observed feeding young on 3 Aug (**BB** 51[1]:46).

It finally was detected in the western half of the state at the end of 1983; a female was seen on the Maryville CBC (DAE, MBR) on 23 Dec. Several additional observations, involving 1–3 birds at scattered localities, were made in 1984. Breeding was established in the Mississippi Lowlands in 1985 when a pair and fledged young were seen on 10 July 1985 at Cape Girardeau (J. Peerman, JW-**BB** 52[4]:20). Additional breeding records were recorded in this region in 1986 at Kennett in Dunklin Co. (H. Schanda-**BB** 53[3]:14) and Charleston in Mississippi Co. (JH-**BB** 53[4]:18). Nesting was also suspected in 1986 at Columbia. Breeding was first recorded in the northwestern corner at St. Joseph (C. Fisher) and Maryville (DAE) in the spring of 1988 (**BB** 55[3]:90–91).

No relatively large concentrations were noted in the state until a flock of 35 was seen at St. Louis on 15 Nov 1987 (SRU-**BB** 55[1]:13). To date, it continues to increase in numbers and solidify its range in the state. Before long it will undoubtedly become a permanent resident in the more open regions of the state; it should remain less common and widespread in the more heavily forested areas of the Ozarks.

Red Crossbill (*Loxia curvirostra*)

Status: Rare and irregular winter resident; casual summer visitant.

Documentation: Specimen: of western, large billed population; male, 1 Nov 1966, west of Maryville (NWMSU, DAE 1102).

Habitat: Almost invariably found in stands of conifers; sunflowers are also an attraction.

Records:

Spring Migration: Following winters when there has been a major influx, birds are regularly seen until mid-Apr and casually into May. Latest dates: 2–15, 5–30 May 1982, Columbia (fide SS-**BB** 49[3]:23); 27 May 1951, Kansas City (D. Hall-**BB** 52[3]:16).

Summer: At least five summer sightings with no evidence of nesting, although late lingering birds should be carefully watched for signs of breeding: female, 14–24 June 1987, east of Butterfield, Barry Co. (F. Abramoritz-**BB** 54[4]:30; photo in ref); 1, captured and released, 12 July 1969,

Stone Co. (BBR-**BB** 38[1]:4); 7, at feeder, 28–30 July 1972, Kansas City (K. Wahl et al.-**BB** 39[4]:10); 1, 29 July 1972, near Hermann, Gasconade Co. (Mrs. Elsenraat); and one at Orchard Farm, without date (Widmann 1907).

Fall Migration: Except during years when there is a major invasion, birds are initially observed about mid-Oct. In falls where there is a large influx (like the falls of 1966, 1972, 1976), birds are seen as early as mid-Sept. During "average" years, the largest numbers are not encountered until late Nov and early Dec. Earliest dates: 2 females or immatures, 5–6 Aug 1966, Table Rock L., Stone Co. (BBR); pair, 21 Aug 1976, along James R., Christian Co. (P. Redfearn-**BB** 44[3]:23); no.?, 1 Sept 1975, St. Louis (R. Laffey-**BB** 44[1]:23). High count: 61, 26 Nov 1969, 10 miles west of Maryville (DAE).

Winter: Irregular. Numbers fluctuate considerably, but at least some are found during the majority of winters. High counts: up to 200, Dec–Jan 1981–82, Columbia (SS-**BB** 49[2]:17); 60, 27 Dec 1955, Parkville, Platte Co. (BK-**AFN** 10:255).

Comments: The taxonomy and systematics of this species is in a state of flux. What names should be applied to various populations is unclear, and there may even be more than one species involved. The vast majority of birds that appear in Missouri are relatively large-billed and have been assigned to the Rocky Mountain race, *L. c. bendirei.* However, at least two other small-billed birds, assigned to the race, *L. c. minor,* of the northwestern coast of North America, have been obtained in the state by DAE: female, 8 Dec 1965, 10 miles west of Maryville (NWMSU, DAE 1042); male, 30 Dec 1973, Maryville (NWMSU, DLD 168). Those interested in the taxonomy of this fascinating species should read treatments by Payne (1987) and Groth (1988). Any birds found dead should be preserved to help clarify what populations (species) occur in the state.

White-winged Crossbill (*Loxia leucoptera*)

Status: Rare and highly irregular winter resident; accidental summer visitant.

Documentation: Specimen: male, 20 Nov 1965, west of Maryville (NWMSU, DAE 1034).

Habitat: Principally found in conifer stands, especially hemlocks.

Records:

Spring Migration: Like the Red Crossbill, some birds linger into this period following winter invasions. With the exception of three records, all observations are during late Feb and the first half of Mar. Latest dates: female, 18 Apr 1907, Shannon Co. (Woodruff 1908; AMNH 229633); 1, caught by cat and released, 17 Apr 1981, Kansas City (M. Jackson; Rising et al. 1978); 2, 7 Apr 1974, St. Joseph (FL).

Summer: One record, but see below: male, apparently starved, 22 July

1969, Maryville (C. Bell-**BB** 38[1]:4; KU 63811). This record and two of the Aug records mentioned below occurred during the same invasion year (see winter).

Fall Migration: With the exception of four records, all observations are from mid-Nov through the remainder of the period. Earliest dates: immature male, photographed (in MRBRC file), 1–9 Aug 1989, Prathersville, Boone Co. (M. Gutsy, BJ et al.-**BB** 57[1]:51); pair, 3 Aug 1969, Kansas City (J. Isenberger et al.-**BB** 38[1]:4); female, 23 Aug 1969, Brickyard Hill WA (F. and H. Diggs-**BB** 37[1]:15; photographed); 5, 17 Oct 1972, Table Rock area, Stone Co. (JC-**BB** 40[1]:9). High counts: 14 (4 males, 10 females), 22 Nov 1981, St. Joseph (FL); 13, 30 Nov 1969, Maryville (DAE).

Winter: The White-winged Crossbill is more irregular in occurrence and is found in much smaller numbers than the Red Crossbill. Although there are observations for all regions of the state, most are from northern Missouri; this may, in part, be a reflection of the ease of finding birds there, as conifer stands are widely scattered and less extensive than in southern Missouri. High count: 9, late Dec 1975, St. Joseph (FL).

The falls and winters of 1969–70, 1975–76, and 1981–82 had especially spectacular invasions. The fall of 1969 was particularly unusual because birds first appeared in late July and early Aug (see above).

Common Redpoll (*Carduelis flammea*)

Status: Rare and irregular winter resident.

Documentation: Specimen: female, 3 Mar 1875, Warrensburg, Johnson Co. (CMSU 34).

Habitat: Open areas with scattered trees (especially birches) and bushes, particularly in fields with sunflowers and ragweed. Roosts in conifers.

Records:

Spring Migration: As with other infrequently occurring northern fringillids, birds remain into Mar following winter invasions. Most observations are during the last week of Feb and the first week of Mar. High counts: 70+, 1 Mar 1970, Maryville (MBR); 50–70, 4 Mar 1978, St. Charles Co. (F. Hallet et al.-**BB** 45[3]:19). Latest dates: 12 Apr 1903, Montgomery Co. (Widmann 1907); 8 Apr 1886, Mt. Carmel, Audrain Co. (Mrs. Musick; Widmann 1907); 1, 3 Apr 1969, St. Louis (PS et al.-**BB** 38[1]:3).

Fall Migration: Even during years when there are large influxes of birds, this species is not seen before mid-Nov. Earliest dates: 4 Nov 1885, Mt. Carmel, Audrain Co. (Mrs. Musick; Widmann 1907); male, 16 Nov 1969, Maryville (MBR-**BB** 38[3]:6). High count: 300, 20 Nov 1981, L. Jacomo (KH-**BB** 49[1]:15).

Winter: Most observations of this species are during Jan and Feb. The winter of 1969–70 saw the most spectacular invasion on record. High counts: 100+, Dec–Jan 1969–70, 10 miles north of Maryville (MBR-**BB** 38[3]:8); 50, 29 Jan 1978, Squaw Creek (FL).

Comments: See Hoary Redpoll.

[**Hoary Redpoll** (*Carduelis hornemanni*)]

Status: Hypothetical winter resident.

Documentation: Photograph: 1–3, 8 Feb–11 Mar 1976, Kansas City (CH, JG, M. Myers et. al-**BB** 44[1]:27).

Comments: There is one additional observation besides the one above: 1, 6–7 Mar 1978, St. Louis (T. Barker et al.-**BB** 45[3]:19). One of the birds at Kansas City was netted and photographed. Opinions by experts who reviewed the slides ranged from stating that it was definitely a "pure" Hoary to stating that it was a hybrid. It seems less likely that the bird was a hybrid, if the recent reinterpretation of purported hybrids between *flammea* and *hornemanni* is taken into account. Apparently, plumage variation within *flammea* (the nominate race) and *hornemanni* (the race *exilipes*) is much greater than originally appreciated (Knox 1988). Knox's summary statement (1988) on redpoll identification emphasizes that knowing the age and the sex of a bird is imperative for identification. Attempts should also be made to obtain vocalizations of any purported *hornemanni,* since the voice may be a diagnostic character for distinguishing these two taxa. Knox, in 1989, examined a series of photos of the 1976 bird and was unable to make a positive identification. Despite recent treatments, the taxonomy of redpolls remains unresolved.

Pine Siskin (*Carduelis pinus*)

Status: Irregular, rare to locally common winter resident; casual summer resident; rare summer visitant.

Documentation: Specimen: female, 13 May 1907, Spring Valley, Shannon Co. (Woodruff 1908; AMNH 229646).

Habitat: Fields, especially with sunflowers; suburban areas, open mixed coniferous–deciduous woodland.

Records:

Spring Migration: There is considerable fluctuation in the number of birds and the length of time in which they remain at this season. Following winters when there has been a sizable population present in the state, birds are commonly seen through Mar and much of Apr. There are many observations for the entire month of May, with relatively large numbers (50+) occasionally seen until mid-May. High counts for May: 300, 3 May 1970, St. Joseph/Squaw Creek (FL); 50, 24 May 1966, Lowry City, St. Clair Co. (SH-**BB** 33[2]:11).

Summer: There are well over twenty observations for this species, primarily during June, from all areas of the state. Most sightings are of 1–2 birds at feeders.

A minimum of five nesting records exist, with four from the Kansas City area; however, breeding has been suspected in other regions of the state. Undoubtedly a few of the other 20+ observations represent breeding birds as well. Definite breeding records are as follows: nested in Mt. Washington Cemetery, Jackson Co. in 1960 and 1961 (JRI-**BB** 28[3]:21); 3 fledglings, 6 June 1982, Cass Co. (JG-**BB** 49[4]:20); pair attempted nesting but abandoned nest, May 1984, Cleveland, Cass Co. (JG-**BB** 51[3]:15); female with brood patch netted and banded, 29 May 1987, Camdenton area, Camden Co. (M. Solomon-**BB** 54[4]:30).

Fall Migration: In average years the first arrivals are seen by early Oct. Larger numbers (30 birds/day) appear at the end of Oct. In falls when there are major incursions it can be quite abundant by the end of Nov and early Dec. High count: 1,500, 8 Dec 1965, St. Louis (SHA-**NN** 38:17).

Winter: Numbers can vary from winters where virtually none are seen all season to winters when literally hundreds may be seen at a single locality. High counts: 2,000, Dec–Jan 1965–66, St. Louis (RA-**BB** 33[1]:15); 2,000, winter 1969–70, St. Louis (m.ob.-**BB** 38[3]:8); 450, 20 Dec 1969, Maryville CBC.

Lesser Goldfinch (*Carduelis psaltria*)

Status: Accidental transient.

Documentation: Photograph: male, at feeder, 4–6 Apr 1971, Kansas City (C. Hicks; N. Edding et al.-**BB** 39[3]:8; Fig 33).

Comments: The above bird had an all black back, which is diagnostic of the nominate race. It represents the only Missouri record.

American Goldfinch (*Carduelis tristis*)

Status: Common permanent resident.

Documentation: Specimen: male, 17 May 1931, Platte Co. (KU 39465).

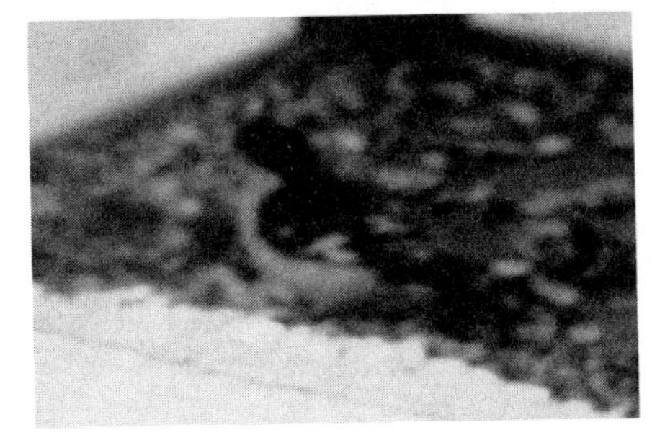

Fig. 33. This male Lesser Goldfinch of the dark-backed nominate population was photographed by Claude Hicks as it visited a Kansas City, Jackson Co., feeder from 4–6 April 1971. It represents the only record for Missouri.

Map 82.
Relative breeding abundance of the American Goldfinch among the natural divisions. Based on BBS data expressed as birds/route.

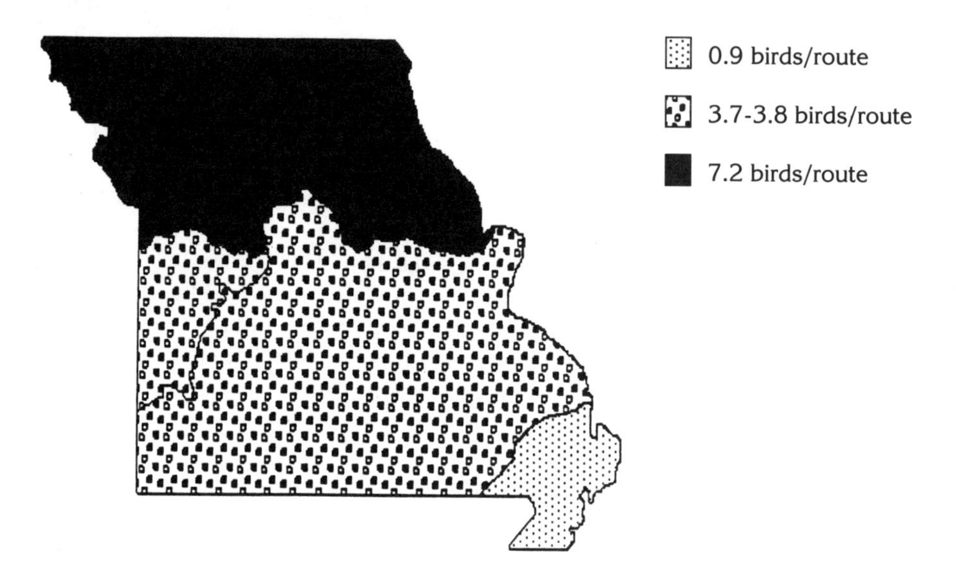

Habitat: Open areas with scattered bushes and trees.

Records:

Spring Migration: During Mar and Apr relatively large-sized flocks are seen throughout the state. Migration continues through early May, but thereafter only pairs or small groups (3–6 birds) are seen. High count: 250+, 27 Mar 1969, Mississippi Co. (JH).

Summer: BBS data indicate that this species is most common in the Glaciated Plains and least common in the Mississippi Lowlands (Map 82).

Fall Migration: Migrants, in relatively small-sized flocks, begin to appear at the beginning of Oct. By the end of the month, larger groups are more frequently encountered. It is commonly seen throughout the remainder of the period. High count: 400, 30 Oct 1937, Joplin (G. Banner-**BB** 4[12]:136).

Winter: The highest average values on CBCs are in the Ozark Border where an overall average of 4.0 birds/hr are recorded. The Ozarks and the Glaciated Plains have a similar average, ca. 3.0 birds/hr, while the Mississippi Lowlands has the lowest average, 1.3 birds/hr. High counts: 962, 29 Dec 1968, Weldon Spring CBC; 914, 17 Dec 1983, Kansas City Southeast CBC.

Evening Grosbeak (*Coccothraustes vespertinus*)

Status: Irregular, rare to locally uncommon winter resident; accidental summer visitant.

Documentation: Specimen: female, 15 Jan 1962, Columbia (MU 2102).

Habitat: Deciduous and coniferous woodland and forest; frequently seen at feeders.

Records:

Spring Migration: Following winters when birds are recorded in the state, small-sized flocks (4–12 birds) are found moving through the state at nonwintering locales during Mar. The largest numbers are encountered during the final two weeks of Apr and the first week of May. Virtually none remain in the state after the second week of May. High count: 60, late Apr 1981, Callaway Co. (RW-**BB** 48[3]:21). Latest date: 4, at feeder, 13 May 1984, Russellville, Cole Co. (JW-**BB** 51[3]:15).

Summer: Two records of late lingering birds: 1, 13 June 1965, near Independence, Jackson Co. (J. Mitchell-**BB** 32[3]:15); 10+, summer 1986, Farmington; female until 15 July 1986, near Farmington (B. Dugal, BRE-**BB** 53[4]:18).

Fall Migration: The first arrivals are seen at the beginning of Nov. Movement continues through the remainder of the period. Earliest dates: 3, 21 Sep–4 Oct 1903, New Haven, Franklin Co. (A. Eimbeck; Widmann 1907); 7, 27 Oct 1985, Jefferson Co. (MP-**NN** 57:85). High count: 15–20, 5 Nov 1977, Maryville (DAE).

Winter: The Evening Grosbeak is encountered more frequently in the Ozarks than in any other region. The highest number of birds/10 pa hrs on CBCs is the 1.5 recorded on the Sullivan count (1971–89). Numbers vary considerably from year to year; some years it is completely absent, whereas hundreds may be locally present during "flight" years. The winters of 1961–62, 1968–69, 1975–76, 1977–78, 1983–84, and 1985–86 were "flight" years. High counts: 75, 13 Jan 1984, Forsyth, Taney Co. (PM-**BB** 51[2]:23); 70, 20 Feb 1962, Columbia (DAE, WG).

Comments: It is unclear where Missouri Evening Grosbeaks are originating, as we fail to find the purported differences that distinguish the western montane race, *brooksi,* from the central and eastern Canadian nominate race. Nevertheless, banding recoveries obtained in Arkansas indicate that birds are of northern instead of western origin (James and Neal 1986).

Family Passeridae: Old World Sparrows

House Sparrow (*Passer domesticus*)

Status: Introduced; common permanent resident.

Map 83.
Relative breeding abundance of the House Sparrow among the natural divisions. Based on BBS data expressed as birds/route.

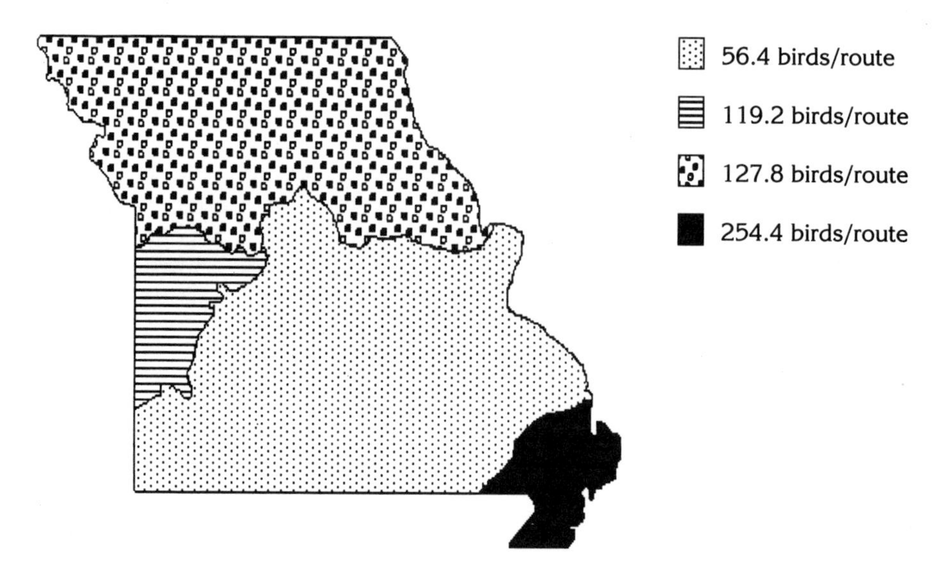

Documentation: Specimen: male, 15 Nov 1956, Columbia (MU 2207).

Habitat: Open areas with scattered trees and bushes, especially in cultivated areas.

Comments: Initially introduced in the United States in 1850 at Brooklyn, New York (*A.O.U. Check-list 1983*). By 1870 it had reached St. Louis (Lever 1987). Widmann (1907) mentioned that it was present wherever there was human habitation. Presently, it is an abundant resident statewide. Both BBS (Map 83) and CBC data indicate that this species is significantly more common in the less forested regions of the state. CBC data, birds/pa hr: 28.4 (Glaciated Plains); 23.5 (Mississippi Lowlands); 13.2 (Ozark Border), and 4.9 (Ozarks).

Eurasian Tree Sparrow (*Passer montanus*)

Status: Introduced; uncommon permanent resident in the St. Louis area, accidental elsewhere.

Documentation: Specimen: male, 26 Dec 1964, St. Louis (NWMSU, RA 6).

Habitat: Open areas with scattered trees and bushes, especially in cultivated areas.

Comments: Between 20 and 34 birds were released in Lafayette Park, St. Louis, in Apr 1870 (Widmann 1907; Jones 1934). Fortunately it has not become abundant, and it has only spread a relatively short distance from St. Louis. Today it is locally common only in St. Louis and St. Charles counties.

By the mid-1930s it had become established in isolated colonies along the Missouri R., as far west as Washington, Franklin Co., and during the 1940s it colonized much of St. Louis and St. Charles counties. To the north at Hannibal, Marion Co., it was first recorded breeding in 1946 (Musselman 1950). Apparently it is still not established at this latter locality. Nevertheless, this species has recently appeared in winter in southeastern Iowa (J. Fuller et al.-**AB** 44:278).

Six birds were seen as far west as Montgomery Co., Missouri, on 15 Nov 1986 (RW-**BB** 54[1]:34), and during the mid-1970s single birds were seen as far south as the Flat River and Farmington areas in St. Francois Co. (B. Samuelson-**NN** 47:2; B. Sutterfield-**NN** 49:17). High counts on CBCs, both on the Orchard Farm CBC: 159, 2 Jan 1983; 151, 2 Jan 1982.

Bibliography

Ahlquist, J. E. 1964. Rufous-necked Sandpiper, *Erolia ruficollis,* in northeastern Ohio. *Auk* 81:432–33.
American Ornithologists' Union. 1886. *The code of nomenclature and check-list of North American birds.*
———. 1895. *Check-list of North American birds.* 2d ed.
———. 1957. *Check-list of North American birds.* 5th ed.
———. 1983. *Check-list of North American birds.* 6th ed.
———. 1985. Thirty-fifth supplement to the American Ornithologists' Union check-list of North American birds. *Auk* 102:680–86.
———. 1987. Thirty-sixth supplement to the American Ornithologists' Union check-list of North American birds. *Auk* 104:591–96.
———. 1989. Thirty-seventh supplement to the American Ornithologists' Union check-list of North American birds. *Auk* 106:532–38.
Anderson, D. 1965. Winter robin roost in Missouri. *Nature Notes* 36:86–87.
Anderson, R., and P. Bauer. 1968. *A guide to finding birds in the St. Louis area.* Webster Groves Nature Study Society.
Ankney, C. D., D. G. Dennis, L. N. Wishard, and J. E. Seeb. 1986. Low genetic variation between Black Ducks and Mallards. *Auk* 103:701–9.
Arbib, R. 1977. Bluelist 1978. *American Birds* 31:1093–94.
Baird, S. F., J. Cassin, and G. N. Lawrence. 1860. *The birds of North America.* Philadelphia: J. B. Lippincott and Co.
Baird, S. F., T. M. Brewer, and R. Ridgway. 1874. *A history of North American birds: Land birds.* 3 vols. Boston: Little, Brown & Co.
Baldwin, R. N. 1911. Notes on rare species in eastern Missouri. *Auk* 28:490–92.
Barksdale, T. 1983. The fourth record of Fulvous-Whistling Duck (*Dendrocygna bicolor*) from Missouri. *Bluebird* 50 (1): 12–13.
Barksdale, T. R., and R. A. Rowlett. 1981. Status of the Gyrfalcon in Missouri,

with notes on field identification. *Continental Birdlife* 2:154–57.

Bennitt, R. 1932. *Check-list of the birds of Missouri.* Univ. of Missouri–Columbia.

———. 1937. Pocket list of Missouri birds. Audubon Society of Missouri.

Betts, N de W. 1909. Brown Creeper's nest in Missouri. *Bluebird* 53 (1): 32–34.

Binford, L. C., and J. V. Remsen, Jr. 1974. Identification of the Yellow-billed Loon (*Gavia adamsii*). *Western Birds* 5:111–26.

Blake, J. G., and J. R. Karr. 1987. Breeding birds of isolated woodlots: Area and habitat relationships. *Ecology* 68:1724-34.

Bock, C. E., and L. W. Lepthien. 1972. Winter eruptions of Red-breasted Nuthatches in North America, 1950–1970. *American Birds* 26:558–61.

———. 1976. Growth in the eastern House Finch population, 1962–1971. *American Birds* 30:791–92.

Bohlen, H. D., and W. Zimmerman. 1989. *The Birds of Illinois.* Bloomington and Indianapolis: Indiana Univ. Press.

Branan, W. V., and H. C. Burdick. 1981. Bird species composition in a Missouri park, 1916 vs. 1973. *Kansas Ornithological Society Bulletin* 32:41–45.

Braun, M. J., and M. B. Robbins. 1986. Extensive protein similarity of the hybridizing chickadees *Parus atricapillus* and *P. carolinensis. Auk* 103: 667–75.

Brown, L. N. 1963. Status of the Roadrunner in Missouri. *Condor* 65:242–43.

Browning, M. R. 1989. The correct name for the Olivaceous Cormorant, "Maiagua" of Piso (1658). *Wilson Bulletin* 101:101–6.

Bryant, J. A. 1895. Clark's Nutcracker in eastern Missouri. *Auk* 12:82.

Burgess, H. 1981. The Trumpeters may again return. *Bluebird* 48 (3): 12.

Burnside, F. L. 1987. Long-distance movements by Loggerhead Shrikes. *Journal of Field Ornithology* 58:62–65.

Bystrak, D. 1981. The North American breeding bird survey. *Studies in Avian Biology* 6:34–41.

Cannon, R. W., and D. W. Christisen. 1984. Breeding range and population status of the Greater Prairie-Chicken in Missouri. *Transactions of the Missouri Academy of Science* 18:33–39.

Cary, D. L. 1984. A stronghold for pheasants. *Nature Notes* 56:37–38.

Childers, L. 1989. Birding the Opolis area, Jasper County, MO. *Western Ozarks Birders Newsletter* 3:12.

Christisen, D. M. 1985. The Greater Prairie-Chicken and Missouri's land-use patterns. Missouri Department of Conservation Terrestrial Series, no. 15.

Clawson, R. L. 1982. The status, distribution, and habitat preferences of the birds of Missouri. Missouri Department of Conservation Terrestrial Series, no. 11.

Collier, J. E. 1955. *Agricultural atlas of Missouri.* Columbia: Univ. of Missouri Press. Bulletin 645.

Collins, C. 1975. Rare Swallow-tailed Kite sighted. *Bluebird* 42 (1): 4.

Comfort, J. E. 1952. Resume for St. Louis area, 1951. *Bluebird* 19 (2): 3–4.

Comfort, J. F. 1972. Otto Widmann's bracketed species. *Bluebird* 39 (3): 2–5.
Confer, J. L., T. Kaaret, and L. Jones. 1979. Effort, location, and the Christmas Bird Count tally. *American Birds* 33:690-92.
Cooke, M. T. 1928. The spread of the European Starling in North America (to 1928). U.S. Department of Agriculture. Circulation 40.
———. 1937. Some returns of banded birds. *Bird-Banding* 8:144–55.
———. 1941. Recoveries of some banded birds of prey. *Bird-Banding* 12: 150–60.
———. 1950. Returns of banded birds. *Bird-Banding* 21:11–18.
Cooke, W. W. 1888. Report on bird migration in the Mississippi Valley in the years 1884 and 1885. *U.S. Department of Agriculture Division Economic Ornithology Bulletin* 2:1–313.
———. 1910. The type locality of *Vireo bellii*. *Auk* 27:342–43.
Cortelyou, R. G. 1975. A Brown Creeper's nest in Nebraska. *Nebraska Bird Review* 43:80–83.
Coues, E. 1872. *Key to North American birds.* Massachusetts: Salem Press.
———. 1874. Birds of the northwest. U.S. Geol. Surv. of the territories. U.S. Department of the Interior, miscellaneous publication 3.
Crosby, G. 1972. Spread of the Cattle Egret in the Western Hemisphere. *Bird-Banding* 43:205–12.
Dierker, W. W. 1979. Birds of the Hannibal, Missouri area. *Transactions of the Missouri Academy of Science* 13:41–51.
Dinsmore, J. J., T. H. Kent, D. Koenig, P. C. Petersen, and D. M. Roosa. 1984. *Iowa birds.* Ames: Iowa State Univ. Press.
Eades, J., D. M. Jones, and D. Symes. 1978. First field identification of a female Barrow's Goldeneye in the St. Louis area. *Bluebird* 45 (1): 9–10.
Easterla, D. A. 1952. New state records. *Bluebird* 19 (9): 2.
———. 1962a. Avifauna of Tucker Prairie. Master's thesis, Univ. of Missouri–Columbia.
———. 1962b. Some foods of the Yellow Rail in Missouri. *Wilson Bulletin* 74:94–95.
———. 1962c. The Bald Eagle nesting in Missouri. *Bluebird* 29 (2): 11–13.
———. 1962d. Grasshopper Sparrow wintering in central Missouri. *Wilson Bulletin* 74:288.
———. 1963. Ruff observed in Missouri. *Wilson Bulletin* 75:274–75.
———. 1964. Summary of tower fatalities at Columbia, Missouri. *Bluebird* 31 (2): 6.
———. 1965a. Range extension of the Fish Crow in Missouri. *Wilson Bulletin* 77:297–98.
———. 1965b. Arctic Loons invade Missouri. *Condor* 67:544.
———. 1966a. Northern fringillids invade northwestern Missouri. *Bluebird* 33 (2): 3–5.
———. 1966b. Short-billed Marsh Wren, Pine Warbler, Palm Warbler and Savannah Sparrow wintering in Missouri. *Bluebird* 33 (2): 26–28.

———. 1967a. The Baird's Sparrow and Burrowing Owl in Missouri. *Condor* 69:88–89.

———. 1967b. A breeding population of Henslow's Sparrows in southwestern Missouri. *Bluebird* 34 (1):18–19.

———. 1969a. 1968 Breeding records of the Vesper Sparrow and Bobolink in Missouri. *Bluebird* 36 (1): 21–22.

———. 1969b. The Snowy Plover in Missouri. *Auk* 86:146.

———. 1970a. First nesting colonies of the Lark Bunting in Missouri. *Wilson Bulletin* 82:465–66.

———. 1970b. Hermit Warbler in Missouri. *Wilson Bulletin* 82:464.

———. 1970c. Long-billed and Short-billed Dowitcher identification. *Birding* 2 (4): insert.

———. 1971. The Hudsonian Godwit at Squaw Creek National Wildlife Refuge, Mound City, Missouri. *Birding* 3 (2): insert.

———. 1975. Grasshopper Sparrow with crossed mandibles. *Kansas Ornithological Society Bulletin* 26:21.

———. 1976. First record of the Little Gull and the White-tailed Kite in Missouri. *Bluebird* 43 (4): 14–16.

Easterla, D. A., and R. A. Anderson. 1967. *Checklist of Missouri birds.* Audubon Society of Missouri.

———. 1969. First Vermilion Flycatcher specimen from Missouri. *Auk* 86:750.

———. 1971. *Checklist of Missouri birds.* Audubon Society of Missouri.

———. 1979. *Checklist of Missouri birds.* Audubon Society of Missouri.

Easterla, D. A., and F. Lawhon. 1971. First specimen of Arctic Loon from Missouri. *Auk* 88:175.

Easterla, D. A., and R. E. Ball. 1973. The Rock Wren in Missouri. *Wilson Bulletin* 85:479–80.

Easterla, D. A., M. B. Robbins, and R. A. Anderson. 1986. *Annotated Check-list of Missouri birds.* Audubon Society of Missouri.

Eddleman, W. R., and R. L. Clawson. 1987. Population status and habitat conditions for the Red-cockaded Woodpecker in Missouri. *Transactions of the Missouri Academy of Science* 21:105–17.

Elder, W. H., and J. Hansen. 1967. Bird mortality at KOMU-TV tower, Columbia, Missouri. Fall 1965 and 1966. *Bluebird* 34 (1): 3–6.

Elliot, J. J., and R. S. Arbib. 1953. Origin and status of the House Finch in the eastern United States. *Auk* 70:31–37.

Evans, K. E. 1980. A recent nesting record of the Chestnut-sided Warbler in Missouri. *Bluebird* 47 (4): 6–7.

Falk, L. L. 1979. An examination of observer's weather sensitivity in Christmas Bird Count data. *American Birds* 33:688–89.

Fleming, J. H. 1912. Sabine's Gull on the Mississippi River. *Auk* 29:388–89.

Fisher, B. 1989. Evolution of a new state record: the 1989 Jackson County Frigatebird. *Bluebird* 56 (3): 79–81.

Fretwell, S. 1978. White-throated Swift at Manhattan, Kansas. *Kansas Orni-*

thological Society Bulletin 29 (4): 31–32.

Funk, J. L., and J. W. Robinson. 1974. Changes in the channel of the lower Missouri River and effects on fish and wildlife. Missouri Department of Conservation Aquatic Series, no. 11.

Garrett, J. 1989. A male and female Mountain Bluebird in Cass County. *Bluebird* 56 (2): 70.

Gaston, A. J., and R. Decker. 1985. Interbreeding of Thayer's Gull, *Larus thayeri,* and Kumlien's Gull, *Larus glaucoides kumlieni,* on Southhampton Island, Northwest Territories. *Canadian Field-Naturalist* 99:257–59.

George, W. 1963. Columbia tower fatalities. *Bluebird* 30 (4): 5.

Giessman, N. F., T. W. Barney, T. L. Haithcoat, J. W. Myers, and R. Massengale. 1986. Distribution of Missouri forests. *Transactions of the Missouri Academy of Science* 20:5–14.

Gill, F. B. 1980. Historical aspects of hybridization between Blue-winged and Golden-winged warblers. *Auk* 97:1–18.

Godfrey, W. E. 1986. *The birds of Canada.* National Museums of Canada, Ottawa.

Goetz, R. E., W. M. Rudden, and P. B. Snetsinger. 1986. Slaty-backed Gull winters on the Mississippi River. *American Birds* 40:207–16.

Goodge, B. 1972. Groove-billed Ani at Columbia. *Bluebird* 39 (4): 11.

Goodge, W. 1977. Laughing Gull at Fountain Grove Waterfowl area. *Bluebird* 44 (2): 5.

Goss, N. S. 1891. *History of birds of Kansas.* Topeka: G. W. Crane and Co.

Griffin, C. R., and W. H. Elder. 1980. Nesting records for the Bald Eagle in Missouri and the lower Mississippi River. *Transactions of the Missouri Academy of Science* 14:5–7.

Griffin, C. R., J. M. Southern, and L. D. Frenzel. 1980. Origins and migratory movements of Bald Eagles wintering in Missouri. *Journal of Field Ornithology* 51:161–67.

Groth, J. R. 1988. Resolution of cryptic species in Appalachian Red Crossbills. *Condor* 90:745–60.

Hallet, D. L. 1988. Statewide Ring-necked Pheasant surveys. Study no. 29. Jefferson City: Missouri Department of Conservation.

———. 1989. Prairie-chickens and pheasants: Do they compete in Missouri? *Missouri Prairie Journal* 11 (1–2): 20–21.

Hanselmann, S. 1963. The warbler and vireo migration in the St. Louis area. *Bluebird* 30 (4): 9–12.

Harford, H. 1959. A flamingo in Missouri. *Bluebird* 26 (6): 1.

Hardin, K. I., T. S. Baskett, and K. E. Evans. 1982. Habitat of Bachman's Sparrows breeding on Missouri glades. *Wilson Bulletin* 94:208–12.

Harris, H. 1919a. Historical notes on Harris's Sparrow (*Zonotrichia querula*). *Auk* 36:180–90.

———. 1919b. Birds of the Kansas City region. *Transactions of the Academy of Sciences, St. Louis.* Vol. 23, no. 8.

Hasbrouck, E. M. 1891. The present status of the Ivory-billed Woodpecker (*Campephilus principalis*). *Auk* 8:174–86.

Hayes, J. 1989a. White River Ozarks Vulture watch. *Bluebird* 56 (1): 19–20.

———. 1989b. Ozarks nightjar surveys. *Western Ozarks Newsletter* 3:15–16.

Hayman, P., J. Marchant, and T. Prater. 1986. *Shorebirds: An identification guide.* Boston: Houghton Mifflin Company.

Heusmann, H. W. 1974. Mallard-Black Duck relationships in the northeast. *Wildlife Society Bulletin* 2:171–77.

Heye, P. L. 1963. Tower fatalities. *Bluebird* 30 (1): 7.

Hobbs, K. 1981. Not all kingbirds are Easterns! *Bluebird* 48 (1): 6–7.

Hoffman, R. 1916. The Pomarine Jaeger and the Purple Gallinule in western Missouri. *Auk* 33:196.

Howell, A. 1911. *Birds of Arkansas.* U.S. Department of Agriculture, Bulletin 38.

Hoy, P. R. 1865. Journal of an exploration of western Missouri in 1854, under the auspices of the Smithsonian Institution. Nineteenth Annual Report Smithsonian Institution.

Hurter, J. 1884. List of birds collected in the neighborhood of St. Louis. *Ornithologist and Oologist* 9:85–87, 95–97.

Jackson, J. 1955. The Least Tern in Missouri. *Bluebird* 22 (8–9): 2–3.

Jacobs, B. 1989. I brake for birds: Missouri's second Lewis' Woodpecker. *Bluebird* 56 (3): 85–87.

James, D. A., and J. C. Neal. 1986. *Arkansas birds: Their status and distribution.* Fayetteville: Univ. of Arkansas.

Jenner, W. 1934. Some bird observations in Howard County, Missouri. *Wilson Bulletin* 46:258–59.

Johnsgard, P. A. 1980. A revised list of the birds of Nebraska and adjacent Plains states. *Occasional Papers of the Nebraska Ornithologists' Union, no. 6.*

Johnson, M. 1975. Maximas on the rocks. *Missouri Conservationist* 36 (7): 8–9.

Johnson, M. S. 1908. Notes on Missouri birds. *Auk* 25:324.

Jones, D. 1978. The House Finch winter invasion reaches Missouri. *Bluebird* 46 (1): 29–30.

Jones, S. P. 1934. The European Tree Sparrow: A bird of Missouri and Illinois. *Bluebird* 1 (2): 9.

Kaufman, K. 1989. Identifying and documenting Clark's Grebe out of range. *Bluebird* 56 (2): 57–62.

———. 1990. *A field guide to advanced birding.* Boston: Houghton Mifflin Company.

Kessel, B. 1953. Distribution and migration of the European Starling in North America. *Condor* 55:49–67.

Kleen, V. M., and L. Bush. 1973. Middlewestern Prairie Region. *American Birds* 27:622–25.

Knox, A. G. 1988. The taxonomy of redpolls. *Ardea* 76:1–26.

Korotev, R. L. 1990. A timetable for the bird spring migration through Forest

Park, St. Louis. Supplement to *Nature Notes* 60:1-4.

Kridelbaugh, A. L. 1981. Population trend, breeding and wintering distribution of Loggerhead Shrikes (*Lanius ludovicianus*) in Missouri. *Transactions of the Missouri Academy of Science* 15:111-19.

Kritz, K. J. 1989. Nesting ecology and nest site habitat of Sharp-shinned and Cooper's hawks in Missouri. Master's thesis, Univ. of Missouri-Columbia.

Kucera, C. L. 1961. The grasses of Missouri. *Univ. of Missouri Studies* 35:1-241.

Kushlan, J. A., and P. C. Frohring. 1986. The history of the southern Wood Stork population. *Wilson Bulletin* 98:368-86.

Lanyon, W. E. 1956. Ecological aspects of the sympatric distribution of meadowlarks in the north-central states. *Ecology* 37:98-108.

Lasley, G. W., and C. Sexton. 1988. The spring season. Texas region. *American Birds* 42:456-62.

Lehman, P. 1988. The changing seasons. *American Birds* 43 (Autumn): 50-54.

Lever, P. 1987. *Naturalized birds of the world.* New York: J. Wiley and Sons.

Lewis, J. B. 1961. Wild Turkeys in Missouri, 1940-1960. *Transactions of the North American Wildlife and Natural Resources Conference* 26:505-12.

Lewis, J. B., J. D. McGowan, and T. S. Baskett. 1968. Evaluating Ruffed Grouse reintroduction in Missouri. *Journal of Wildlife Management* 32:17-28.

Ligon, J. D. 1967. Relationships of the Cathartid vultures. Univ. Michigan Museum of Zoology, Occasional Papers, no. 651.

Lundberg, C. 1990. Rare loons at Table Rock Lake. *Western Ozark Birders Newsletter* 5:11-12.

McKinley, D. 1960a. The Carolina Parakeet in pioneer Missouri. *Wilson Bulletin* 72:274-87.

———. 1960b. A history of the Passenger Pigeon in Missouri. *Auk* 77:399-419.

———. 1961. History of the Canada Goose as a breeding bird in Missouri. *Bluebird* 28 (3): 6-12.

———. 1962. The Trumpeter Swan in Missouri. *Bluebird* 29 (1): 6-11.

———. 1964. History of the Carolina Parakeet in its southwestern range. *Wilson Bulletin* 76:68-93.

Mayes, E. A. 1937. Forest restoration in Missouri. Univ. of Missouri-Columbia, College of Agriculture Research Bull. 392. Agricultural Experimental Station.

Murray, B. G., Jr. 1989. A critical review of the transoceanic migration of the Blackpoll Warbler. *Auk* 106:8-17.

Musselman, T. E. 1937. American Magpie in Missouri and Illinois. *Auk* 54:393.

———. 1950. European Tree Sparrow at Hannibal, Missouri. *Auk* 67:105.

Myers, J. P., R. F. Cardillo, and F. B. Gill. 1986. VIREO: procedures for the ornithological community. *Condor* 88:115-17.

Neff, J. A. 1923. Some birds of the Ozark region. *Wilson Bulletin* 35:202-15.

———. 1930. The starling in the Missouri ozarks. *Wilson Bulletin* 42:66.

Nelson, P. W. 1985. *The terrestrial natural communities of Missouri.* Jefferson

City: Missouri Natural Areas Committee.

Newton, E. T. 1942. Western Sandpiper in western Missouri and eastern Kansas. *Auk* 59:109–10.

Norris, D. J., and W. H. Elder. 1982a. Distribution and habitat characteristics of the Painted Bunting in Missouri. *Transactions of the Missouri Academy of Science* 16:77–83.

———. 1982b. Decline of the Roadrunner in Missouri. *Wilson Bulletin* 94: 354–56.

Oberholser, H. C. 1974. *The bird life of Texas.* Ed. E. B. Kincaid, Jr., Univ. of Texas, Austin.

Payne, R. B. 1987. Population and type specimens of a nomadic bird: Comments of the North American crossbills *Loxia pusilla* Gloger 1834 and *Crucirostra minor* Brehm 1845. Occasional Papers of the Museum of Michigan, no. 714.

Paynter, R. A., Jr. 1970. Cardinalinae. In *Check-list of the birds of the world,* 13:216–45. Cambridge: Museum of Comparative Zoology.

Peters, M. 1988. Fall 1987 hawk migration—St. Louis area. *Nature Notes* 60: 55–56.

Phillips, A. R. 1986. *The known birds of North and Middle America.* Denver, Colorado.

Pitocchelli, J. 1990. Plumage, morphometric, and song variation in Mourning (*Oporornis philadelphia*) and MacGillivray's (*O. tolmiei*) warblers. *Auk* 107: 161–71.

Prater, T., J. Marchant, and J. Vuorinen. 1977. *Guide to the field identification and ageing of Holarctic Waders.* Tring, England: British Trust for Ornithology.

Prevett, J. P., and C. D. MacInnes. 1972. The number of Ross' Geese in central North America. *Condor* 74:431–38.

Rafferty, M. D. 1982. *Historical Atlas of Missouri.* Norman: Univ. of Oklahoma Press.

Raynor, G. S. 1975. Techniques for evaluating and analyzing Christmas Bird Count data. *American Birds* 29:626–33.

Reffalt, W. C. 1985. A nationwide survey: Wetlands in extremis. *Wilderness* 49:28–41.

Rising, J. D. 1965. Townsend's Solitaire and Pine Grosbeak in Missouri. *Auk* 82:275.

Rising, J. D., T. R. Anderson, M. L. Myers, and S. R. Leffler. 1964. Extreme dates of occurrence of birds in northeastern Kansas and adjacent Missouri. *Kansas Ornithology Society Bulletin* 15:17–18.

Rising, J. D., T. Pucci, N. Johnson, and R. Dawson. 1978. *Birds of the Kansas City area.* Burroughs Audubon Society and Shawnee Mission High School. Kansas City, Kansas.

Robbins, C. S. 1979. Effects of forest fragmentation on bird populations. In *Proceedings of the Workshop on Management of North-central and North-*

eastern Forests for Nongame birds. Eds. R. M. DeGraff and K. E. Evans, 198–212. U.S. Forest Service General Technical Report NC-51.

Robbins, C. S., D. Bystack, and P. H. Geissler. 1986. The breeding bird survey: Its first fifteen years, 1965–1979. *U.S. Fish and Wildlife Service. Publ.* 157.

Robbins, M. B. 1975. Long-tailed Jaeger at Maryville. *Bluebird* 42 (1): 4.

———. 1986. Rediscovery of nesting Brown Creepers in Missouri. *Bluebird* 53 (1): 32–34.

———. 1989. Swainson's Hawk migration through the upper Missouri river valley of Missouri. *Bluebird* 56 (1): 21–22.

Robbins, M. B., and D. A. Easterla. 1981. Range expansion of the Bronzed Cowbird with the first Missouri record. *Condor* 83: 270–72.

———. 1985. First Missouri record of the Varied Thrush. *Bluebird* 52 (2): 30–31.

———. 1986. Range expansion of the Great-tailed Grackle into Missouri, with details of the first nesting colony. *Bluebird* 53 (4): 24–27.

Robbins, M. B., D. A. Easterla, and F. Lawhon. 1986a. Notes on Missouri's first breeding record of the Burrowing Owl (*Athene cunicularia*). *Bluebird* 53 (2): 21–22.

Robbins, M. B., M. J. Braun, and E. A. Tobey. 1986b. Morphological and vocal variation across a contact zone between the chickadees *Parus atricapillus* and *P. carolinensis. Auk* 103:655–66.

Roberson, D. 1989. More on Pacific versus Arctic loons. *Birding* 21:154–57.

Rudden, B. 1980a. First St. Louis area sighting of a McCown's Longspur. *Nature Notes* 52:1–2.

———. 1980b. There is a jaeger at the dam. *Nature Notes* 52:14–16.

Say, T. 1823. *Fringilla grammaca.* In *Account of an expedition [S. Long expedition] from Pittsburgh to the Rocky Mountains,* ed. E. James, 1:139–140. (Philadelphia ed.).

Schoen, E. 1955. Red-breasted Nuthatch nesting. *Bluebird* 22 (7): 3.

Schroeder, W. A. 1981. Presettlement Prairie of Missouri. Natural History Series, no. 2. Missouri Department of Conservation.

Schulenberg, T. 1989. More on Pacific vs. Arctic Loons. *Birding* 21:157–58.

Schwartz, C. W. 1945. The ecology of the prairie-chicken in Missouri. *Univ. of Missouri Studies* 20:1–99.

Scott, W. E. D. 1879. Notes on birds observed during the spring migration in western Missouri. *Bulletin of the Nuttall Ornithological Club* 4:139–47.

Shirling, A. E. 1920. *Birds of Swope Park.* Kansas City, MO.: McIndoo Publ. Co.

Sibley, C. G., and J. E. Ahlquist. 1985. The relationships of some groups of African birds, based on comparisons of the genetic material, DNA. In *Proceedings of Symposium on African Vertebrates.* Ed. K. L. Schuchmann, 115–161. Bonn: Zoologisches Forschungsinstitut und Museum A. Koenig.

Sidle, J. G., J. J. Dinan, M. P. Dryer, J. P. Rumancik, Jr., and J. W. Smith. 1988. Distribution of the Least Tern in interior North America. *American Birds* 42:195–210.

Smith, J. W. 1985. Improving the status of endangered species in Missouri (Interior Least Tern habitat and nest survey). Endangered species project no. SE-01-12. Jefferson City: Missouri Department of Conservation.

———. 1987. Improving the status of endangered species in Missouri (Interior Least Tern habitat and nest survey). Endangered species project SE-01-12. Jefferson City: Missouri Department of Conservation.

Smith, K. G. 1986. Winter population dynamics of Blue Jays, Red-headed Woodpeckers, and Northern Mockingbirds in the Ozarks. *American Midland Naturalist* 115:52-62.

Smith, N. G. 1963. Evolution of some Arctic Gulls (*Larus*): A study of isolating mechanisms. Ph.D. diss., Cornell Univ., Ithaca, New York.

———. 1966. Evolution of some Arctic Gulls (*Larus*): An experimental study of isolating mechanisms. *Ornithological Monographs* 4.

Snell, R. R. 1989. Status of *Larus* gulls at Home Bay, Baffin Island. *Colonial Birds* 12:12-23.

Snyder, D. E. 1961. First record of the Least Frigate-bird (*Fregata ariel*) in North America. *Auk* 78:265.

Sparks, F. W. 1891. Fulvous Tree Duck in Missouri. *Forest and Stream* 36:476.

Street, P. B. 1948. The Edward Harris collection of birds. *Wilson Bulletin* 60: 167-84.

Tanner, J. T. 1942. *The Ivory-billed Woodpecker.* New York: National Audubon Society. Research Report No. 1.

Tate, J., Jr. 1986. The Blue list for 1986. *American Birds* 40:227-36.

Teachenor, D. 1940. Western Burrowing Owl in western Missouri. *Auk* 57:573.

Thom, R. H., and J. H. Wilson. 1980. The natural divisions of Missouri. *Transactions of the Missouri Academy of Science* 14:9-23.

Thompson, M. C., and C. Ely. 1989. *Birds in Kansas.* Vol. 1. Lawrence: Univ. of Kansas Museum of Natural History.

Tingley, S. J. 1983. The 83rd Audubon Christmas bird count. Atlantic provinces. *American Birds* 37:376.

Toland, B. 1986. Evidence of an increasing nesting population of Swainson's Hawks in Missouri. *Bluebird* 53 (4): 28-32.

Tulenko, P. Q. 1950. *Avis rara. Bluebird* 17 (12): 1.

United States Fish and Wildlife Service. 1985. Interior population of the Least Tern determined to be endangered. Federal Register 50:50720-50734.

Vanden Berge, J. C. 1970. A comparative study of the appendicular musculature of the order Ciconiiformes. *American Midland Naturalist* 84: 289-364.

Veit, R. R., and L. Jonsson. 1984. Field identification of smaller sandpipers within the genus *Calidris. American Birds* 38:853-76.

Walkinshaw, L. 1983. *Kirtland's Warbler, the natural history of an endangered species.* Bloomfield Hills: Cranbrook Institute of Science.

Walsh, T. 1988. Identifying Pacific Loons. Some old and new problems. *Birding* 20:12-28.

Walters, D. L. 1939. Lazuli Bunting in Missouri. *Oologist* 56:117.
Warner, C. A. 1966. Breeding range expansion of the Scissor-tailed Flycatcher into Missouri and in other states. *Wilson Bulletin* 78:289–300.
Watkins, L. 1968. Missouri's summer Bobolinks. *Bluebird* 35 (1): 7.
Whitcomb, R. F., C. S. Robbins, J. F. Lynch, B. L. Whitcomb, M. K. Klimkiewicz, and D. Bystrak. 1981. Effects of forest fragmentation on avifauna of the eastern deciduous forest. In *Forest island dynamics in man-dominated landscapes.* Eds. R. L. Burgess and D. M. Sharpe, 123–205. New York: Springer-Verlag.
Widmann, O. 1885. Note on the capture of *Coturniculus lecontei* and *Dendroeca kirtlandi* within the city limits of St. Louis *Auk* 2:381–82.
———. 1895a. Swainson's Warbler an inhabitant of the swampy woods of southeastern Missouri. *Auk* 12:112–17.
———. 1895b. An hour with Baird's and LeConte's Sparrows near St. Louis, Missouri. *Auk* 12:219–25.
———. 1897. The summer home of Bachman's Warbler no longer unknown. *Auk* 14:305–10.
———. 1907. A preliminary catalog of the birds of Missouri. Academy of Science, St. Louis.
———. 1908. Another Clark's Crow taken in Missouri. *Auk* 25:222.
———. 1928. Chimney Swifts in November, 1925. *Wilson Bulletin* 40:151–54.
Wilhelm, E. J. 1958. *Birds of the St. Louis area.* St. Louis Audubon Society.
Williams, F. 1973. Southern Great Plains. *American Birds* 27:633–37.
———. 1988. Southern Great Plains. *American Birds* 42:284.
Williams, H. C. 1913. White Ibis (*Guara alba*) in Missouri. *Auk* 30:268.
Williamson, V. 1946. Arctic 3-toed Woodpecker. *Bluebird* 13 (5): 1.
Wilson, J. D. 1986. Missouri's first record of Roseate Spoonbill. *Bluebird* 53 (3): 16–18.
———. 1989. First record of White-throated Swift in Missouri. *Bluebird* 56 (3): 82–84.
Wilson, S. S. 1896. Notes from Missouri. *Wilson Bulletin* 8:8.
Wood, D. S., and G. D. Schnell. 1984. *Distributions of Oklahoma birds.* Norman: Univ. of Oklahoma Press.
Woodruff, E. S. 1907. Some interesting records from southern Missouri. *Auk* 24:348–49.
———. 1908. A preliminary list of the birds of Shannon and Carter counties, Missouri. *Auk* 25:191–214.
Wywialowski, A. 1988. Long-term trends in Missouri's Greater Prairie-Chicken population. *Missouri Prairie Journal* 10 (2): 3–5.
Wywialowski, A., and D. M. Christisen. 1989. Greater Prairie-chicken populations of Missouri: An overview. *Missouri Prairie Journal* 11 (1–2): 8–11.

Index

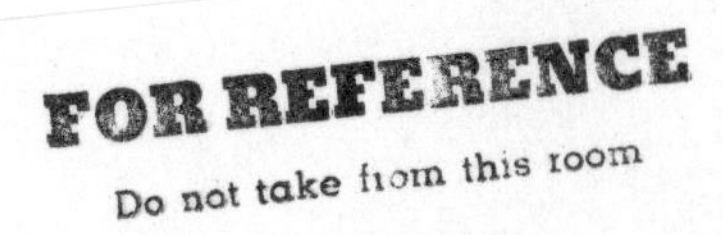
FOR REFERENCE
Do not take from this room